In Situ Tests in Geotechnical Engineering

Series Editor
Gilles Pijaudier-Cabot

In Situ Tests in Geotechnical Engineering

Jacques Monnet

ISTE

WILEY

First published 2015 in Great Britain and the United States by ISTE Ltd and John Wiley & Sons, Inc.

ISTE Ltd
27-37 St George's Road
London SW19 4EU
UK

www.iste.co.uk

John Wiley & Sons, Inc.
111 River Street
Hoboken, NJ 07030
USA

www.wiley.com

Library of Congress Control Number: 2015952403

British Library Cataloguing-in-Publication Data
A CIP record for this book is available from the British Library
ISBN 978-1-84821-849-9

Contents

Foreword

Jacques Monnet is a very incisive researcher, who does not accommodate with the classical approaches of *in situ* testing techniques or interpretation approaches, innovating with utmost originality in equipment, procedures and analysis. However, what could be seen as a pure speculative attitude, which frequently characterizes some of our researchers seeking for new publications, often complicating the objectiveness of ground characterizations, he has always proved to have a practical attitude giving solutions to problems that other more common test procedures do not solve. That was the case of some of the innovative pieces of equipment like the *Géomécamètre* described along with others in this book.

The result of his work expressed now in this book surpassed my expectations since I was not envisaging such an enriching document, plenty of objective solutions, together with the description of advanced techniques that he has been pursuing during his professional activity, allowing the enrolment of new solutions in "non-textbook" ground conditions and special projects.

As Jacques refers to, at beginning of the book, *in situ* tests for ground characterization have been suffering from a lack of credibility to some well known scientists, who prefer well controlled laboratory tests, pointed out as fundamental where the test conditions are well known and the measures are realiable and can allow the direct determination of the parameters that feed the increasing numerical codes available in geotech practice, with complex models behind them. This may be a consequence of the inertia of *in situ* testing specialists to diffuse the potentiality of *in situ* data, as well as being well interpreted and obtained from the most adapted techniques.

Jacques Monnet decided not to follow the usual pattern developed in many textbooks where the distinct technologies (equipment and procedures) are displayed sequentially. This practice is somehow an invertion of the natural process of geoengineering, conception/design of geotechnical problems, looking at the properties to be determined in each big group of grounds (soils or rocks – or intermediate material), and then looking for the tools/methods that can help that process.

This book starts with the techniques and processes that allow the measurement of the essential physical indices and the sampling processes to obtain representative samples to deduce the best information in laboratory tests, where we forget that the soil has its own history before arriving at the laboratory. After detailing the processes of drilling, preparation, transportation of soils, through sampling, the chapter concludes with an approach to rock mechanics, finishing the natural step to the following, most important chapters related to the mechanical and hydraulic properties of soils. The approaches of each chapter are ruled by the properties that geotechnical practitioners are seeking, for which distinct solutions in terms of techniques (equipment, procedures and methods) are proposed, including common/classical tests, or advanced techniques. This comes as being quite logical as, for "direct" shear strengths, shear tests in boxes, vane tests, and the "Philonnat *phicomètre*" are suggested; for stress-strain relationships, and therefore for the evaluation of deformability and compressibility, the surface load tests (PLT), dilatometers (DMT), penetrometers (SPT, DP, CPT...), pressuremeters (PMT, SBPT) and dilatometer (DMT) tests as well as some new techniques, like "Géomécamètre", are presented and their essential interpretation methods are developed. Then a very important chapter, currently forgotten in handbooks of site investigation, deals with permeabilty tests in soils, executed in localized points (from boreholes), but also the more generalized solutions for the evalaution of transmissivity, like the pumping tests with multi-checking points. The books ends with an interesting resumé of the essential properies that can be derived and how they can be deduced from sustained correlations.

In summary, this book is a highly recommendable document for students and researchers, accademics and practitioners in engineering and will contribute to constructing one more pillar to allow for a better understanding of the geotechnical “terroirs”, enhanced by the advantage and the potentiality of using good *in situ* equipment, adapted for the purpose of our projects and interpreted with competence.

António Viana da Fonseca
Chair of ISSMGE-TC1202
Ground Characterization from In Situ Testing
September 2015

Symbols & Notations

Symbols	Designations	Units
a	Side of the square plate	m
a	Dilatation coefficient of the pipes (PMT)	cm^3/MPa
a	Bore radius (PMT)	m
A	Transverse section of the pile	m^2
A	Reading of the detachment pressure of the membrane (DMT test)	kPa
A	Length of penetration measured (drilling parameter)	m
A_c	Straight section of the cone (penetrometer)	m^2
A_n	Straight section of the pile at the level of n	m^2
A_u	Straight section of the cone (penetrometer) above the cylindrical part of the cone	m^2
α	Angle of attack of the bevel of case of the core drill, in degrees	degree
α	Slope of the unitary drawdown curve (well)	m^{-2}/h
α	Empirical drilling parameter (drilling parameter)	
b	Radius of the first plastic zone (PMT)	m
B	Diameter of the plate or width of the foundation or the pile	m
B	Reading of the pressure for a membrane advancement of 1.1 m (DMT)	kPa

B	Diameter of the injection or pumping cavity (Lefranc)	m
B_c	Semi-thickness of the pressure sensor	m
$B\theta$	Normalized interstitial pressure ratio (CPTU)	
β	Empirical bore parameter (drilling parameter)	
c	Radius of the second plastic zone	m
c'	Effective cohesion of the soil	kPa
c_u	Apparent undrained cohesion of the soil	kPa
c_r	Residual cohesion of the soil	kPa
C	Coefficient of the shape of the cavity (Lefranc)	
C	Empirical coefficient of the distribution of constraints (sensor)	
C_a	Surface index of a core drill, expressed as a percentage	
$C_{\square}$	Creep coefficient	
C_c	Compression index (odometer; Log scale)	
C_i	Internal play index of a core drill expressed as a percentage	
C_0	External play index of a core drill expressed as a percentage	
C_R	Calculated rotational torque of the drill head (drilling parameter)	kN.m
C_{Rmax}	Maximal rotational torque of the drill head (drilling parameter)	kN.m
C_s	Elastic swelling index (odometer; Log scale)	C_c
C_v	Vertical consolidation coefficient	m^2/s
C_{vh}	Horizontal consolidation coefficient	m^2/s
d_c	Diameter of the sensitive surface of the pressure sensor	m
d_i	Interior diameter of the calibration tube (PMT)	m
d_{ij}	Distance between two data sets (cluster)	
d_s	Diameter of the pressuremeter probe (PMT)	m
d_t	Diameter of the borehole (PMT)	m
D	Granularity, all parameters linked to it	m

D	Bore diameter (drilling parameter)	m
D_c	Diameter of the pressure sensor	m
DG	Degradability coefficient of the rocky material expressed as a percentage	
D_1	Internal diameter of the case or the crown of the core drill	m
D_2	External diameter of the case, crown at the end of the core drill	m
D_3	Internal diameter of the sheath or tube of the core drill, in millimeters	m
D_4	External diameter of the coring sheath or tube	m
$\delta\chi$	Deformation of the pressure sensor	m
$\delta\sigma$	Deformation of the ground at the level of the pressure sensor	m
$\delta\chi$	Deflection of the sensitive surface of the pressure sensor	m
ΔA	Reading of the detachment pressure of probe membrane in open air (DMT)	kPa
ΔB	Reading of the pressure for a membrane advancement of 1.1 m in open air (DMT)	kPa
$\Delta\sigma\iota$	Stacking increment of the plate	m
$\Delta\sigma\iota$	Pressure increment on the plate	kPa
e	Void ratio	
e_0	Initial void ratio	
e_1	Total stacking of the first load of the plate	m
E	Elastic modulus of the soil; Young's modulus	kPa
E^e	Elastic modulus of the soil (PMT)	kPa
E_{oed}	Odometric modulus of the soil	kPa
E_u	Undrained elastic modulus of the soil; Young's modulus	kPa
E	Drilling energy (drilling parameter)	kW
E_D	Elastic modulus of the soil (DMT)	kPa
E_M	Pressiometric modulus of the soil (PMT)	kPa

E_n	Elastic modulus at level n of the pile	kPa
E_v	Vertical elastic modulus of the soil	kPa
F	Safety coefficient	
F_{max}	Maximal push force (drilling parameter)	MN
F	Hammer impact frequency (drilling parameter)	Hz
FR	Fragmentation coefficient of a rock, expressed as a percentage	
Φ	Friction angle of the soil	degree
Φэ	Effective friction angle of the soil	degree
$\Phi_{\chi\varpi}$	Angle at the critical or residual state	degree
Φ_{μ}	Intergranular friction angle	degree
Φ_{ρ}	Residual friction angle	degree
Φ_{υ}	Undrained friction angle of the soil	degree
Γ	Elastic shearing modulus	kPa
γ	Specific weight of the soil	kN/m^3
γ	Empirical drilling parameter (drilling parameter)	
γ_{δ}	Dry specific weight of the soil	kN/m^3
γ_{M}	Specific weight of the drilling fluid (drilling parameter)	MN/m^3
$\gamma_{O\Pi N}$	Specific weight of the soil at the Proctor optimum	kN/m^3
γ_{σ}	Specific weight of the grains of the soil	kN/m^3
$\gamma_{\sigma\alpha\tau}$	Saturated specific weight of the soil	kN/m^3
h	Load variation due to sampling or injection (Lefranc)	M
h_p	Height of the water in the well (well)	M
H	Length of penetration of the core drill, from the bottom of the bore	M
H	Distance from the injection cavity to the impermeable surface (Lefranc)	M
H_{ini}	Initial overload or discharge (Lefranc)	M

H_i	Load at injection ("*Géomécamètre*" test)	M
H_{max}	Maximal retaining force (drilling parameter)	MN
H_p	Load at pumping ("*Géomécamètre*" test)	M
H_w	Distance from the injection cavity to the water table (Lefranc)	M
i	Hydraulic gradient	
I_A	Calculated alteration index (drilling parameter)	
I_c	Consistency index	
i_{crit}	Terzaghi critical hydraulic gradient	
IC	Net coring index	
I_D	Compactness index	
I_D	Material index (DMT)	
I_f	Fracturing index	
I_p	Plasticity index	
k	Permeability coefficient of the soil	m/s
k_h	Horizontal permeability coefficient of the soil	m/s
K_c	Sensor influence coefficient	
K_D	Horizontal constraint index (DMT)	
K_s	N_s/M_s ratio	
k_s	Reaction modulus of the soil	kPa/m
K_W	Westergaard modulus of the soil	
K_0	Coefficient of ground at rest	
k_0	Empirical parameter based on the drilling results of the calibration test (drilling parameter)	
K_1	Empirical parameter based on the drilling results of the calibration test (drilling parameter)	
κ	Slope of the compression curve–elastic decompression (compression.isotrope; Ln scale)	λ
l_g	Length of the guard cell of the pressure meter probe (PMT)	M

l_s	Length of the central cell of the pressure meter probe (PMT)	M
L	Length of the injection cavity, of the sieve (Lefranc ; well)	M
LA	Los Angeles coefficient	
L_b	Length of the modified or lost sample at its inferior part	M
L_e	Length of the shortening of the sample during coring	M
l_e	Length of the probe ("*Géomécamètre*" test)	M
L_g	Gross length of the sample, after retrieval of the core drill, including the sections modified or lost at the extremities	M
L_n	Net length of the sample after conditioning	M
L_t	Useful length of the coring tube	M
λ	Lamé coefficient	kPa
λ	Slope of the blank compression curve (compression, isotrope; Ln scale)	
m	Coefficient of the shape of the cavity (Lefranc)	
m_0	Corrected coefficient of the shape of the cavity (Lefranc)	
m_{td}	Total dry mass	Kg
m_{th}	Total wet mass	Kg
M	Mass of the hammer (dynamic penetrometer)	Kg
M	Drained vertical elastic modulus of the soil (DMT)	kPa
M'	Mass struck: anvil + string of rods (dynamic penetrometer)	Kg
M_c	Elastic modulus of the sensor	kPa
M_E	Soil modulus during VSS plate test	kPa
M_s	Elastic modulus of the soil	kPa
M_{DE}	Micro Deval coefficient	
μ	Lamé coefficient	kPa
n	Porosity	Kg
n	Dilatancy ratio (PMT)	
N	Friction ratio (PMT)	

N_s	Deformation coefficient of the soil, pressure sensor	
N_{dh}	Number of blows to produce a penetration of h (dynamic penetrometer)	m^{-1}
$N\ni_{70}$	Number of blows (SPT)	
ν	Poisson's coefficient	
ν_υ	Undrained Poisson's coefficient of the soil	
OCR	Overconsolidation ratio	
p	Pressure applied by the pressuremeter probe on the terrain (PMT)	kPa
p	Hydraulic pressure measured in the feed motor or in the jack (drilling parameter)	MPa
p_{CR}	Hydraulic pressure in the rotation motor (drilling parameter)	MPa
p_e	Inherent resistance of the pressuremeter probe, after correction (PMT)	kPa
p_e	Inherent resistance of the packers (Lefranc)	kPa
p_{el}	Inherent resistance limit of the pressuremeter probe (PMT)	kPa
P_E	Calculated net push applied to the drilling tool (drilling parameter)	MPa
p_f	Creep pressure of the pressuremeter test (PMT)	kPa
P_F	Pressure of the drilling fluid measured at the level of the pump exit (drilling parameter)	MPa
p_g	Inflation pressure of the packers (Lefranc)	kPa
p_H	Back pressure measured (drilling parameter)	MPa
P_H	Back pressure measured (drilling parameter)	MPa
p_{Hmax}	Maximal back pressure measured (drilling parameter)	MPa
p_i	Water pressure at injection ("*Géomécamètre*" test)	kPa
P_I	Pressure of the drilling fluid calculated at the level of the exit of the drilling tool (drilling parameter)	MPa
p_j	Water pressure at the injection threshold j (Lugeon)	kPa
P_M	Power of the hammer (drilling parameter)	MW

p_{max}	Maximum water pressure at injection (Lefranc)	kPa
p_{max}	Maximum push pressure in the feed motor or the jack for a force of F_{max} (drilling parameter)	MPa
P_{max}	Maximum push on a tool measured in the terrain (drilling parameter)	MPa
P_O	Gross calculated push applied to the drilling tool (drilling parameter)	MPa
p_p	Water pressure at pumping ("*Géomécamètre*" test)	kPa
p_0	Membrane detachment pressure (DMT)	kPa
p_1	Pressure for a membrane advancement of 1.1 mm (DMT)	kPa
p_1	Pressure at the start of the pseudo elastic range (PMT)	kPa
p_2	Pressure at the end of the pseudo elastic range (PMT)	kPa
p_{LM}	Pressiometric pressure limit of the terrain (PMT)	kPa
p_{LM}^*	Net pressiometric pressure limit of the terrain (PMT)	kPa
p_m	Inherent resistance of the membrane of the central measurement cell (PMT)	kPa
p_r	Pressure read on the pressure volume controller (PMT)	kPa
q_c	Resistance to the static penetration of the cone (CPT)	kPa
q_d	Resistance to the dynamic penetration of the cone (PDA, PDB)	kPa
q_{net}	Ultimate resistance of the soil under a foundation	kPa
q_{sn}	Mean unitary lateral friction of section n of the pile	kPa
Q	Water flow injected or pumped into the bore (Lefranc)	m^3/s
Q_a	Constant water flow pumped or injected into the bore (Lefranc)	m^3/s
Q_C	Pile creep critical load	kN
Q_e	Maximum load applied during the pile control test	kN
Q_i	Water flow at injection ("*Géomécamètre*" test)	m^3/s
Q_p	Water flow at pumping ("*Géomécamètre*" test)	m^3/s
Q_t	Normalized resistance to penetration (CPT)	

Q_{LE}	Conventional rupture load of the pile	kN
Q_{ELS}	Load at the service limit state of the pile	kN
Q_{ELU}	Load at the ultimate limit state of the pile	kN
Q_G	Elastic limit load of the constitutive materials of the pile	kN
Q_j	Flow injected at threshold j (Lugeon)	m^3
Q_L	Limit load of the pile	kN
Q_{max}	Maximum load for the pile loading test	kN
Q_n	Effort in the straight section n of the pile	kN
Q_N	Nominal load of the pile	kN
Q_T	Total effort on the cone of the penetrometer	kN
Q_u	Effort exerted by the interstitial pressure on the superior part of the cone (penetrometer)	kN
Ψ	Dilatancy angle	degree
r_p	Well radius (well)	m
R	Radius of the plate	m
R	Radius of influence of the well (well)	m
R_c	Resistance to simple compression	kPa
R_e	Recording ratio σ_c/σ_s, pressure sensor	
R_f	Friction ratio (CPT)	
R_p	Resistance to penetration (drilling parameter)	s
R_{SR}	Calculated soil-rock resistance (drilling parameter)	MPa/m/s
RQD	Cracking coefficient of a rocky sample	kPa
R_{TB}	Resistance to Brazilian traction	Pa
ρ	Wet density	kg/m^3
ρ_δ	Dry density	kg/m^3
$\rho_{\iota\varphi}$	Correlation coefficient between variables i and j	
$\rho_{O\Pi N}$	Specific weight of the soil at the Proctor optimum	kg/m^3

ρ_σ	Specific weight of the grains of soil	kg/m^3
$\rho_{\sigma\alpha\tau}$	Saturated specific weight of the soil	kg/m^3
S	Packing or deflection	m
S	Well drawdown (well)	m
S_{air}	Matrix sucking of air intake or output	kPa
S	Straight section of the pile	m^2
S	Straight section of the bore (Lefranc)	m^2
S_d	Calculated Somerton index	
S_e	Surface area of the vents at the head of the core drill	m^2
S_O	Section of soil removed by the tool, considered equal to the section of the drilling tool (drilling parameter)	m^2
S_r	Degree of water saturation	
$S_{lat,n}$	Lateral surface of section n of the pile	m^2
σ_χ	Total constraint on the sensor	kPa
σ_χ	Preconsolidation constraint	kPa
σ_0	Total horizontal constraint at rest	kPa
$\sigma\ni_0$	Effective horizontal constraint at rest	kPa
σ_π	Vertical preconsolidation constraint	kPa
σ'_π	Effective vertical preconsolidation constraint	kPa
σ_ρ	Total radial constraint (PMT)	kPa
$\sigma\ni_\rho$	Effective radial constraint (PMT)	kPa
σ_σ	Total constraint on the soil at the level of the sensor	kPa
$\sigma_{\varpi 0}$	Total vertical constraint at rest	kPa
σ_ϖ	Total vertical constraint	kPa
$\sigma'_{\varpi 0}$	Effective vertical constraint at rest	kPa
σ'_ϖ	Effective vertical constraint ("*Géomécamètre*")	kPa
σ_θ	Total circumferential constraint (PMT)	kPa

$\sigma \ni_\theta$	Effective circumferential constraint (PMT)	kPa
σ_ζ	Total vertical constraint (PMT)	kPa
$\sigma \ni_\zeta$	Effective vertical constraint (PMT)	kPa
T	Time	s
T	Tare	kg
T	Hydraulic transmissivity (well)	m^2/s
TC	Coring percentage	
τ_f	Resistance to shearing, in kilopascal	kPa
u	Interstitial pressure	kPa
u_0	Interstitial pressure at ground rest	kPa
U_L	Permeability of the rock in Lugeon units	m^2/s
V_A	Advancement velocity (drilling parameter)	m/h
V_{BS}	Value of methylene blue	
V_c	Volume injected to ensure contact with the calibration tube (PMT)	cm^3
V_{max}	Theoretical maximal advancement velocity measured in the field, allowing normalization of the real advancement velocity (drilling parameter)	m/h
V_r	Volume read on the pressure volume controller (PMT)	cm^3
V_s	Volume of the measurement cell (PMT)	cm^3
W	Water content	
w_L	Liquidity limit	
w_{nat}	Natural water content	
w_p	Plasticity limit	
w_r	Removal limit	
$W_{drill\ head}$	Weight of the drill head (drilling parameter)	MN
W_{rod}	Weight of a drilling rod (drilling parameter)	MN
ξ	Length of the plastic deformation	

z_c	Altimetric level of point zero of the pressure volume controller (PMT)	m
z_m	Altimetric level of the injection pump (Lefranc)	m
z_M	Level of the water table (drilling parameter)	m
z_s	Altimetric level of the probe (PMT)	m
z_2	Stacking of the second plate loading	m
Z_f	Depth of the extremity of the core drill after coring and before retrieval of the tool, in relation to the natural terrain	m
Z_i	Depth of the bottom of the borehole before coring and depth at the start of sampling in relation to the natural terrain	m
Z^0	Decimetric description of the sample	m
Z^+	Centimetric description of the sample, rough description of the sample	m

Introduction

> To be a man is, precisely, to be responsible. It is to feel shame at the sight of what seems to be unmerited misery. It is to take pride in a victory won by one's comrades. It is to feel, when setting one's stone, that one is contributing to the building of the world.
>
> Antoine de Saint Exupéry
> Terre des Hommes, Chapter 2

For a long time, *in situ* geotechnical tests have suffered from a lack of credibility among elite scientific academics who have always preferred laboratory tests in which the test conditions are well known and the measures are well controlled. This vision might suggest a misleading perfection of laboratory tests for which we forget that the soil has its own history before reaching the laboratory. It is subjected to sampling, transport and conservation conditions that transform it so much so that it arrives more or less altered on the triaxial press or the oedometer.

This work is a plea to rehabilitate *in situ* geotechnical tests and we will show that they not only allow us to understand all the mechanical quantities that are measured in the laboratory, but that they also allow us to go further and find data that are inaccessible to laboratory tests.

1

Measuring Water Content and Density

1.1. Sample collection method

1.1.1. *Measuring water content using the frying pan method*

This method of measuring water content is carried out on site. It is subject to a standard [AFN 96c]. The procedure is as follows:

– weigh an empty and dry container: tare T;

– yake the soil sample to be studied;

– weigh the container filled with the soil to be tested: the total wet mass m_{th};

– heat the soil enough to dry it out completely but without breaking it down (for clay). A Bunsen burner, camping gas or methylated spirits can be used…;

– weigh the container with the dry soil: the total dry mass m_{td}.

The water content of the soil w is then given by the formula:

$$w = \frac{m_{th} - m_{td}}{m_{td} - T} \quad [1.1]$$

1.1.2. *Measuring water content using the oven-dry method*

This method is carried out in a laboratory and is subject to a standard [AFN 95c]. The approach is the same as before, but the soil is heated using an electric oven at 105°C for a minimum of 4 and 24 h if possible. This method enables:

– the prevention of clay decomposition;

– the evaporation of all the free water contained within the soil.

It is also possible to use a microwave oven. This process is subject to a standard [AFN 96b].

1.3.1. *Measuring dry density using a membrane densitometer*

This measurement is carried out in situ with a membrane densitometer or a sample of soil on a core taken from the soil. In both cases, a wet density is measured and then converted into a dry density as follows:

– the water content of the soil taken is measured using one of the previous methods, with w as the water content;

– a sample of soil of mass m_h is taken, which then breaks down into a grain mass m_s and a mass of water m_w [1.2]. If the water content w is known, the water mass of the sample can be determined [1.3]. If the volume of the sample V is known, the wet density can be determined [1.4] and the dry density deduced [1.6]. Measuring the volume is done following the membrane densitometer method or using the external volume of the core.

$$m_h = m_s + m_w \qquad [1.2]$$

$$m_w = w.m_s \qquad [1.3]$$

$$\rho = \frac{m_h}{V} = \frac{m_s}{V} + w.\frac{m_s}{V} \qquad [1.4]$$

$$\rho = \rho_d.(1 + w) \qquad [1.5]$$

$$\rho_d = \frac{\rho}{(1+w)} \qquad [1.6]$$

The densitometer is a device that measures cavity volume (Figure 1.1). To use it, a hole is dug in the soil whose volume is to be determined. The device is first filled with water and a membrane is placed at its base. The procedure is as follows:

– the densitometer is positioned at the surface of the ground to be tested after some surface stripping;

– the piston is pressed so that the flexible membrane is applied onto the ground surface and an initial volume reading is taken;

– the superior part of the device is removed, leaving the perforated disk (on the left in Figure 1.1);

– a hole is made in the ground to match the diameter of the densitometer;

– the extracted soil is weighed and its water content determined;

– the densitometer is repositioned back onto the perforated disk;

– the membrane is pushed into the hole by pushing on the densitometer's piston, and a final volume reading is taken;

– the volume of the hole is determined by the difference between the two volumes;

– the wet density is determined following the formula [1.4] and the dry density following formula [1.6].

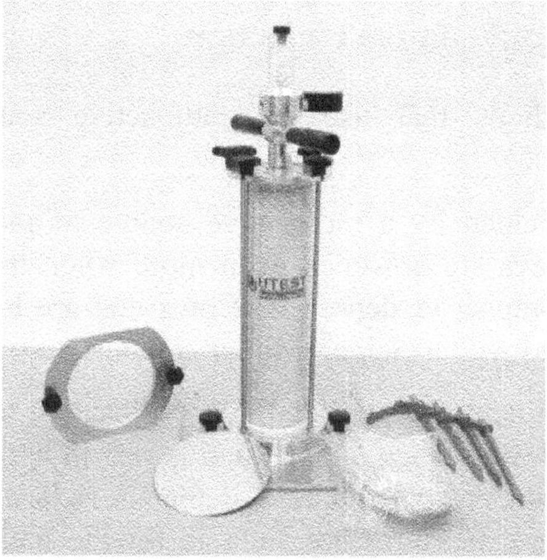

Figure 1.1. *Membrane densitometer (taken from www.lindqvist-international.com)*

1.1.4. *Measuring dry density using the excavation method*

This method consists of making a hole on the surface of the ground, measuring the total wet weight and the water content of the extracted soil. The hole is then lined and filled with water, measuring the volume introduced, giving the volume occupied by the soil. Formula [1.6] gives the dry density.

Figure 1.2. *Measuring the density of surface waste a) excavation by shovel b) filling the excavated and lined pit using a water tank, ADEME [ADE 05]*

1.2. Method without sample collection

These are methods that use the radioactive sources listed below [AGE 03]:

– γ rays are provided by a radioactive source of radium, beryllium or cesium 137. The rays are absorbed even more when the medium traveled through is more compact or denser. The rays that are backscattered by the ground are measured with a Geiger–Müller counter.

– Rays from fast neutrons provided by a radioactive source of americium 241, beryllium is slowed when the medium is rich in hydrogen atoms. Since water is the material present in the soil that is richest in hydrogen, it is considered that these measures are suitable for the measurement of water content in the soil. Slow neutron detectors enable the measurement of the moisture in the environment.

Measurements are taken from backscatter or surface transmission, from drilling, or between two boreholes. The device is calibrated beforehand for reference materials whose compactness or water content is known. The difference in receiving γ rays or neutrons between reference materials and the soil is put into a chart that then provides the density or water content of the soil. The volume tested is in the order of dm^3.

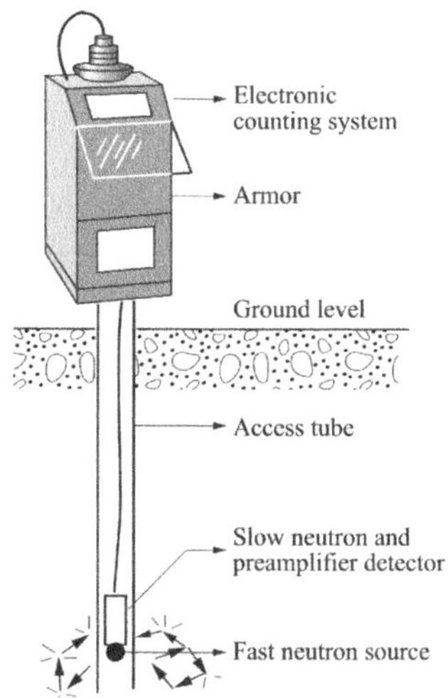

Figure 1.3. *Deep neutron probe (from Campbell Pacific Nuclear)*

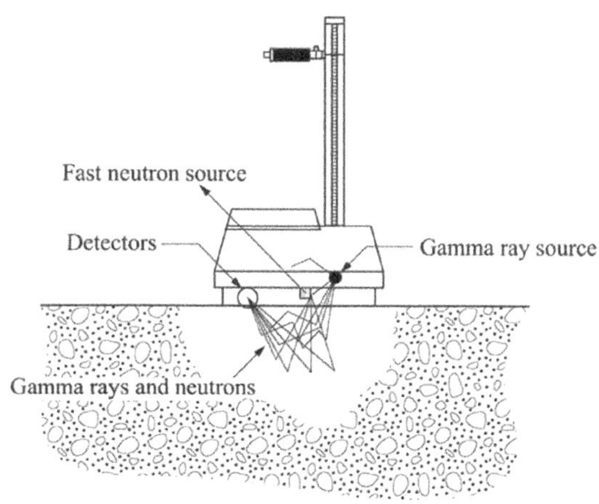

Figure 1.4. *Probe in position for measuring the water content and density of the soil surface (from CPN MC-3 Portaprobe)*

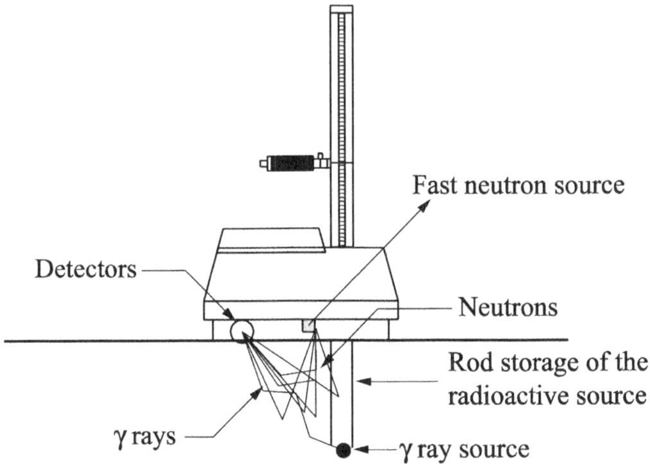

Figure 1.5. *Probe in position for measuring the water content and density of the layer between the radioactive source and the detector (from CPN MC-3 Portaprobe)*

2

Soil and Rock Sampling Methods

2.1. Sampling classes and nomenclature

Sampling classes define, *a priori*, geotechnical characteristics that are measurable from samples. The main elements that determine the quality of a sample depend on:

– the nature of the material (clay, sand, granular soil, heterogeneity and rock);

– the consistency of the material, whether it is soil (Table 2.1) or rock (Table 2.2);

– the technique used to collect the sample, taking into account the drilling machine, the sampler and the use of its;

– the competence and expertise of the personnel.

Soil	Simplified test	Undrained cohesion c_u (kPa)
Very soft	Escapes through fingers with some pressure	< 20
Soft	Can be kneaded with a little pressure from fingers	20 to 40
Firm	Can be kneaded with strong pressure from fingers	40 to 75
Stiff	Cannot be kneaded with fingers, thumb leaves an imprint	75 to 150
Very stiff	Scratched with a fingernail	150 to 300
Hard		> 300

Table 2.1. *Classification of cohesive soils [AFN 95d]*

Description	Unconfined compression strength R_c (MPa)
Very weak resistance	< 6
Weak resistance	6 to 20
Average resistance	20 to 60
High resistance	60 to 200
Very high resistance	> 200

Table 2.2. *Classification of rocks [AFN 95d]*

The classes are numbered from 1 to 5 in decreasing order of the number of measurable geotechnical parameters (Table 2.3) with the previously mentioned notations.

Classes	Main parameters obtained from the sample	
	Nature characteristic	State characteristic
1	D, w_L, w_P, e_{max}, e_{min}, ρ_s, MO, C_c, C_s, Z^+	Z^+, I_D, e, W_{nat}, ρ_d, τ_f, [ϕ', c', c_u, ϕ_{uu}, c_{uu}], E, R_c, RTB, RQD, σ'_P, k, c_v
2	D, w_L, w_P, e_{max}, e_{min}, ρ_s, MO, Z^+	Z^+, I_D, e, w_{nat}, ρ_d, RQD, k
3	D, w_L, w_P, e_{max}, e_{min}, ρ_s, MO, Z^0	Z^0, w_{nat}
4	D, W_L, W_P, e_{max}, e_{min}, ρ_s, MO, Z^0	Z^0
5	Incomplete sampling, Z^-	Z^-

Table 2.3. *Definition of material sampling classes*

2.2. Sampling techniques

Sampling techniques use several methods:

– wells, trenches, excavation and tunnels: mechanically or manually dug;

– manual drilling: done by hand with an auger;

– punch drilling: the core barrel is plunged without rotation, either by percussion (frequency greater than 2 Hz, compression and extension effort), by beating (frequency less than 2 Hz) or by drilling at a constant speed (greater than 2 cm/s);

– rotation drilling: the core drill can be cooled by a drilling fluid (water, air, bentonite slurry, etc.).

Drilling can be classed into three categories:

- core drilling;
- semi-destructive drilling;
- destructive drilling.

Left to itself, a borehole in the ground will always close after a certain amount of time. Boreholes must, therefore, be made by maintaining the walls, either with metal casing or with the introduction of sludge, the static pressure of the mud and then balancing the horizontal pressure of the soil.

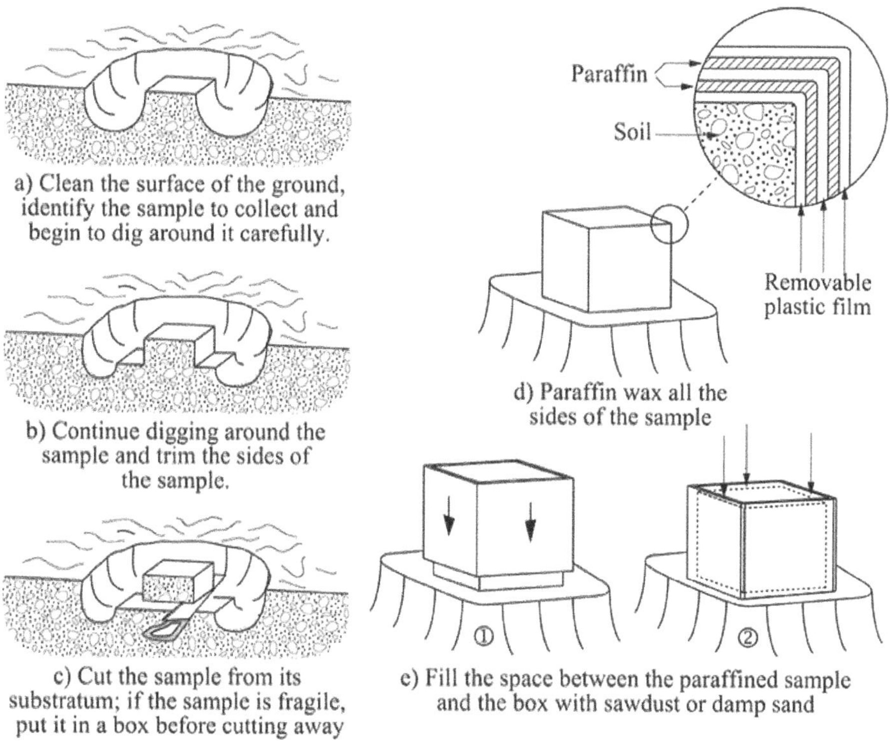

Figure 2.1. *Example of sampling blocks by hand [AFN 95d]*

2.2.1. *Manual samples*

Blocks of soil or firm rocks can be collected by hand in excavations. If all the necessary care is given to sizing, packaging, transport and storage, these blocks can provide samples of very high quality (class 1). An example of collecting a sample is shown in Figure 2.1).

The shape of the box is cubic or parallelepiped, of decimetric dimension and is often removable in order to enable the extraction of samples in the laboratory.

The box can consist of two semi-cylindrical shells with sharp lower edges. If it is plunged under pressure, its area index is such that C_a<=15%.

During transport and storage, the removable bases are sealed and unaffected by water (Figure 2.2).

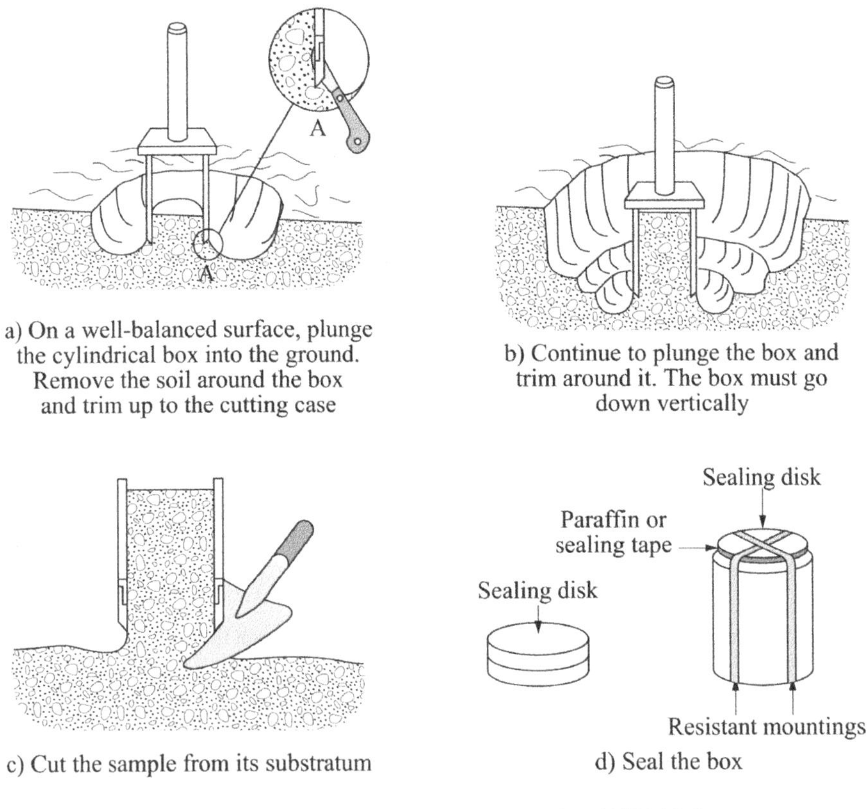

Figure 2.2. *Example of manual sampling [AFN 95d]*

The characteristic symbols of samplers are shown in Figure 2.3, while the characteristic symbols of samples are shown in Figure 2.4.

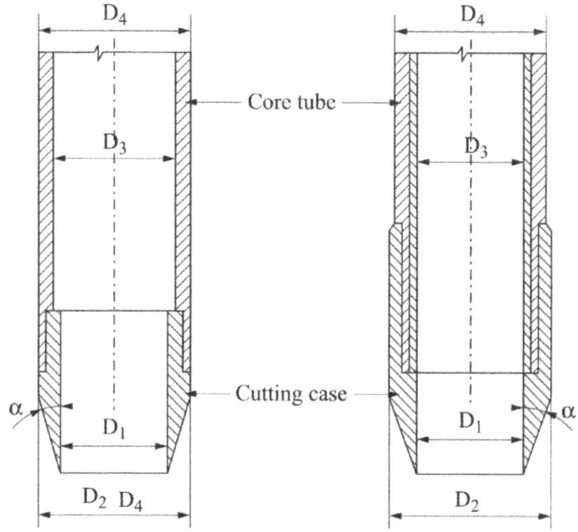

Figure 2.3. *The notations used for drilling diameters [AFN 95d]*

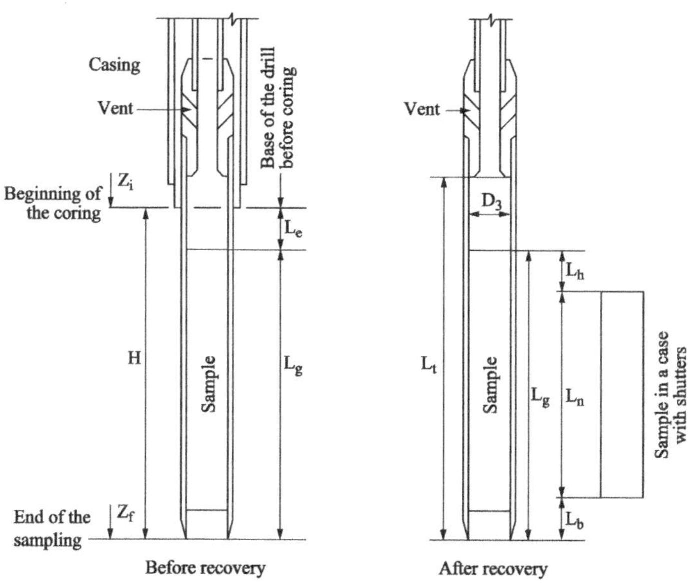

Figure 2.4. *The notations used for length of the samples [AFN 95d]*

2.2.2. Core drilling

2.2.2.1. *Punch core drilling*

The punch drill variety is shown in Figure 2.3. Drilling is done by jacking, beating or vibro sinking. The core of soil is maintained in place during the ascent of the tool.

Punch drills are characterized by the following parameters:

– area index of a core drill C_a [2.1];

– backlash index inside a core drill C_i [2.2];

– backlash index outside a core drill C_0 [2.3];

$$C_a = \frac{D_2^2 - D_1^2}{D_1^2} \quad [2.1]$$

$$C_i = \frac{D_3 - D_1}{D_1} \quad [2.2]$$

$$C_0 = \frac{D_2 - D_4}{D_4} \quad [2.3]$$

– the angle of the bevel of the cutting case α (Figure 2.3);

– the section S_e of superior exhaust ducts for air and water;

– the maximum ratio of the net length of the core drill by the inner diameter;

$$\frac{L_t}{D_1} \quad [2.4]$$

– the possible presence of an extractor in the inferior part of the core drill with partial (fingers and flexible blades) or full closure (profiled plugs and inflatable pressure sleeve);

– net core drill index IC [2.5];

– the core drill percentage TC [2.6].

The values imposed by these parameters are shown in the from Tables 2.4 to 2.15.

$$IC = \frac{L_n}{H} \qquad [2.5]$$

$$TC = \frac{L_g}{D_1} \qquad [2.6]$$

The punch core drill with thin walls and no inner casing (Figure 2.5) is designed to meet the requirements of a geotechnical sample. It must have an area index C_a of less than or equal to 15%. The tube serves as a casing during transport and storage. In the laboratory, the sample is extracted either by cutting the tube following two generators, or by jacking in the direction of the existing ingress. Furthermore, the cutting head must have water evacuation vents that provide minimal pressure loss and a special valve ensuring absolute tightness, which is essential in certain cases, such as for the collection of soil boiling.

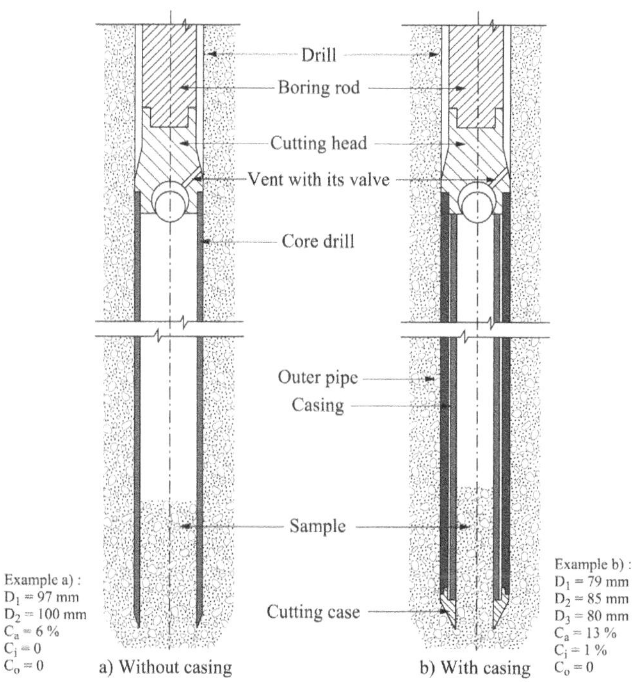

Figure 2.5. *Example of a punch drill with thin walls [AFN 95d]*

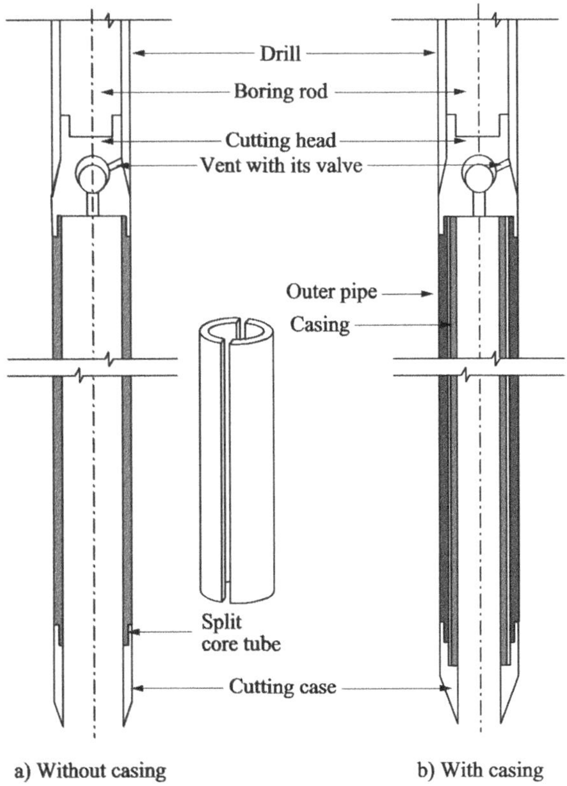

Figure 2.6. *Example of a punch core drill with thick walls [AFN 95d]*

The core drill with thin walls and an inner casing (Figure 2.5) is designed with the same principles as the previous core drill. It must have an area index C_a less than 15%. The pipe houses a cylindrical metal or polymer casing for transport and storage.

The core drill with thick walls (Figure 2.6) is a robust ordinary steel device that has walls 8 mm in thickness. It can have an area index C_a more than 15%. It can be equipped with a simple thin casing that provides it with a high wall coefficient. In some cases, it is possible to equip it with a thin casing calculated from current standards. It is put in place by beating into sandy or gravelly soils.

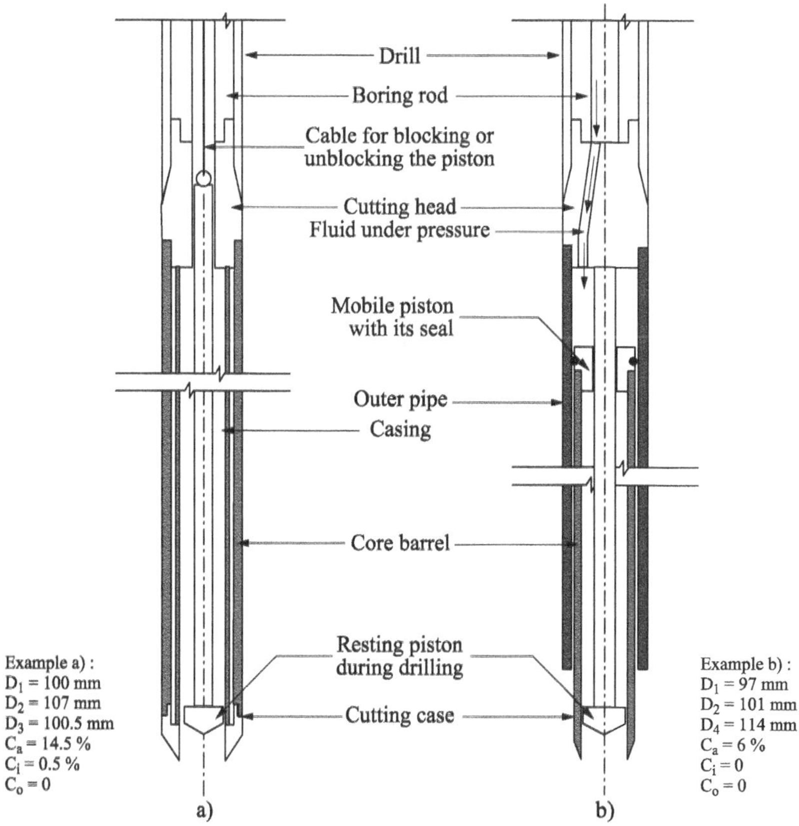

Figure 2.7. *Example of a stationary piston core drill [AFN 95d]*

The stationary piston corer (Figure 2.7) is a sampler with thin walls and, in general, an inner casing in which there is a piston. The latter is locked into a low position and obstructs the entrance of the drill during the installation phase up until the coring level. The piston is then disengaged and kept fixed from the surface, during the core drilling phase. When the desired depth is reached, the piston and the sampling tube are secured again and the core is recovered with the sample. Sealing between the piston and the casing is ensured by the piston seal.

Bell samplers (Figure 2.8) are suitable for collecting samples in sand and granular waterlogged soils. The difficulty with sampling these drills is the lack of cohesion. However, it is possible to create conditions such that under capillary force, the ground acquires a noticeable cohesion. The collection of a sample is done by classic hammering using sampling tubes that meet the current regulations. After sinking the tube, it is recovered in an air pocket that removes the sample from water contact. It is necessary to take sufficient precautions to avoid any shock during its recovery and handling.

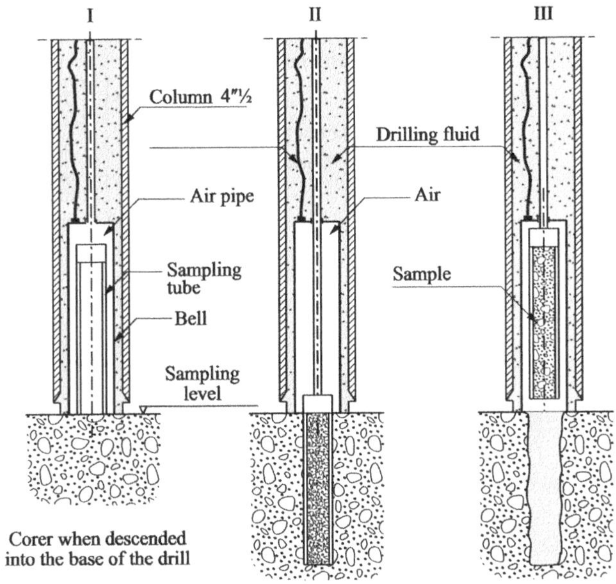

Figure 2.8. *Example of a bell core drill [CAS 81]*

2.2.2.2. *Rotary core drilling*

Twist samplers are tubular tools ending in a crown that cuts through soil or rock by rotation. An interstitial fluid lubricates the tool and removes debris. It is injected continuously through the head and leads to the crown. Twist samplers allow the collection of samples from lands of the marl or compact clay variety.

The simple industrial corer (Figure 2.9(a)) has a length less than or equal to 500 mm. It does not have an extractor. The fluid injected circulates between the core and the tube. However, samples thus obtained are sometimes far from correct because during drilling the core is in contact with the lubrication water, especially between the inner pipe which is found further back relative to the cutting plate of the crown and the extremity of the corer.

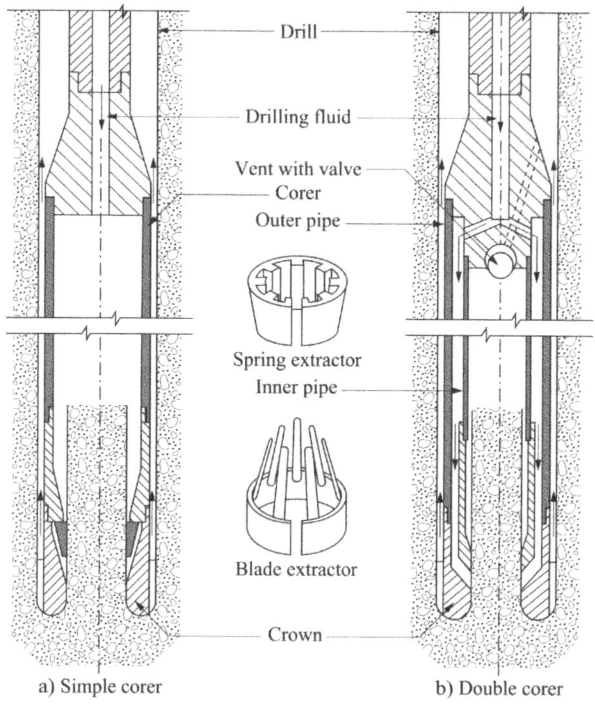

Figure 2.9. *Example of a twist drill [AFN 95d]*

The simple mining corer is fitted at its base with a conical extractor for sectioning the core and maintaining it as it is recovered.

The double corer (Figure 2.9(b)) comprises not only a rotating outer pipe carrying the crown, but also a free inner pipe mounted on a pivot in which a sample is stored during its collection. The inner pipe does not turn during coring. The injected fluid circulates between the two pipes.

The triple corer is a double corer whose inner pipe is equipped with a removable casing used for collecting the sample gradually during extraction. A double corer can be changed into a triple corer by adding a casing and replacing the original crown by a crown of smaller diameter to take into account the thickness of the casing.

The triple corer with a protruding cutting case (Figure 2.10) is a sampler whose inner pipe is equipped with a cutting case that extends past the crown. When drilling, the case collects the sample by punching in the same manner as a corer with a thin wall and the crown removes the material at the back of the cutting case. Due to a spring located in the cutting head, regulating the overrun of the case is done automatically according to the resistance of the ground. Beyond a fixed limit, the case retracts and places itself in abutment with an overrun of around 2 cm. The geometric characteristics of punch drills with thin walls apply to the entire case-inner pipe as well as the vents. For lands of consistent hardness ranging from plastic clay to the chalk of Parisian regions, this core drill is successfully used. This device has a negligible surface index.

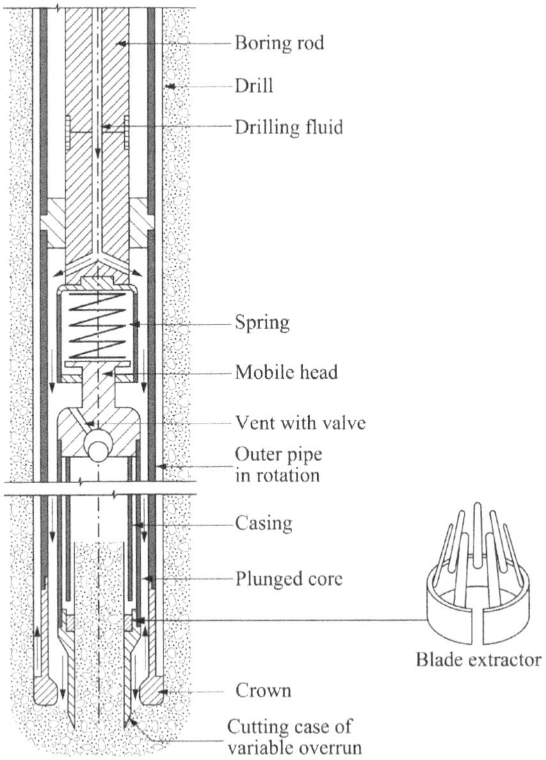

Figure 2.10. *Example of a triple rotary core drill [AFN 95d]*

More generally, double corers are recommended for loose ground, soft rock, altered rock and in soils likely to be disrupted by drilling fluid (water or slurry).

Simple core drills, without inner corers, are recommended for sound rock.

2.2.3. *Semi-destructive drills*

The process that allows the extraction of soil in a continuous manner in an amended form is known as semi-destructive.

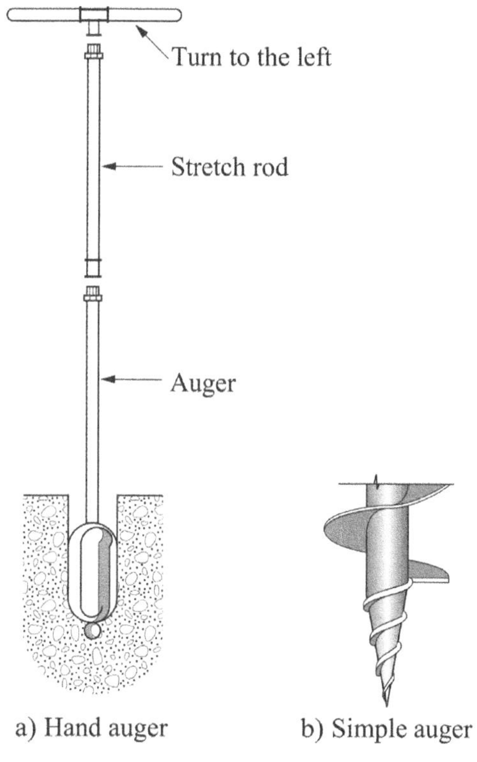

Figure 2.11. *Hand auger and simple auger [AFN 95d]*

2.2.3.1. *Hand augers*

Drilling is done by hand with an auger driven by a turn to the left. Deep reconnaissance is limited to several meters in fine soils. The presence of gravel makes it impossible to use.

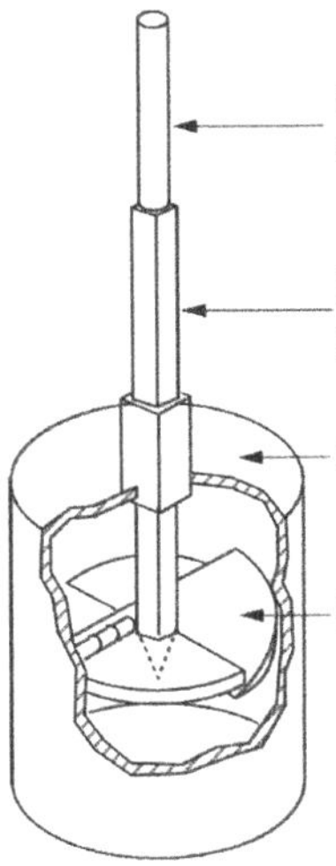

Figure 2.12. *Rotary bucket with mobile auger [AFN 95d]*

2.2.3.2. *Simple augers and rotary buckets*

The tool is driven by a rod operated from the surface. They are limited in depth to a reconnaissance of several meters in fine soils.

2.2.3.3. *Drilling with continuous flight auger*

This drilling is done with a motorized auger that drills into the ground using a helical screw system, of the Archimedes screw variety (Figure 2.13). The ground, therefore, remains within the turns of the helix and can be collected when the auger is recovered. The reconnaissance depth can reach several dozen meters.

2.2.3.4. *Valve drilling*

This is a punch corer, with a valve system (Figure 2.14) for retaining soil.

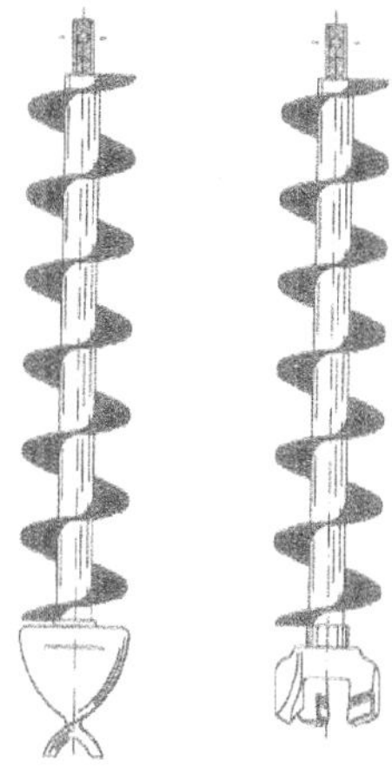

Figure 2.13. *Continuous flight auger [AFN 95d]*

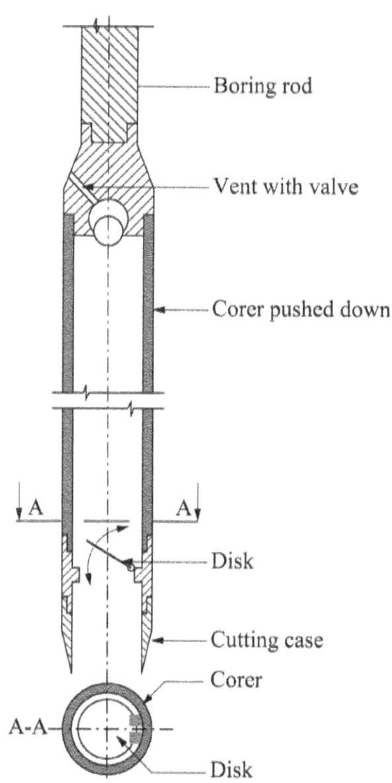

Figure 2.14. *Core drill with valve [AFN 95d]*

2.2.3.5. *Clamshell grab and spoon sampler*

The spoon sampler (Figure 2.15) is equipped with a simple auger that disintegrates the soil in order to introduce it into the sampler.

The clamshell bucket (Figure 2.16) is driven by a crane. It collects soil with two jaws enabling the ascent of the disturbed soil to the surface.

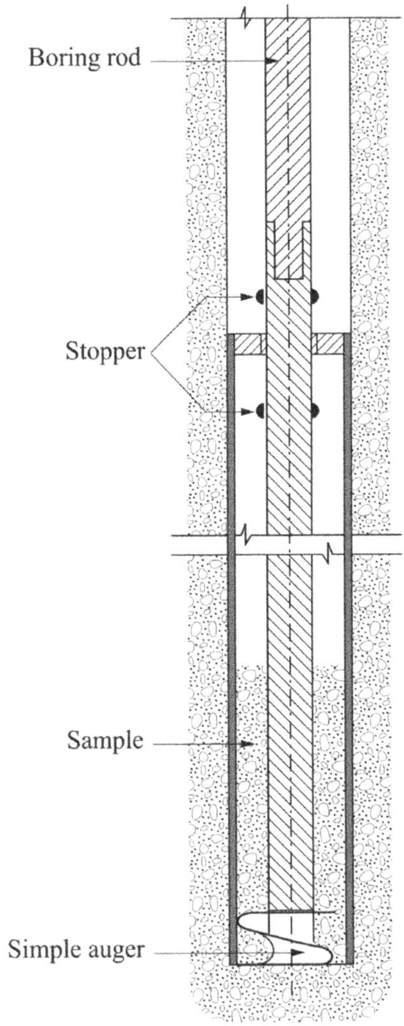

Figure 2.15. *The spoon core drill [AFN 95d]*

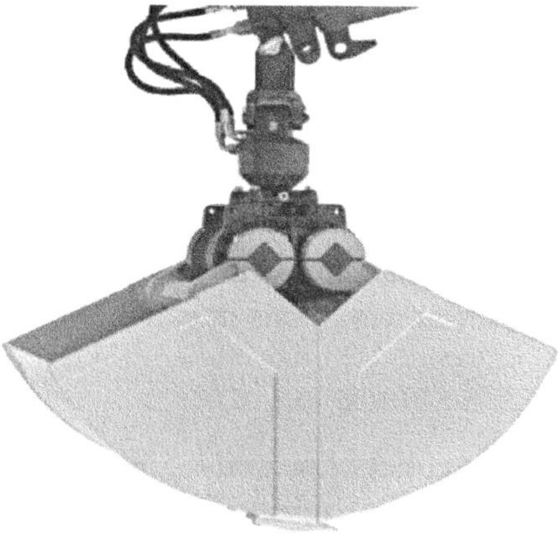

Figure 2.16. *Example of a clamshell grab*

2.2.4. *Destructive drills*

This drilling is done by hammering, rotation or vibro sinking of a drill bit. Drilling fluid, water, air or slurry is injected into the drill and transports the sediment to the surface. It is possible to measure the drilling parameters in order to try and connect them to the geology encountered. In this case, the parameters measured include:

- the torque of the tool;
- the forward speed;
- the pressure of the lubricating fluid;
- the vertical force on the tool;
- the rotation speed of the tool.

Unless otherwise indicated, the holes must be resealed with a sounder.

2.3. The procedures

The procedures followed are not arbitrary. They are now defined by a standard [AFN 95d]. To guarantee the quality of samples obtained, the recommended procedures must be adhered to.

2.3.1. *Sample collection methods*

2.3.1.1. *Ground sample collection*

Previous examples have been given (Figure 2.1 and 2.2). In the case of the box, the free space between the sample and the internal walls must be filled with paraffin or any other material to ensure the conservation of the sample and preserving it from variations in water content.

2.3.1.2. *Punch core drill sample collection*

This method is used to collect fine soils. Sinking under continuous force, without interruption or rotation, results in the collection of a sample of superior quality to that collected by hammering. The speed of sinking must be greater than or equal to 2 cm/s. The length of penetration H must always be lower than the length H_t of the corer.

2.3.1.3. *Sample collection with a casing*

The extremities of the sample are trimmed and the disturbed areas removed. For an efficient filling of the extremities, a mix of paraffin and Vaseline showing no withdrawal can be used, or any other O-ring solution (Figure 2.17) or an interlocking cap (Figure 2.18). When the casing becomes deformable, it should be blocked into a rigid shell.

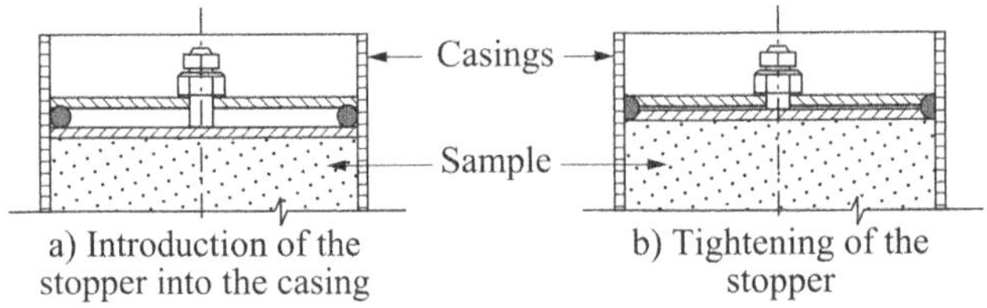

Figure 2.17. *Stopper cap of the O-ring casing [AFN 95d]*

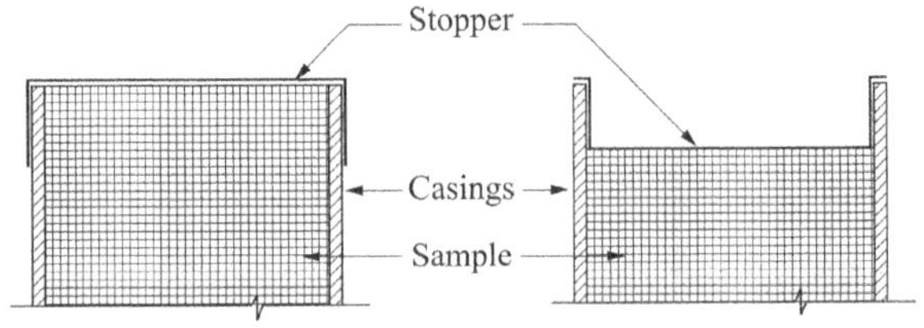

Figure 2.18. *Closing of the casing with the interlocking stopper [AFN 95d]*

2.3.1.4. *Sample collection without a casing*

For class 1 and 2 samples, the core (coherent) is wiped off, scraped, coated in a waterproof film (for example, a mix of paraffin and Vaseline), and then jammed into a rigid cylindrical shell.

For class 3, samples are placed into waterproof bags.

2.3.1.5. *Sample collection with a disintegration tool*

Class 3 samples are placed into waterproof bags. Class 4 and 5 samples are placed into bags that prevent the dispersion of its constituents.

2.3.2. *The choice of sampling technique*

Sampling techniques must be selected according to the soil to be tested. Table 2.4 relates to very soft fine soil, Table 2.5. relates to firm fine soil, Table 2.6 relates to stiff fine soil, Table 2.7 relates to hard fine soil, Table 2.8 relates to clean coarse soil, Table 2.9 relates to clayed sand and loam soil as well as sandy clay and Table 2.10 relates to very heterogeneous soil and rocks.

The use of sampling tools must be determined according to the nature of the soil to be collected and in correlation with the recommended sampling techniques. Table 2.11 relates to ground sampling, Table 2.12 relates to punch core drill sampling, Table 2.13 relates to stationary piston drill sampling, Table 2.14 relates to punch core and twist core drill sampling and Table 2.15 relates to semi-destructive sampling methods.

Sampling material (see Table 2.1)	Sampling techniques				Sampling class	
	Tool name	Use	Fluid	Technique efficiency	Possible optimal class	Comments
Very soft to soft fine soil	- Stationary piston core drill	- Continuous sinking under force ($v \geq 2$ cm/s)	No	Recommended	1	- $D_1 \geq 75$ mm
	- Thin walled core drill with casing	- Continuous sinking under force ($v \geq 2$ cm/s)	Slurry	Adapted	1 2 (if extractor)	- $D_1 \geq 75$ mm - Tubed, cleaned drilling with slurry
	- Auger and rotary bucket	- Rotation	No	Inaffective under water table	3	Class 4 for very soft fine soils

Table 2.4. *Sampling techniques recommended for soft fine soil [AFN 95d]*

Sampling material (see Table 2.1)	Sampling techniques			Sampling class		
	Tool name	Use	Fluid	Technique efficiency	Possible optimal class	Comments
Firm fine soil	- Stationary piston core drill	- Continuous sinkings under force (v ≥ 2 cm/s)	No	Recommended	1	- $D_I \geq 75$ mm
	- Thin walled core drill with or without casing	- Continuous sinkings under force (v ≥ 2 cm/s)	Slurry	Adapted	1	- $D_I \geq 75$ mm
		- Hammering	No	Adapted	2	- $D_I \geq 75$ mm
	- Triple corer with overrunning case	- Rotation	Water or slurry	Adapted	1	- $D_I \geq 75$ mm
	- Triple corer	- Rotation	Water or slurry	Tolerated	2	- $D_I \geq 85$ mm
	- Double corer	- Rotation	Water or slurry	Tolerated	2	- $D_I \geq 85$ mm
	- Auger and rotary bucket	- Rotation	No	Adapted	3	- Simple augers are lifted every 0.3m at most and continuous augers every 1m at most
	- Sampling blocks by hand	- By hand	No	Recommended	1	
	- Sampling blocks by box	- By hand or sinking under force	No	Recommended	1	
	- Loose	- Mechanical shovel	No	Tolerated	3	

Table 2.5. *Sampling techniques recommended for firm fine soil [AFN 95d])*

Sampling material (see table 1)	Sampling techniques			Sampling class		
	Tool name	Use	Fluid	Technique efficiency	Potential optimal class	Comments
Stiff to very stiff fine soils	- Thin-walled corer	- Continuous sinking under force	No	Adapted	1	- $D_I \geq 75$ mm
		- Hammering	No	Adapted	2	
	- Thick-walled corer	- Continuous sinking under force	No	Tolerated	3	- $D_I \geq 90$ mm
		- Hammering	No	Tolerated	3	- $D_I \geq 90$ mm
	- Triple corer with overrunning case	- Rotation	Water or slurry	Recommended	1	- $D_I \geq 75$ mm
	- Triple corer	- Rotation	Water or slurry	Adapted	1	- $D_I \geq 85$ mm
	- Double corer	- Rotation	Water or slurry	Adapted	2	- $D_I \geq 85$ mm class 1
	- Simple corer	- Rotation	No	Tolerated when dry and short run	3	- $D_I \geq 85$ mm class 2 and exceptional use
	- Auger and rotary bucket	- Rotation	No	Adapted	3	
	- Block sampling by hand	- Manual	No	Adapted	1	- To obtain a Z^+ cut, simple augers are lifted every 0.3m and continuous augers every 1m at most.
	- Block sampling by box	- By hand or forced sinking	No	Adapted	1	
	- Loose	- Mechanical shovel	No	Tolerated	3	

Table 2.6. *Sampling techniques recommended for stiff to very stiff fine soil [AFN 95d]*

Sampling material (see Table 2.1)	Sampling techniques				Sampling class	
	Tool name	Use	Fluid	Technique efficiency	Potential optimal class	Comments
Hard fine soils	- Triple corer-with overrunning case	- Rotation	Water	Unadapted to the hardest soils	1	- $D_1 \geq 75$ mm The limitation is due to the non-penetration of the case
	- Triple corer	- Rotation	Water	Recommanded	1	- $D_1 \geq 85$ mm
	- Double corer	- Rotation	Water	Adapted	2	- $D_1 \geq 85$ mm - Class 1
	- Simple corer	- Rotation	Water	Adapted	3	- Class 2
	- Auger and rotary bucket	- Rotation	Water	Adapted	3	Simple augers a raised every 0.3m at most and continuous augers every 1m at most.
	- Block sampling by hand	- Manual	Water	Adapted	1	
	- Block sampling by box	- By hand or forced sinking	Water	Adapted	1	
	- Loose	- Mechanical shovel	Water	Tolerated	3	

Table 2.7. *Sampling techniques recommended for hard fine soil [AFN 95d]*

Sampling materials (see Table 2.1)	Sampling techniques				Sampling class	
	Tool name	Use	Fluid	Technique efficiency	Potential optimal class	Comments
Clean coarse soil Sand	- Stationary triple piston corer	- Forced sinking		Recommended for loose sand	2	- $D_1 \geq 75$
	- Thin-walled corer with casing	- Forced sinking		Recommended	2	- $D_1 \geq 75$ mm
		- Hammering		Recommended	3	- $D_1 \geq 75$ mm
	- Triple corer	- Rotation	Slurry	Recommended	2	- $D_1 \geq 85$ mm
	- Valve corer	- Hammering	No	Adapted	5	
	- Auger-rotary bucket	- Rotation	No	Tolerated (out of water) Adapted (under water)	3 5	Simple augers a raised every 0.3m at most and continuous augers every 1m at most
Gravel	- Thick-walled corer	- Hammering or vibro-sinking	No	Recommended	4	Tubing during advance $D_1 \geq 3.D_{max}$
	- Valve corer	- Hammering	No	Adapted	5	
	- Auger-rotary bucket	- Rotation	No	Tolerated (out of water) Adapted (under water table)	3 5	

Table 2.8. *Sampling techniques recommended for clean seedy soil [AFN 95d]*

Sampling materials (see Table 2.1)	Sampling techniques			Sampling class		
	Tool name	Use	Fluid	Technique efficiency	Potential optimal class	Comments
Clayey sand Loamy sand	- Stationary piston corer	- Continuous forced sinking	No	Recommended	1	
	- Thin-walled corer with casing	- Continuous forced sinking	No	Recommended	1	
		- Hammering	No	Recommended	2	
	- Thick-walled corer (and casing)	- Hammering or percussion	No	Tolerated	3	
	- Auger and rotary bucket	- Rotation	No	Adapted	3 away from water 5 under water table	- Simple augers a raised every 0.3m at most and continuous augers every 1m at most.
Gravelly clay (gravel in a matrix of fine soils)	- Thin-walled corer and casing	- Hammering	No	Adapted	3	- For soil away from water $D_1 \geq 3.D_{max}$
	- Double corer	- Rotation	Slurry	Tolerated	3	$D_1 \geq 3.D_{max}$
	- Auger and rotary bucket	- Rotation	No	Tolerated	3 away from water 5 under water table	$D_1 \geq 3.D_{max}$ - Simple augers a raised every 0.3m at most and continuous augers every 1m at most

Table 2.9. *Sampling techniques recommended for sand-clay or loam soil and sandy clays [AFN 95d]*

Sampling material (see Table 2.1)	Sampling techniques			Sampling class		
	Tool name	Use	Fluid	Technique efficiency	Potential optimal class	Comments
Sandy clay	(See fine soil)					
Very heterogeneous and coarse soil prior to strong granularity (coarse alluvium, scree, moraine, embankments)	- Thick-walled corer	- Hammering or percussion or roto-percussion	No	Recommended	3	- Risk of early refusal
	- Clamshell bucket tool	- Hammering	No	Recommended	4	
	- Auger	- Rotation	No		3 away from water 4 under water table	- $D_1 \geq 3.D_{max}$ - Simple augers are raised every 0.3m at most and continuous augers every 1m at most
Rock	- Simple corer	- Rotation	Water Water + Slurry additive	Adapted	1	
	- Double corer	- Rotation	Water Water + Slurry additive	Adapted	1	
	- Triple corer	- Rotation	Water Water + Slurry additive	Adapted	1	
	- Disintegrator tool	- Rotation or roto-percussion	Water Water + Slurry additive	Adapted	5	- if the shavings are recovered at surface

*) Eventual, if the disturbed lateral parts are eliminated and immediately conditioned with a waterproof envelope.

Table 2.10. *Sampling techniques recommended for heterogeneous soil and rocks [AFN 95d]*

Sampling tools			Drilling procedures			Sampling class		
Designation	Characteristics	Sample dimensions	Penetration mode	Cutting and extraction	Fluid	Collectable materials	Potential optimal class	Comments
Ground sampling - Loose		- Several dm^3		- Shovel	No	- Fine soils - Seedy soil away from water - Seedy soil in water	3 3 5	
- In blocks, by hand		- Several dm^3	- Hand cutting	- Hand tool, saw	No	- Fine soils firm to hard fine soils, away from water	1	- Special care to be taken during wrapping and transport
- By box		- Several dm^3	- Hand cutting		No	- Firm to hard fine soils, away from water	1	
With a thin-walled laboratory corer	- $C_a \leq 15$ % - $\alpha \leq 10°$ - $C_i = 0{,}5$ % to 1,5 % - $C_o = 0$	- D_1 = following lab tests - $L_t/D_1 = 2$ to 3	- Forced sinking - Hammering		No No	- Firm to hard fine soils, away from water - Soft to firm fine soils, away from water	1 (sinking) 2 (hammering)	- Internal backlash C_i=0 acceptable for short corers ($L_t/D_1 \leq 6$)

Table 2.11. *Ground sampling method of use [AFN 95d]*

Sampling tools			Drilling procedures			Sampling class		
Designation	Characteristics	Sample dimensions	Penetration mode	Découpage extraction	Fluid	Collectable materials	Potential optimal class	Comments
Punch core drills - With thin walls (with or without inner casing)	- $C_a \leq 15\%$ and $\alpha \leq 10°$ - $C_i = 0{,}5\%$ to 1,5% - $C_o = 0\%$ to 2% - $S_e \geq 0{,}1.D^2{}_1$	- $D_1 \geq 75$ mm and $H/D_1 \leq 10$	- Continuous forced sinking ($H/L_t < 9$) - Hammering - Vibro-percussion	- Cutting case + sometimes an extractor	No	- Very soft to stiff fine soils - Clayey sand - Gravelly clay - Clean coarse soils	1 (sinking) 2 (Hammering) 2 2	- Possible increase by 1 point in the class if an extractor is present - See Tables 2.4 to 2.10 for more details on collectable soils and the appropriateness of the process - Internal backlash $C_i = 0$ acceptable for short corers ($L_t/D_1 \leq 6$)
- With thick wall (and inner casing))	- $C_a \leq 15\%$ $10 < \alpha < 40°$ - $C_i = 0\%$ to 4% - $C_o = 0\%$ to 4% - $S_e \geq 0{,}1.D^2{}_1$	- $D_1 \geq 90$ mm $H/D_1 \leq 15$	- Forced sinking ($H/L_t < 0{,}95$) - Hammering - Vibro-percussion	- Cutting case + sometimes an extractor	No No No	- Stiff to hard fine soils and sandy clay - clayey sand - gravelly clay -Gravelly sands and pebbles	3 3 3 4	- $D_1 \geq 3.D_{max}$
- Standard Penetration Test (SPT) corer	- $C_a = 112\%$ and $\alpha = 20°$ - $C_i = C_o = 0$ (see standard NF P 94-116)	- $D_1 = 35$ mm	- Hammering	- Case	No		4	-This tool is foremost a test material for dynamic penetration

Table 2.12. *Punch core drill method of use [AFN 95d]*

Sampling tools			Drilling procedures			Sampling class		
Designation	Characteristics	Sample dimensions	Penetration mode	Cutting extraction	Fluid	Collectable materials	Potential optimal class	Comments
- Stationary piston core drill	- $C_a \leq 15\%$ and $\alpha \leq 10°$ - $C_i = 0{,}5\%$ to 1,5% - $C_o = 0$ to 2%	- $D_l \geq 75$ mm and $H/D_l \leq 10$	- Continuous forced sinking ($H/L_t < 0{,}9$)	- Case	No	- Very soft to firm fine soils	1	- High sinking speed in soft fine soils is favorable - Internal backlash $C_i = 0$ acceptable for short corers ($L_t/D_l \leq 6$)

Table 2.13. *Stationary piston core drill method of use [AFN 95d]*

2.3.3. *Labeling samples*

When a sampling tube is forced into the ground, depending on the nature of the ground, two possibilities are presented:

– The ground is forced into the corer, it is therefore compromised and not intact.

– The ground is pulled lower and therefore not cored.

These phenomena appear during extraction when the percentage of coring is measured. To obtain an ideal sample, a coring percentage close to or equal to 100% is needed. In these conditions, a corer that does not, or barely, disturbs the soil, must meet three criteria:

- limit or even eliminate deformation of the ground during sinking, so that the influence of the corer volume is reduce as much as possible;

- limit the friction of the soil on the exterior of the corer;

- limit the friction of the corer on the interior of the casing.

It should be noted that even when a 100% coring is obtained, a geotechnical laboratory can only use up to 60% of it, taking into account the losses from the high parts due to decompression and the disturbance caused by drilling water, and losses from the low parts due to decompression and the disturbance from the rotation of the corer tube during the shearing of the sample.

The standardization aims to eliminate the drawbacks mentioned above and take samples in optimal conditions.

Sampling tools			Drilling procedures			Sampling class		
Designation	Characteristics	Sample dimensions	Penetration mode	Découpage extraction	Fluid	Collectable materials	Potential optimal class	Comments
Twist core drill - Simple industrial corer			- Rotation	- Crown	Water, Water + slurry additive	- (Treated paved layers)	1	- The maximum class is relative to the diameter of the core and the nature of the ground (see Tables 2.4 to 2.10)
- Simple mining corer			- Rotation	- Crown	Water, Water + slurry additive	- Sound rocks	1	- Can be used rarely without fluid for a short while.
- Double corer		- $D_1 \geq 85$ mm	- Rotation	- Crown	Water, Water + slurry additive	- Firm to hard fine soils - Gravelly clay, rock	1	- Class 1 samples can be obtained with stiff to hard fine soils with scraped and paraffined cores, as of collection
- Double corer with overrunning case	- Case $C_a \leq 15\%$ $\alpha < 10°$ $C_i = 1,5\%$ to 3,5%	- $D_1 \geq 75$ mm $H/D_1 \leq 10$	- Rotation and sinking from pressure on the case $H/L_t < 0,9$	- Crown + cutting case	Water, Water + slurry additive	- Stiff to hard fine soils - Rocks	1	
- Triple corer		- $D_1 \geq 85$ mm $H/D_1 \leq 10$	- Rotation	- Crown	Water, Water + slurry additive	- Firm to hard fine soils - Rocks	1	
- Triple corer with overrunning case	- Case $C_a \leq 15\%$ $\alpha < 10°$ $C_i = 1,5\%$ to 3,5%	- $D_1 \geq 75$ mm $H/D_1 \leq 10$	- Rotation and sinking from pressure on the case $H/L_t < 0,9$	- Crown + cutting case	Water, Water + slurry additive	- Firm to hard fine soils - Tender rocks	1	- The use of an extractor can increase the class by a point.

Table 2.14. *Twist core drill method of use [AFN 95d]*

Sampling tools			Drilling procedures			Sampling class		
Designation	Characteristics	Sample dimensions	Penetration mode	Découpage extraction	Fluid	Collectable materials	Potential optimal class	Comments
Other tools - Hand augers		- Drill diameter ≥ 40 mm	- Rotation		No	- Fine soils, sand away from water	3 (4 if peaty and very soft soil)	
- Continuous augers	- $D_{axis} \leq 0,33$ d_{ext} pitch of the helix = external diameter	- ≥ 60 mm	- Rotation		No	- Fine soils, - Coarse soil away from water	3 (4 if peaty and very soft soil)	- Simple augers are raised every 0.3m at most and continuous augers every 1m at most.
- Valve corer						- Graves	5	
- Clamshell bucket and curette tools						- Very heterogeneous soils	3 (away from water) 5 (under the water table)	
- Simple auger and rotary bucket		- ≥ 300 mm	- Rotation		No	- Fine soils, - Coarse soil away from water	3	- Simple augers are raised every 0.3m at most and continuous augers every 1m at most.

Table 2.15. *Semi-destructive tool method of use [AFN 95d]*

To eliminate the above mentioned drawbacks and take samples in optimal conditions, it has been necessary to standardize the dimensions of core drills. After sampling, the samples are placed in a chest labeled with:

– the site reference and its name;

– the name of the drilling company that took the sample;

– the sample's reference (survey number and sample number);

– the sampling method (tool type and usage conditions);

– sampling date;

– the lower and upper levels of the sample compared to the natural ground, with the top and bottom of the coring.

2.3.4. *Transporting the samples*

The following conditions must be adhered to on the site and during the transport to the laboratory:

– keep it from the freeze;

– no strong heat or direct exposure to the sun;

– manipulation without impact and minimal vibrations (for classes 1 and 2).

For classes 1 and 2, the casings are transported well wedged in a case carpeted with damping material. The case is blocked into the transport vehicle. It is recommended that the procedure presented in Figure 2.19 be followed for all soils transported vertically and that presented in Figure 2.20 for all soils transported horizontally.

2.3.5. *Storage before testing of class 1 and 3 samples*

The samples must be stored at a constant temperature away from freeze. If a long storage period (more than 2 months) is necessary, a device preserving the water content should be used (humidity chamber, vacuum packing, etc.).

2.3.6. *Competence of the providers*

The competence and know-how of the borehole team is one of the best factors determining the quality of the sample. The borehole team organization must support the skill of its personnel.

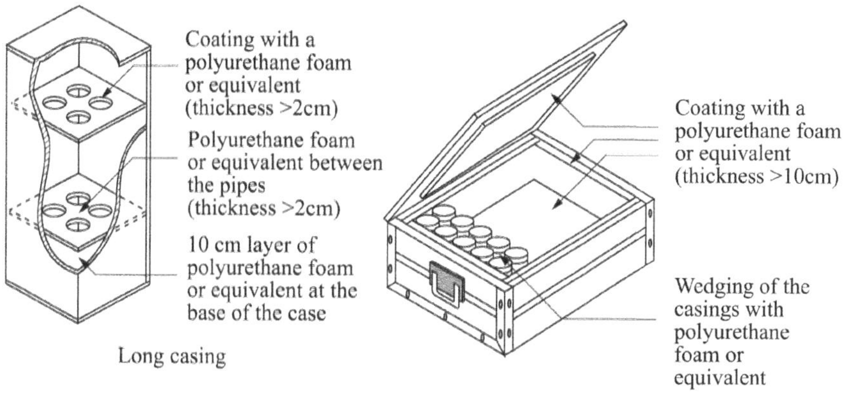

Figure 2.19. *Vertical transport case for class 1 and 2 samples [AFN 95d]*

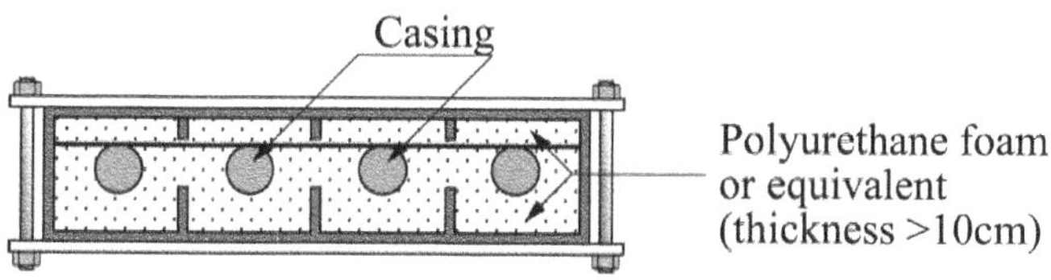

Figure 2.20. *Horizontal transport case for class 1 and 2 samples [AFN 95d]*

2.3.7. *Checking: controls*

The correct function of drilling equipment must be checked before any use. With punch core drills, the cutting case must well tapered and not blunt. Only the cutting extremity can be rounded (radius less than 0.5 mm).

With twist core drills, check that:

– the crown is in a good state and adapted for the material;

– the rotation of the inner pipe is free for double or triple corers.

In all cases, the internal surface of the corer or the casing must be clean (but can be greased or oiled), smooth, without any ridges or irregularities.

2.3.8. *Sampling written record*

The written records include the following information:

– reference to the standard NF P 94-202;

– the geographical location of the sampling site;

– the name of the organization that carried out the sampling as well as the name of the team leader and the type of machine used.

Identification and specifications of the drilling that was carried out will include, on the written drilling record:

– the technical cut;

– possibly a statement of the drilling parameters noted or recorded based on the depth (speed of advance, force pushing down on the tool, etc.);

– the conditions of refill carried out according to the contracted specifications (height of refill and type of filler).

The sample cam be identified through:

– the class of sample requested;

– the sampling date;

– the position of the drill (depth of the high and low points of the sample compared to the natural terrain at the level of the drill);

– the type of sampling tool used, its specifications (C_0, C_i, C_a, L_t) and the method of use;

– the sample diameter (D_l), the length of the sample (L_g), net length (L_n), and possibly the eliminated lengths from the top (L_h), and from the base (L_b), the depth of penetration (H), the net coring index (IC), the coring percentage (TC);

– a summary description of the sample (reduced to the two extremities if the latter is under a casing), nature, consistency and color.

The observations linked to the execution of the drilling and obtaining the sample, as well as the unintentional operational details, are susceptible to influencing the results. An example of a datasheet is shown in Table 2.16.

Borehole company			Data sheet Class 1 sample collected in conformance with the standard P 94-202		
Site :			Collection method		N° document :
Specification of the corer			Casing from: to m		Drilling n°
Type :	Inner diameter D_1 : mm	Usable length L_t : mm			Collection date :
			Sample n° :		
C_o :	C_i :	C_a :	code		
Sample specification			Depths		Inner diameter of the casing D_3 = mm
Penetration length H : mm			Z_i = m	Z_f = m	Net length L_n = mm
Gross length of the sample L_g : mm			Nature of the soil :	High :	L_h = mm / Survey
Net coring index $IC = L_n/H$ =	Coring percentage $TC = L_g/H$ =			Base :	L_p = mm / Signature

Table 2.16. *Data sheet from a sample [AFN 95d]*

2.4. The drilling section

A section type is shown in Figure 2.21. It shows the following:

– the name of the borehole;

– the X, Y, Z coordinates of the borehole tip;

– the kind of machine used;

– the name of the drilling company;

– the drilling dates;

– the sample collection rates;

– the piezometric rates;

– drilling run references;

– the synthetic geological cut;

– the name of all the different soils encountered.

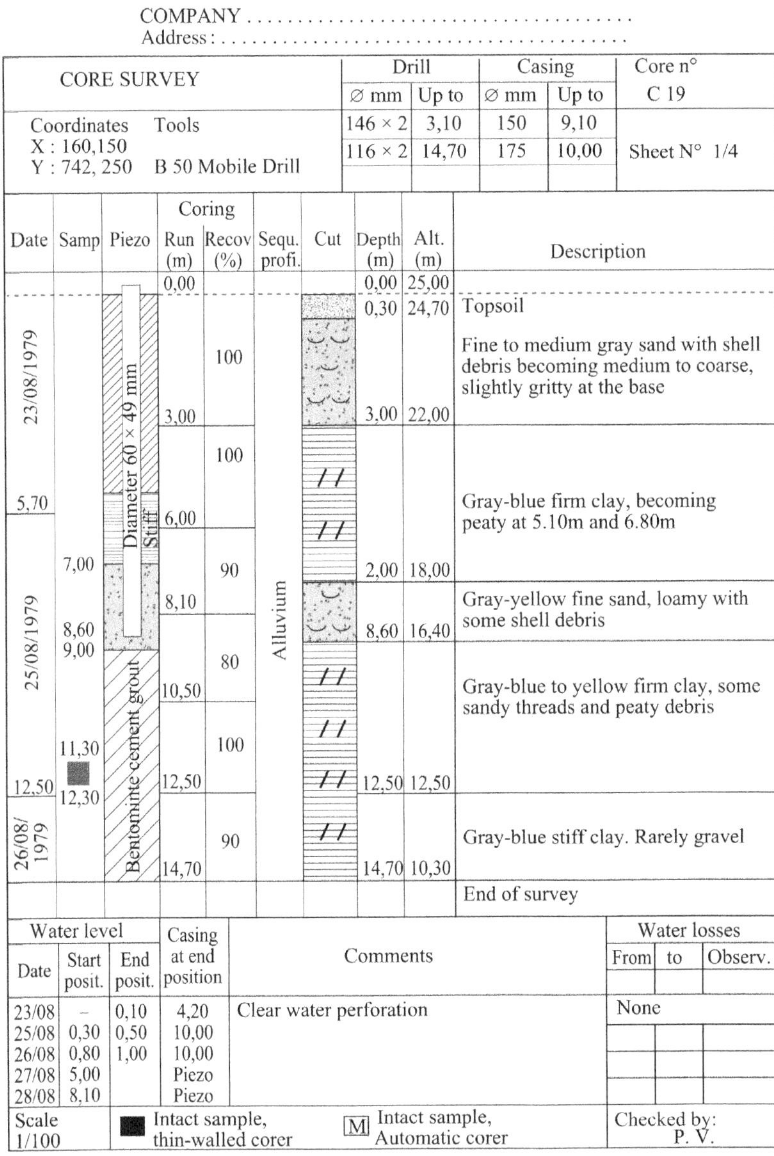

Figure 2.21. *Example of a survey profile*

A summary of the different symbols used to represent the soils and rocks is also shown in Figure 2.22.

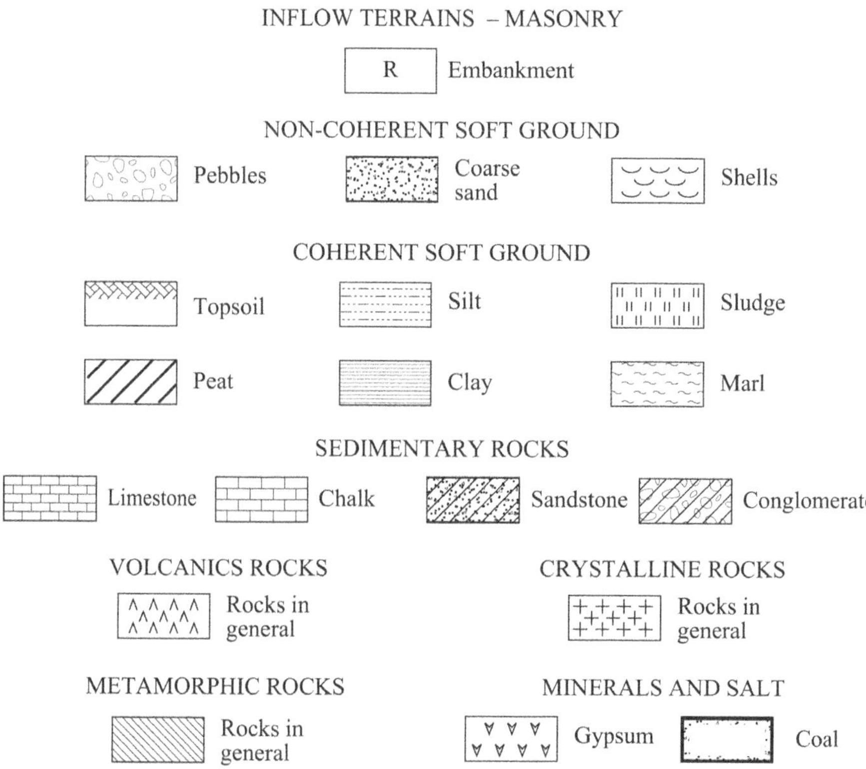

Figure 2.22. *The symbols allocated to the different layers of soil and rock*

2.5. The rock-quality designation (RQD)

For rock drillings, the fracturing density is stated and called the rock-quality designation (RQD) [2.7]. This allows the classification of rocks based on their weathering (Table 2.17). The RQD value can be usefully completed by estimates of the fracturing index I_f. This corresponds to the median length of the cores collected over the core run (Figure 2.23).

$$RQD = \left(\frac{\sum \text{Length of core>10cm}}{\text{Length of run}} \right).100 \qquad [2.7]$$

Classe	RQD	Description
RQD 1	>90%	Very weak fracturing
RQD 2	90 to 75%	Weak fracturing
RQD 3	75 to 50%	Average fracturing
RQD 4	50 to 25%	Strong fracturing
RQD 5	<25%	Very strong fracturing

Table 2.17. *Classification of rocks by fracturing*

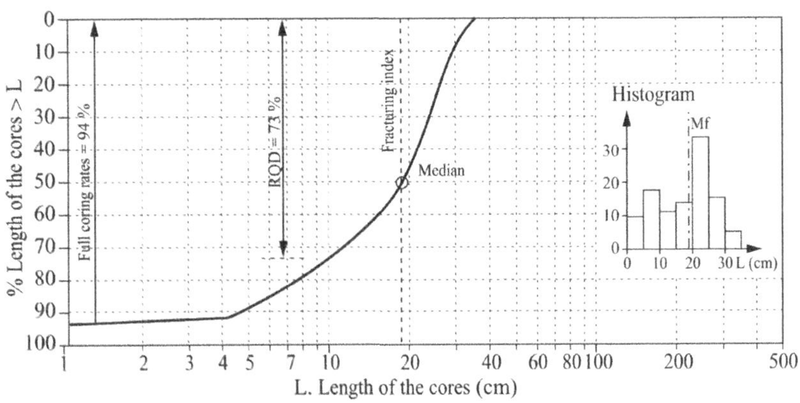

Figure 2.23. *Calculating the median fracturing length*

3

Measuring the Total Pressuring, the Interstitial Pressure and the Groundwater Table Rating

3.1. Measuring the total pressure within the soil

3.1.1. *Measurement principles*

Measuring the total pressure in the ground is done through the intermediate of a pressure sensor buried in the ground, which transfers the required information to the surface. The ground pressure is changed into a measurable phenomenon that is:

– either the displacement of a membrane;

– or the displacement of a piston.

This implies movement that disturbs the distribution of stresses in the soil. This influence of the sensors on the soil presssures is poorly understood. The devices are generally presented as a flat disk, or sometimes as a square or rectangular plate.

3.1.2. *Mode of action of the sensors*

The three potential actions are shown in Figure 3.1:

– direct action;

– indirect action;

– opposite action.

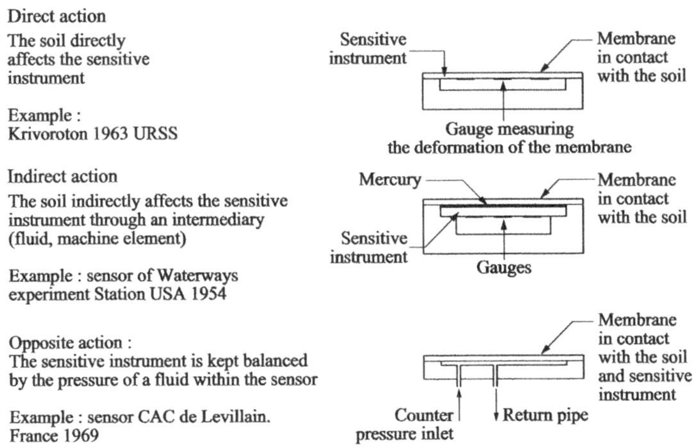

Figure 3.1. *The different modes of action of sensors*

3.1.3. *Deformation measuring systems*

These measuring systems are based on various principles:

– mechanical: this follows the friction dynanometer principle. Its use is restricted to the laboratory;

– hydraulic: this uses a system of liquid-filled pipes for pressure transmission. The static pressure is thus transmitted from the sensor toward the surface of the ground. Its running is simple, but caution must be taken with losses of charge and remaining within hydrostatic conditions;

– acoustic: this uses the variations in resonance frequency of a chord based on its tension. This system is specific, but relatively cumbersome, and sensitive to temperature;

– resistant wire gauges: resistant wires are stuck onto a deformable insulating support. The deformation of the support causes changes in the section and wire length, all of which results in a change in resistance. This system is used more often, but care must be taken with the adhesive connection between the deformation gauges and the membrane whose deformation is to be measured;

– capacitive: this system is based on variations in capacity of a capacitor by the relative displacement of its plates. In this case, the sensor is quite thick.

– inductive: this system is based on variations of impedance of an inductor by the displacement of a metal core in a coil;

– piezoelectric: this system is based on the production of an electrical charge by quartz crystals when they are compressed. It is poorly adapted to the measurement of static pressures that do not vary much.

3.1.4. *Soil-sensor interaction*

The sensor affects the stress field that it is meant to measure. It therefore measures a constraint tainted by error. Current work seeks to evaluate this recording error of the sensors.

3.1.4.1. *The case of sensors subject to a uniaxial constraint*

This type of sensor is shown in Figure 3.2.

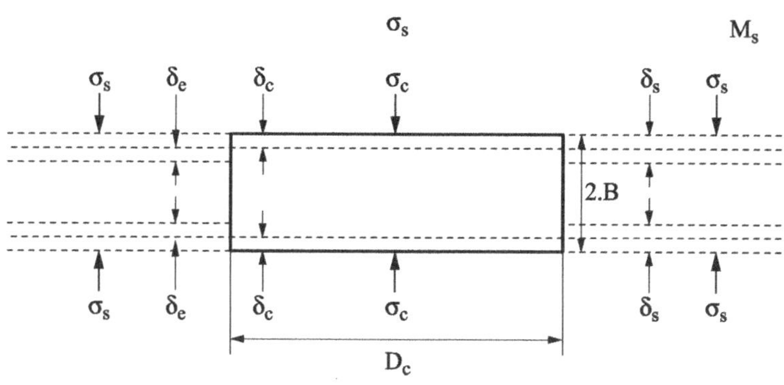

Figure 3.2. *Diagram of a uniaxial sensor [MEM 80]*

The differential deformation δ_e between the sensor and a layer of ground of the same thickness needs to be evaluated. The deformation of the sensor δ_c is given by [3.1]. The deformation of the ground δ_s at the sensor's level is given by [3.2], with B_c the halved thickness of the sensor, M_c the elastic modulus of the sensor, M_s the elastic modulus of the ground, σ_c the total pressure acting on the sensor and σ_s the total pressure acting on the ground at the sensor level.

The deformation of the sensor is given by equation [3.1] and that of the ground by [3.2]:

$$2.\delta_c = 2.B_C.\frac{\sigma_C}{M_C} \quad [3.1]$$

$$2.\delta_s = 2.B_C.\frac{\sigma_S}{M_S} \quad [3.2]$$

The assumptions of the theoretical calculation are:

– the sensor is considered a flat disk of uniform compressibility and radially rigid;

– the ground is elastic;

– the horizontal constriction of the ground does not influence the vertical constriction ($\sigma_X = \sigma_Y = 0$);

– the friction on the walls of the sensor is nil;

– there is one single vertical constriction σ_Z.

The recorded ratio R_e is the relationship of the total pressure applied on the sensor with the pressure in the soils at the same level [3.3]. This term is the same as statement [3.4], with N_S a coefficient dependent on the deformations of the ground, K_S a ratio defined by [3.5].

$$R_e = \frac{\sigma_C}{\sigma_S} \quad [3.3]$$

$$R_e = \frac{\sigma_C}{\sigma_S} = \frac{(D_C/B_C)+(N_S/M_S)}{(D_C/B_C)+(N_S/M_C)} = \frac{(D_C/B_C)+K_S}{(D_C/B_C)+K_S.(M_S/M_C)} \quad [3.4]$$

$$K_S = \frac{N_S}{M_S} \quad [3.5]$$

For the uniform displacement of a sensor membrane, the elastic calculation [TIM 51] enables the determination of the coefficient K_s [3.6], with ν as the Poisson coefficient. For the non-uniform displacement of the flexible membrane variety, formula [3.7] is used, with C as a function coefficient of the distribution of pressures on the membrane.

$$K_S = \frac{4}{\pi.\left(1-\upsilon^2\right)} \quad [3.6]$$

$$K_S = \frac{1}{C.\left(1-\upsilon^2\right)} \quad [3.7]$$

The evolution of the recorded ratio according to the relation of the sensor moduli and the ground (Figure 3.3) for a uniaxial vertical field of constraint corresponds to formula [3.3]. For an ideal sensor, when the sensor modulus is equal to the soil modulus, there will be no variation on the pressure distribution. Compare to this the fact that the sensor modulus is higher than the soil modulus and the recording ratio is larger than one. The theory shows that this ratio trends toward a limiting value that can be accurately known.

Table 3.1 shows the evolution of C and K_S for different constraint distributions.

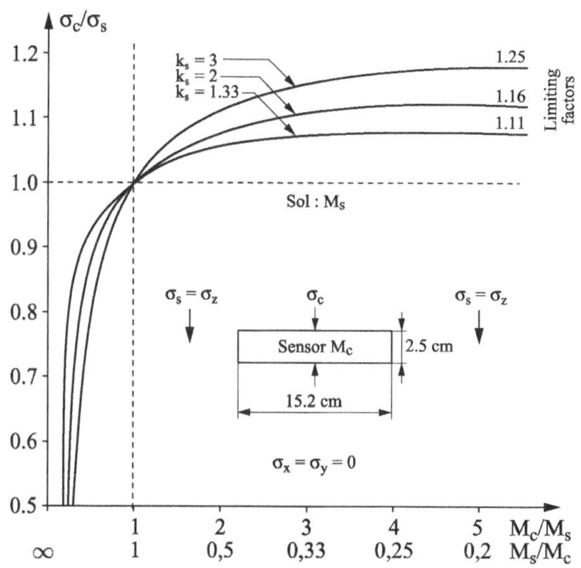

Figure 3.3. *Evolution of the recorded ratio based on the ratio M_C/M_S for a horizontal sensor in a uniaxial field of constraint [MEM 80]*

	Rigid membrane	Flexible membrane	
Charge	Uniform	Parabolic (max in center)	Parabolic (zero in center)
Constraint distributions	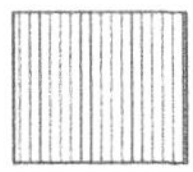	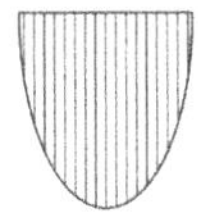	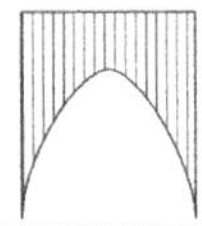
Coefficient C	0.849	0.903	0.793
k_s (for $v = 1/3$)	1.33	1.24	1.42

Table 3.1. *C and K_S values for different constraint distributions*

3.1.4.2. *The case of the sensor subjected to a triaxial state of constraint*

In this case, the recorded ratio is given by relation [3.8] for a sensor placed horizontally (sensitive side perpendicular to σ_Z) and by relation [3.9] for a sensor placed vertically (sensitive face parallel to σ_Z):

$$R_e = \frac{\sigma_c}{\sigma_s} = \frac{\sigma_c}{\sigma_z} = \frac{\left(D_c / B_c \right) + K_s . \left(1 - \upsilon . \frac{\sigma_x + \sigma_y}{\sigma_z} \right)}{\left(D_c / B_c \right) + K_s . \left(M_s / M_c \right)} \qquad [3.8]$$

$$R_e = \frac{\sigma_c}{\sigma_s} = \frac{\sigma_c}{\sigma_x} = \frac{\left(D_c / B_c \right) + K_s . \left(1 - \upsilon . \frac{\sigma_y + \sigma_z}{\sigma_x} \right)}{\left(D_c / B_c \right) + K_s . \left(M_s / M_c \right)} \qquad [3.9]$$

In the case of a triaxial field of constraint, the evolution of the recorded ratio is shown in Figure 3.4 for a horizontal sensor and in Figure 3.5 for a vertical sensor. It is noted that a sensor placed horizontally will give an over-recording, whereas a sensor placed vertically will give an under-recording.

3.1.5. *Other measuring errors*

To protect it, as well as to obtain a better distribution of constraints on their sensitive sides, sensors are often surrounded by a pocket of fine sand. This could potentially be considered an inclusion.

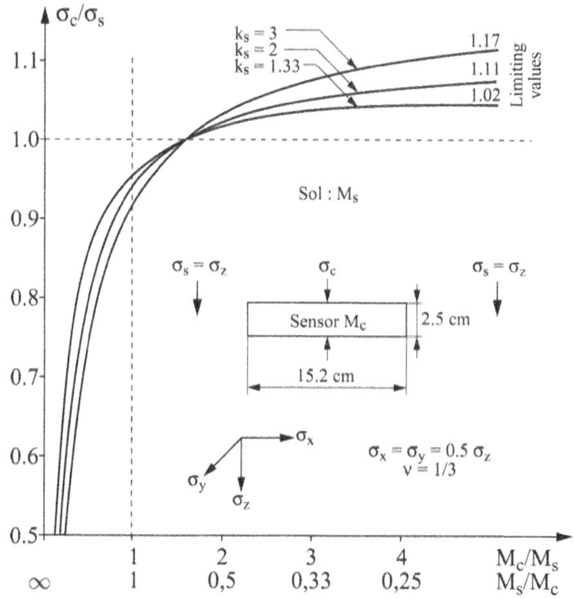

Figure 3.4. *Evolution of the registration report according to the M_C/M_S ratio for a horizontal sensor placed in a triaxial field of constraint*

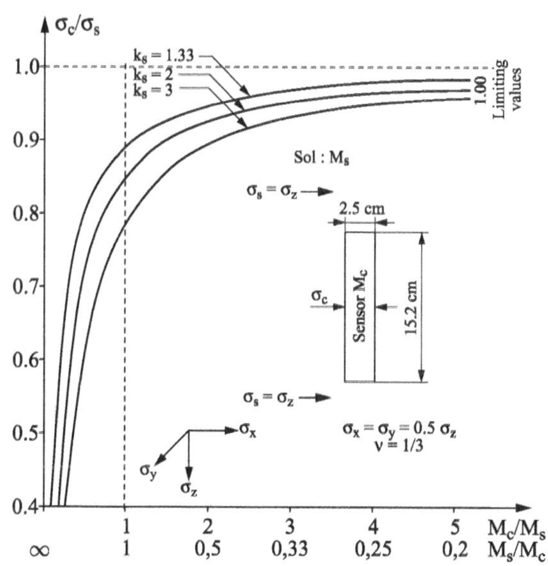

Figure 3.5. *Evolution of the registration report according to the M_C/M_S ratio for a vertical sensor placed in a triaxial field of constraint*

3.1.5.1. *Influence of interstitial pressure*

A variation Δu of interstitial pressure does not significantly affect the deformation of the ground in the layer of the same thickness. However, it does change the total pressure and the deformation of the sensor becomes [3.10]. This results in a variation of the recorded ratio [3.11].

$$2.\delta_C = 2.B_C.\frac{\sigma_C+\Delta u}{M_C} \qquad [3.10]$$

$$\Delta\frac{\sigma_C}{\sigma_S} = -\frac{K_S.\frac{M_S.\Delta u}{M_C.\sigma_S}}{D_C/B_C+K_S.M_S/M_C} \qquad [3.11]$$

An increase in the interstitial pressure increases the total constraint, diminishes the over-recording or increases the under-recording. By introducing the M_C' equivalent modulus compliant to formula [3.12] into formula [3.1], the initial expression of the recorded ratio can be kept.

$$M'_C = M_C.\frac{\sigma_C}{\sigma_C+\Delta u} \qquad [3.12]$$

3.1.5.2. *Influence of temperature*

Changes in temperature can have an effect on:

– the electric circuit of the deformation gauges;

– the expansion of the liquid inside the sensor;

– the global expansion of the sensor which will be different from that of the neighboring ground;

– the vibrating wires if present.

3.1.5.3. *Redistribution of constraints in the vicinity of the sensor*

For piston sensors, the movement of the sensitive surface is uniform. Several authors [TAY 47, MON 50] suggest the model shown in Figure 3.6 and show the distribution of pressures on the sensor, represented here by a piston that moves in the ground.

For membrane sensors, Allwood [ALL 56] suggests the distribution of constraints as shown in Figure 3.7, which have been calculated by finite elements [MEM 80]. The results of the finite element calculation are also covered. In the center of the sensor, where there is the greatest membrane displacement, the contact pressure with the blade is minimal.

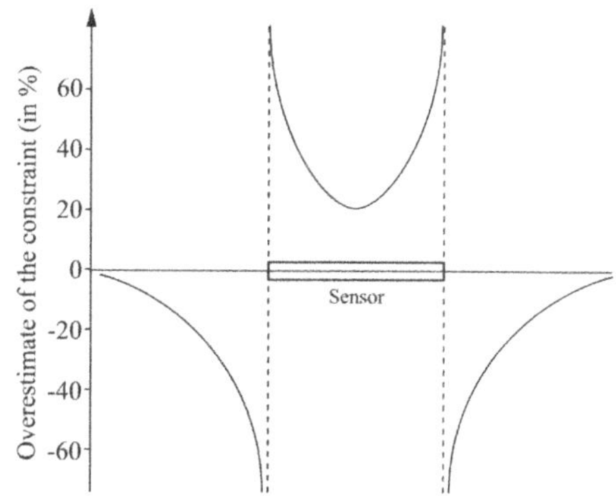

Figure 3.6. *Distribution of constraints around the piston sensor [TAY 47, MON 50]*

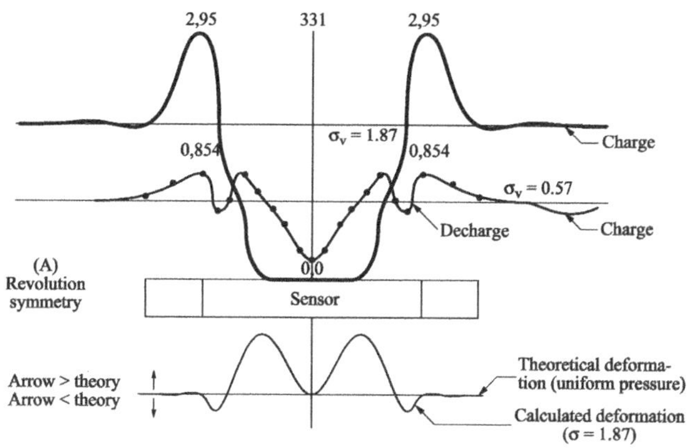

Figure 3.7. *Distribution of constraints around the membrane sensor. From [ALL 56] and [MEM 80]*

3.1.5.4. *Ratio of height to diameter*

Recorded errors are denoted as σ_e/σ_S. This can be expressed with the $2B_c/D_c$ ratio through the relation:

$$\frac{\sigma_e}{\sigma_s} = \frac{\sigma_c - \sigma_s}{\sigma_s} = K_c.\frac{2.B_c}{D_c} \qquad [3.13]$$

with the influence coefficient K_c which depends on the sensor (modulus and sensitive surface) and on the ground (Poisson coefficient and modulus).

If the $2B_c/D_c$ ratio tends toward zero, then the recorded error tends toward zero. However, an infinitely thin sensor is a figment of the imagination. A good compromise is suggested by Hvorslev [HVO 76] with $D_c/2B_c > 10$, as well as Peattle and Sparrow [PEA 54] suggesting $D_c/2B_c > 30$.

An example of the evolution of constraints for a piston sensor and for different values of the B_c/D_c ratio is shown in Figure 3.8.

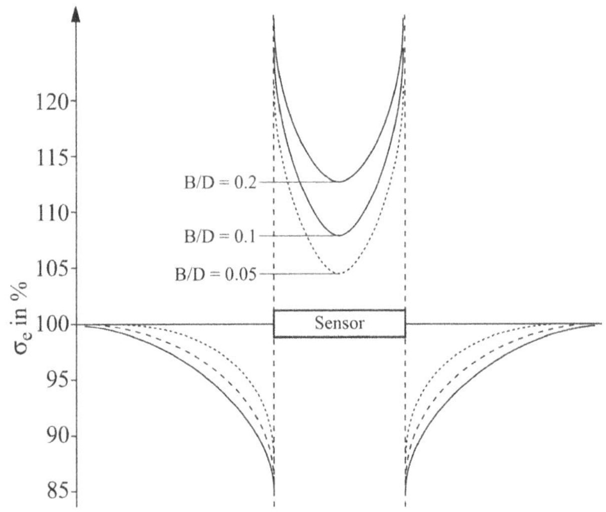

Figure 3.8. *Distribution of constraints for a piston sensor, for different values of the B/D ratio and for M_C/M_S = 2*

3.1.5.5. *Link between the sensor modulus and the ground modulus*

If the sensor modulus is equal to the ground modulus, the recorded error is nil. This condition is difficult to achieve. It is preferable to choose the biggest possible M_C/M_S ratio, in order to reach the asymptote found (Figures 3.4 and 3.5). Some authors [PEA 54] suggest $M_C/M_S > 10$ and even $M_C/M_S > 30$. The evolution of the influence coefficient K_c in equation [3.13] depending on the M_C/M_S ratio is shown in Figure 3.9. It can be seen that a ratio of 10 is enough to reach the recorded ratio asymptote.

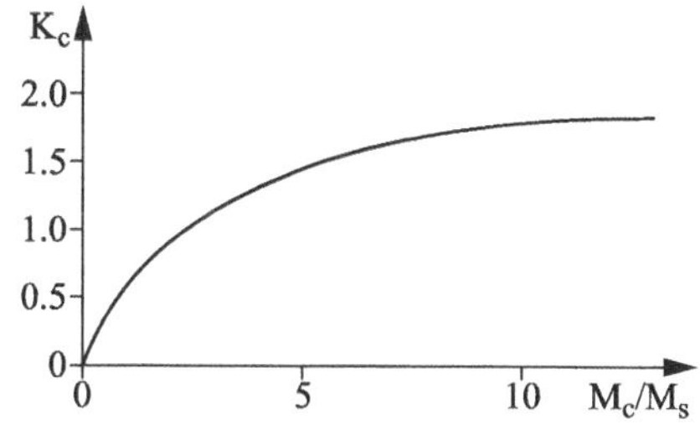

Figure 3.9. *Evolution of the coefficient K_c depending on the M_C/M_S ratio [PEA 54]*

3.1.5.6. *Link between the sensitive surface and the total surface*

Membrane sensors seem preferable to piston sensors since they cause less disturbance to the field of constraint in the ground (Figures 3.6 and 3.7). However, the deflection of the membrane generates a vault effect delaying the constraints on the edge of the sensor. A good compromise is obtained through the ratio between the sensitive surface A1 over the total surface A2 as shown in [3.14].

$$\frac{1}{3} < \frac{A_1}{A_2} < \frac{1}{2} \tag{3.14}$$

Sensor	Mode of action	Measuring system	Height to diameter ratio 2B/D	Membrane deflection	Ease of use	Reliability	Temperature influence	Observation	Schematic diagram	
Mini Glötzl	Opposite	Pneu-matic	Very good (0.05)	Comes and goes Amplitude ?	Long, very delicate	Good	Negligible	– Reduced size – Pipes too supple	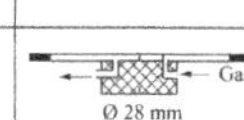	Very simple concept
INSA Glötzl prototype	Opposite and indirect	Pneu-matic	Average (0.20)	Comes and goes	Long, very delicate	Average	Unknown	+ Reduction of membrane movement – Unsuitable assembly	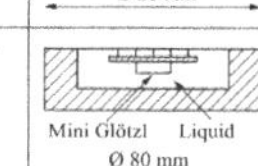	
New Glötzl	Opposite and indirect	Pneu-matic	Very good (0.05)	Comes and goes	Long and delicate	Very good	Unknown	+ No movements coming down from the membrane + Robustness	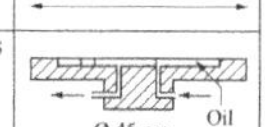	
LRPC Angers prototype	Opposite	Pneu-matic	Very good (0.08)	Comes and goes	Long and delicate	Good	Negligible	– Significant membrane movement	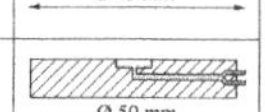	
Schaevitz	Indirect	Gauges	Average (0.23)	Big (1/3500)	Easy and fast	Very good	Weak	+ Robustness and stability of measurements – δ/D important	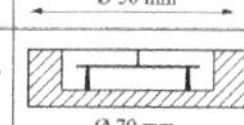	Rapid measurement
INSA electric prototype	Indirect	Gauges	Good (0.07)	Very weak (1/200.000)	Easy, very fast	Bad	Unknown	+ δ/D very weak – Unsuitable as-sembly causing variations in the response	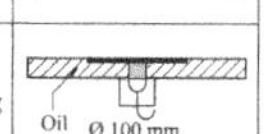	

Table 3.2. *Current examples of total pressure sensors [MEM 80]*

3.1.5.7. *Membrane deflection*

This must be as small as possible. It is calculated by the relationship between the membrane deflection δ_c and the diameter of the sensitive surface d_c. Troloppe and Lee [TRO 61] suggest [3.15] and others recommend [3.16].

$$\frac{\delta_c}{d_c} < \frac{1}{2000} \qquad [3.15]$$

$$\frac{\delta_c}{d_c} < \frac{1}{10000} \qquad [3.16]$$

3.1.5.8. *Granularity*

Assuming a cylindrical piston and spherical grains, Kallstenius and Bergau [KAL 56] calculated the error due to the number N of grains in contact with the piston [3.17]. There must be over 3,000 grains to have an error of less than 3%, so a diameter D_c of the sensor greater than 3,000 D_{max} (D_{max} maximum size of the grains of soil). This stipulation allows the sizing of the grains of sand that surround the sensor.

$$\frac{\sigma_e}{\sigma_s} = \frac{\sigma_c - \sigma_s}{\sigma_s} = \frac{\pi}{\sqrt{2.N}} \qquad [3.17]$$

3.1.6. *Examples of total pressure sensors*

The different sensors that are used are shown in Table 3.2. An example of a flat sensor is shown in Figure 3.10.

3.2. Measuring the interstitial pressure and the level of the water table

Interstitial pressure is measured in piezometers.

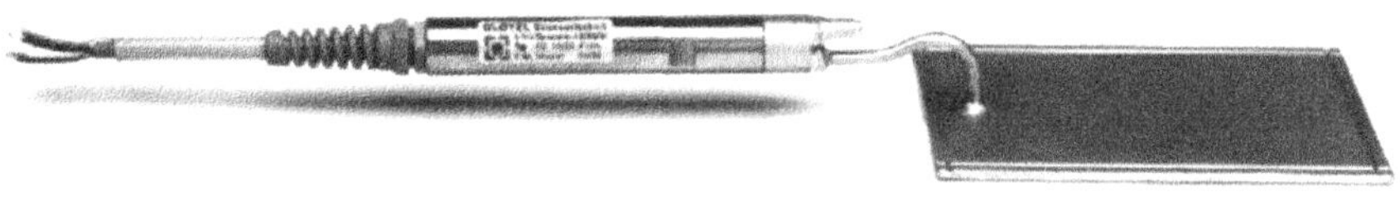

Figure 3.10. *Example of a flat total pressure Glötzl sensor for the ground (Doc.Glötzl)*

3.2.1. *Open-tubed piezometer*

Water table levels are measured in a piezometric tube made by a cased wellbore whose lower part is a strainer. The measurement is either done directly, or by the descent of an electric contact that closes when it reaches the surface of the water. Pressure at a depth h (Figure 3.11) is then given by equation [3.18].

$$u = \gamma_W . h \quad [3.18]$$

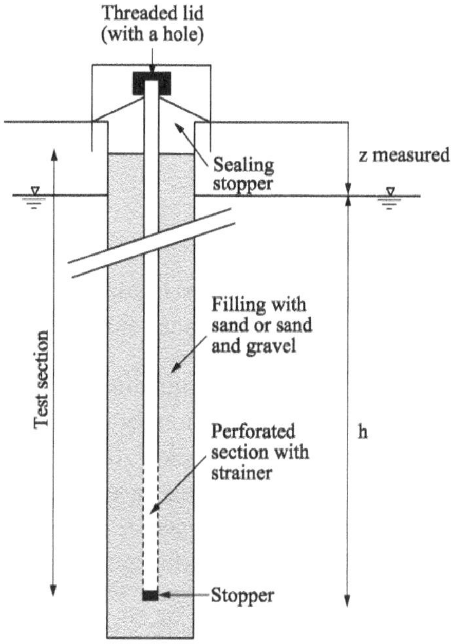

Figure 3.11. *Open-tubed piezometer*

The water table level is relative to the ground compared to the strainer. It can be overpressurized compared to the surface (artesian ground), just as it can be underpressurized compared to a layer above in another level of ground (perched water table).

The open-tubed piezometer does not work in clay because the flow is too weak. The water level in the piezometer is measured manually and possibly with an electric probe (Figure 3.12).

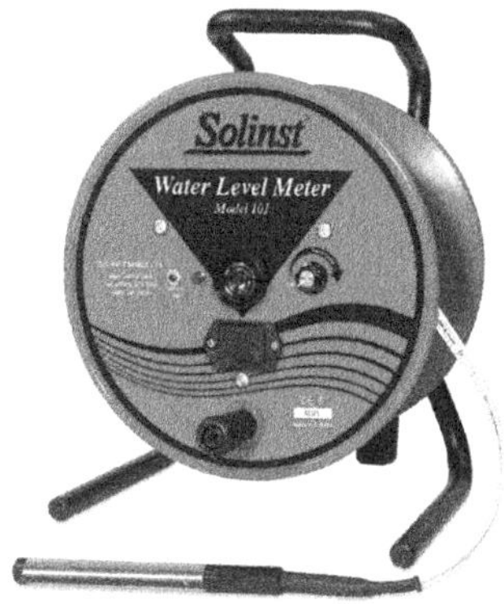

Figure 3.12. *Example of an electric probe*

3.2.2. *Closed-tube piezometer*

This kind of piezometer is shown in Figure 3.13. The ceramic of the piezometer is saturated before it is introduced into the ground.

With pneumatic piezometers, the piezometer communicates with a chamber separated into two parts by a membrane (Figure 3.14). When the gaseous pressure at the entrance is weaker than the water pressure, the membrane does not detach. In the opposite case where the gaseous pressure is equal to or greater than the water pressure, the membrane detaches and the volume of the returning gas is recorded.

A volumeter helps determine the system change that corresponds to the interstitial pressure. Several examples of interstitial pressure sensors are shown in Figure 3.15.

This operating system is also used by total pressure systems. Its measurement is always punctual.

To correctly use this system, it needs:

– the saturation of the cell and the porous stone before its introduction into the ground;

– limitation of the drilling speed to 1–10 cm/s in the case of a sunken sensor;

– to respect the rest period before taking the measurement.

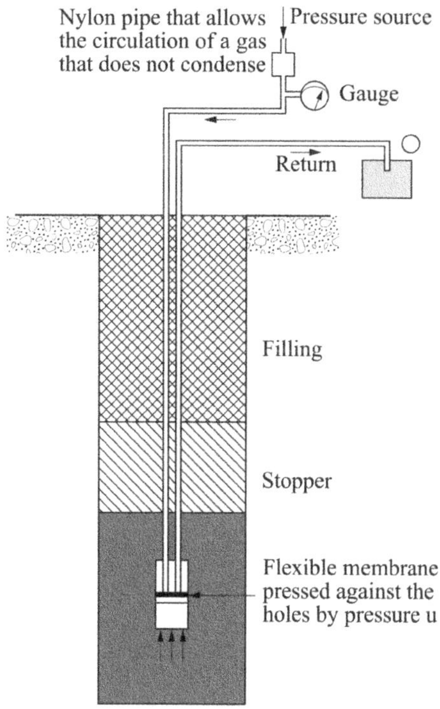

Figure 3.13. *Closed-tube piezometer, pneumatic*

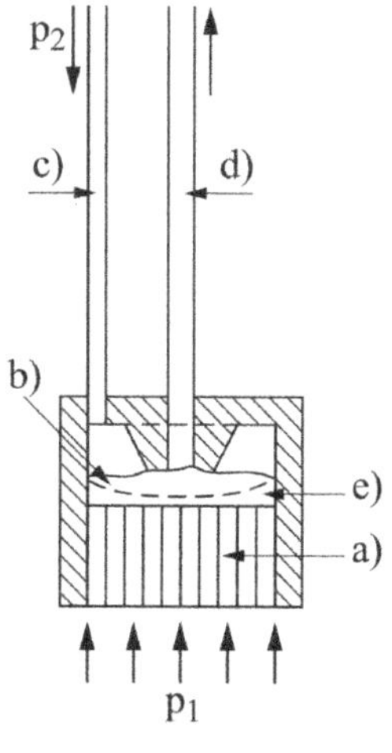

Figure 3.14. *Schematic diagram of the Glötz piezometer*

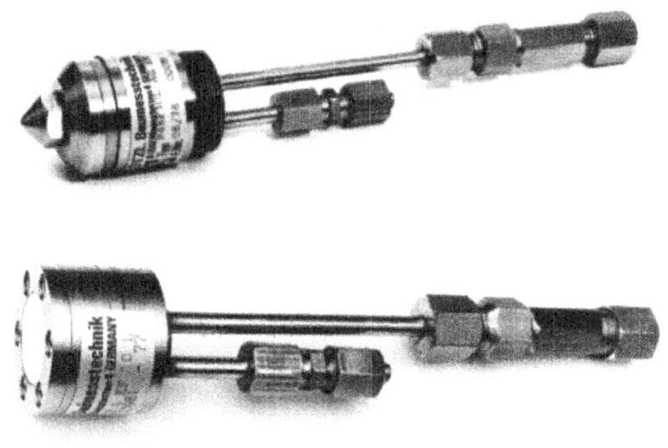

Figure 3.15. *Examples of a Glötz piezometer*

4

Measuring Movement, Settling and Force

4.1. Measuring movement

4.1.1. *Topography*

Topographic measurements are used to appraise landslides, the movement and settling of buildings, etc. They make use of different techniques, which we will highlight here.

At any given time, points on the ground's surface are determined using coordinates (X, Y, Z). Comparing a point's coordinates at different times indicates the point's movement.

4.1.1.1. *Ground surveying using a theodolite*

Points are identified on the ground's surface using a sight or a rod equipped with a reflecting prism. The target point is identified by a system of spherical coordinates with a horizontal angle, a vertical angle and a sloped distance, with respect to a fixed station where the theodolite is situated (Figure 4.1).

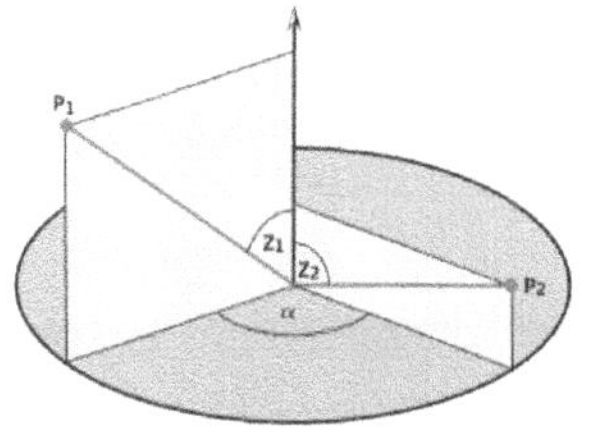

Figure 4.1. *Measurement of spherical coordinates of survey points using a theodolite*

The reference direction for the horizontal angle is another fixed station, or a point with known coordinates. The reference direction for vertical angles is the zenith.

The horizontal angle thus determined is the dihedral angular deviation between the vertical planes passing through the reference direction and the target point.

The vertical angle is the angular deviation between the direction of the zenith and the target point.

The accuracy of the survey is in the order of centimeters, and, with special precautions, of millimeters or even tenths of millimeters.

Using this method, the positions of points can be identified on all three dimensions (Figure 4.2).

The comparison of the same point's coordinates on different dates indicates, to the geotechnician, the point's movement in X, Y, Z.

One example of this is shown in Figure 4.4, where the coordinates' progression of points on the rocky landslide of Séchilienne, near Grenoble in France, can be observed. This progression serves to determine the speed and range of the movement linked to the landslide.

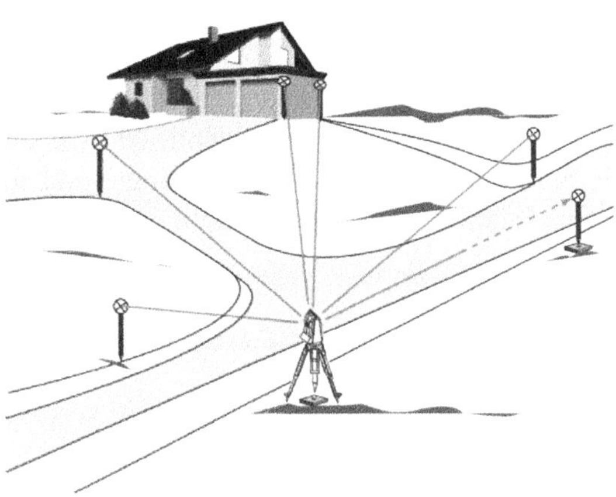

Figure 4.2. *Ground surveying using a theodolite*

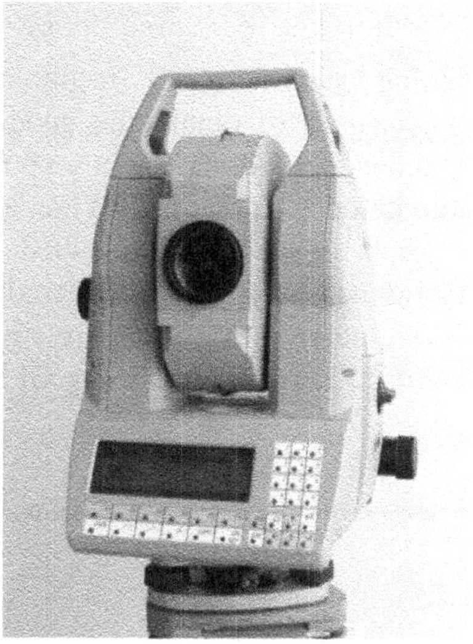

Figure 4.3. *Example of theodolite*

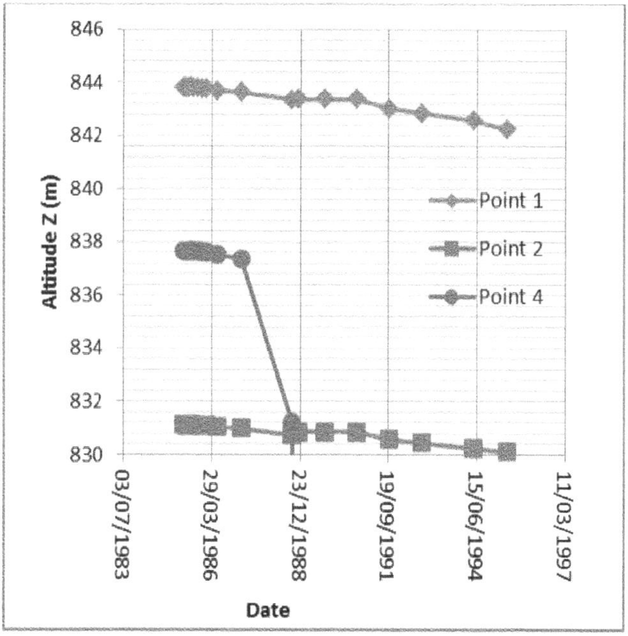

Figure 4.4. *Progression of coordinates for points 1, 2 and 4 at the top of the Ruins of Séchilienne from 1983 to 1997*

4.1.1.2. *Ground surveying using GPS technology*

The global positioning system (GPS technology), a fast-growing technology (Figure 4.5), is used to obtain the geodetic coordinates for a stationed point (one antenna per point) directly, but with poor accuracy [ASH 03] and is unusable in topography (dm). When using two antennas, the distance between two points is accurately obtained (mm), as well as the position of the vector being measured. This is referred to as measurement by differential GPS, that is one point's information relative to another. The gap in coordinates (ΔX, ΔY and ΔZ) is measured using GPS.

Figure 4.5. *Example of a Leica Geosystem Total system*

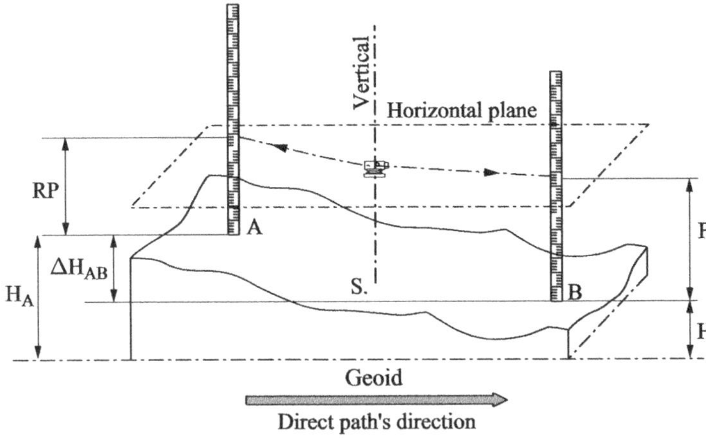

Figure 4.6. *Direct or geometric leveling*

4.1.1.3. *Leveling*

The points' vertical position is determined in relation to a reference level (usually, the average sea level), and along a lengthwise beam. This is done using a level machine (Figure 4.7).

Figure 4.7. *Example of a level (Leica documentation)*

The level sets a horizontal plane. This horizontal plane's intersection with a sight gives, along the direction of the path, the forward playback (FP) and, in the opposite direction, the reverse playback (RP). The height difference (ΔH) between the FP and the RP is given by formula [4.1].

$$\Delta H = RP - FP \qquad [4.1]$$

The accuracy of the leveling is to a few millimeters, and, with special precautions, to a tenth or even a hundredth of a millimeter. Comparing the same point's altitude at different dates indicates its movement (Z), during the settling of foundation structures, for example.

4.1.1.4. *Photogrammetry and LiDAR*

Photogrammetry consists of taking aerial photographs of an object on the ground (Figure 4.8, point A) from an aeroplane, a helicopter or a drone. The same point is thus seen from two different locations, and appears in two different images (a and a'). Transferring these two images, after correcting their base and altitude, to a photogrammetry evaluation device, will give for a trained operator a three-dimensional image, or can be seen in a computer. The points on the ground are identified by measuring the angles at peaks S1 and S2. It is then possible to monitor the object's outlines, or to dismiss curves that have an equal altitude value. The level curves can, therefore, be defined from stereoscopic vision.

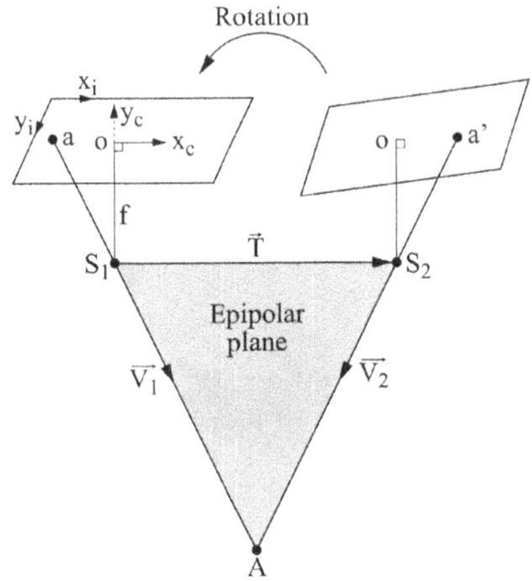

Figure 4.8. *Photogrammetry's basic geometric principle*

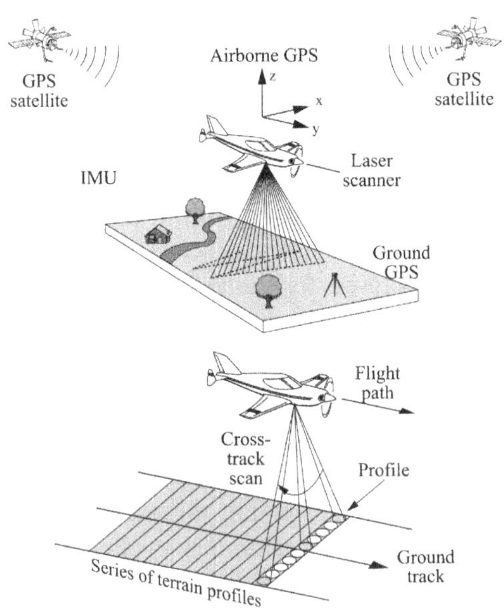

Figure 4.9. *The principle of LiDAR mapping [MAL 10]*

Light detection and ranging (LiDAR) mapping combines three proven technologies: the compact and robust laser range finder (LiDAR), an inertial measuring unit (IMU) and the GPS. Using a vantage point's position, and measuring, via laser, the distance to the point being measured, the latter's position is calculated. By integrating the three aforementioned technologies into a single instrument, installed on board of an aeroplane, a helicopter or a truck, it becomes easy to produce topographic, digital, three-dimensional and accurate maps depicting the terrain, the structures and the objects being scanned. The measuring frequency can reach 500,000 points per second.

With the stronger technological tools available to us, it is now possible to process these large amounts of data.

4.1.2. *Measuring distance directly*

The distance between two points can be measured directly and with great accuracy using a device equipped with an INVAR wire (distomatic or distinvar). The concept of this means of measuring distance is to control the variation in length between two points at different times, using an INVAR wire (which is temperature-resistant) subjected to a constant horizontal force.

The INVAR wire takes the form of a small chain from the influence of its own weight (Figure 4.10).

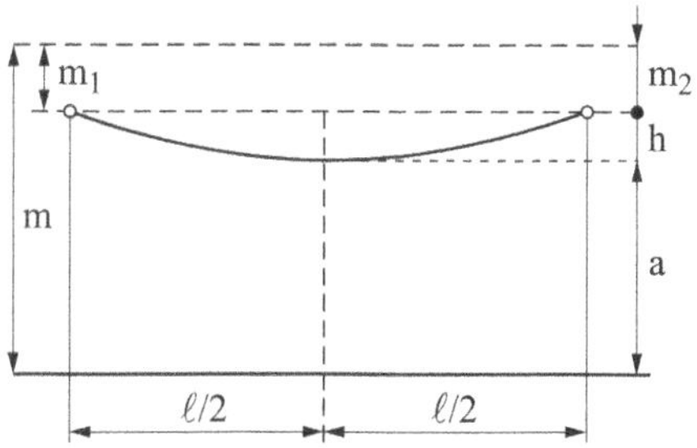

Figure 4.10. *Catenary shape of an INVAR wire*

The constant horizontal force generates an elastic elongation of the wire, which compensates for the small chain form (Figure 4.11). The usual force used on the wires is 50 kN.

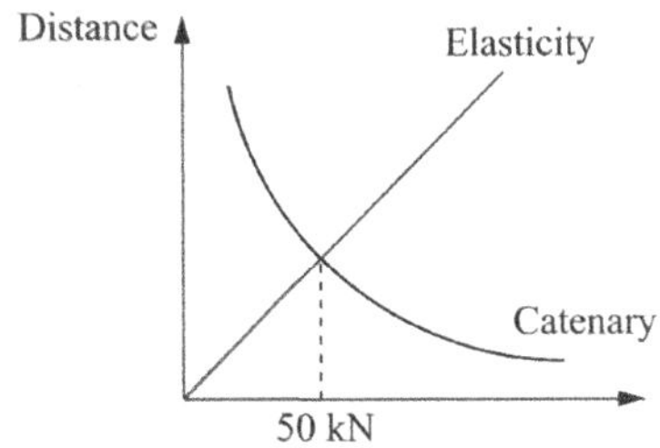

Figure 4.11. *Calculation of the operating voltage of a chain or INVAR wire*

It is clear that when this technique is used at the ground level, no tension force should be applied to the INVAR wire, as the small chain form will not occur.

This method of directly measuring distance is very accurate (2/100 mm) for short distances.

However, it provides only a relative value (the distance between two points) using the INVAR wire, and cannot be used to determine the absolute direction for movement.

It also necessitates a fixed wire length between each pair of measured points.

This technique is used:

– in tunnels, to measure their convergence (Figure 4.12);

– in rock-slope stability, to determine the relative movement between the two edges of a crack at the top of landslides;

– between the stable slope and the unstable boulder, in the case of boulder falls.

To measure longer distances, with lower precision (in the order of millimeters), a modulated electromagnetic wave is used (Figure 4.13), through a white light, infrared or laser-type carrier wave.

These types of distance meters can be associated with opto-mechanic or opto-electronic the dolites.

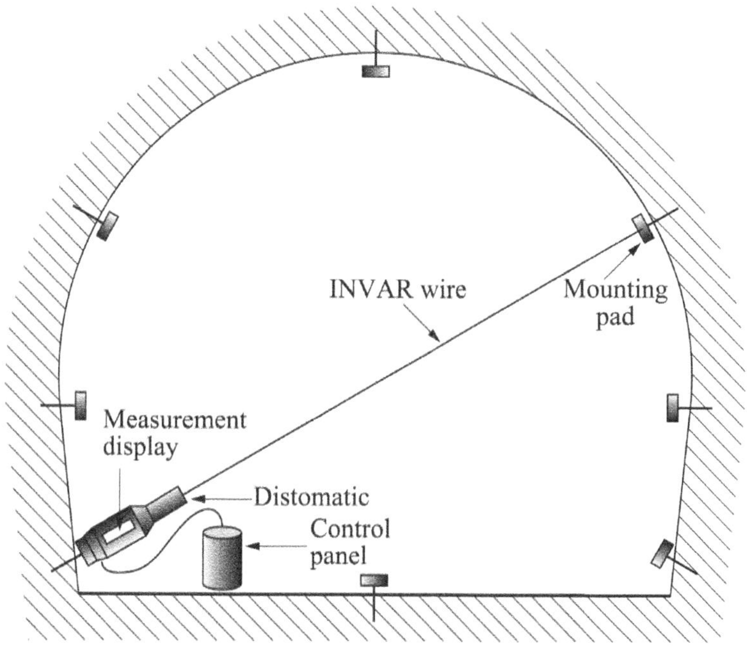

Figure 4.12. *Measurement of a tunnel's convergence using a distomatic*

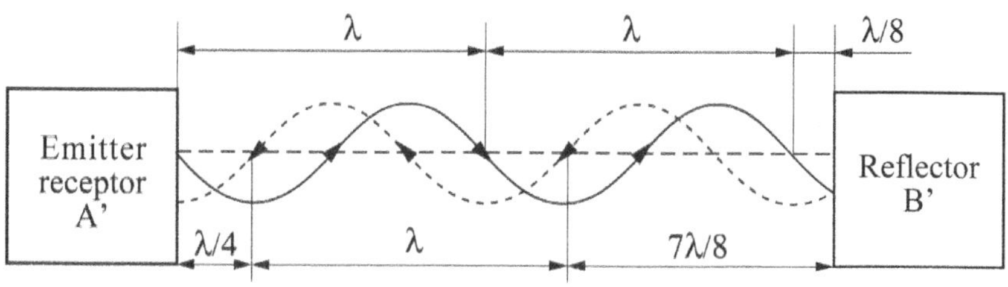

Figure 4.13. *The principle of opto-electronic distance measurement*

4.1.3. *The inclinometry technique*

4.1.3.1. *Inclinometry measurement principle*

The inclinometer is a device widely available in civil engineering (companies). It is used to determine the deformation, over time, of a tube placed vertically on the ground. If possible, the tube will be anchored to a healthy and motionless substratum, and deformed by the movement of the terrain (Figure 4.14). The inclinometer probe (Figure 4.15) is in the form of a metallic tube equipped with four retractable wheels, and is around 50 cm in length. It is inserted into the inclinometer tube, which is itself equipped with four lengthwise grooves, into which the probe's wheels slide. The probe is held into place at a level determined by its electrical wire. This device is based on the same principle as the accelerometer's. A floating mass is held into a constant position by a magnetic servo-system.

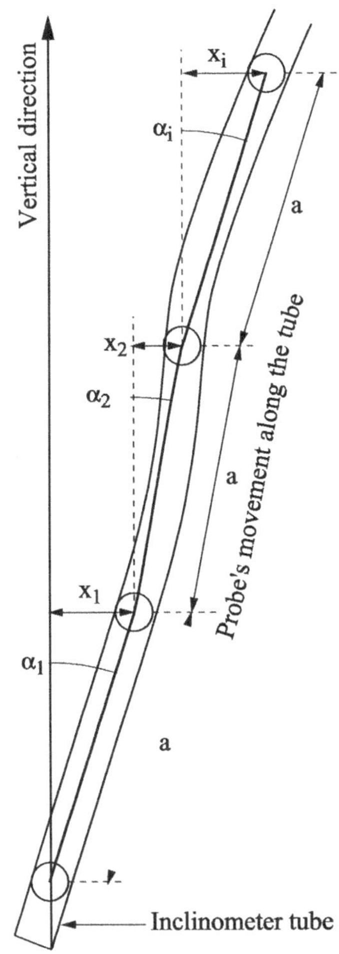

Figure 4.14. *An inclinometer measuring system*

Figure 4.15. *The Glötzl probe*

Two inclinometer sensors (servo-accelerometers) are placed on perpendicular planes A+ A- and B+ B-. The tube can be placed vertically in order to measure horizontal movement, or horizontally to measure vertical movement.

The test procedure is subject to a standard [AFN 95b].

The probe's tilt angle is measured relative to the following two axes:

– to the precision of 10^{-5} radian for one measurement in the plane of the four wheels for one measurement;

– to the precision of 3.10^{-5} radian and for one movement in the plane perpendicular to the plane of the four wheels.

and for a moderate inclination not exceeding 1 radian. The tilt-measuring device has a square base and is inside the probe that can slide into a grooved tube of 50 mm in diameter.

It should be noted that the value of the movement (x) at a given level (i) in relation to the tube's bottom end is provided by formula [4.2], under the assumption that the toe of the inclinometer tube is stationary:

$$x_m = a.\Sigma_{i=1}^{n} \sin(\alpha_i) \qquad [4.2]$$

– a: step of measure (distance between two measuring levels, 50 cm on average);

– α_i: probe's inclination at level (i).

The GLÖTZL device presents the angle's radian angle, or its sinus value.

The interpretation of these measurements provides the localized movement, that is the values of a . sin α_i between two distinct levels, or the cumulative movement, that is the value of x_m. An example of inclinometer measurements results is shown in Figure 4.16.

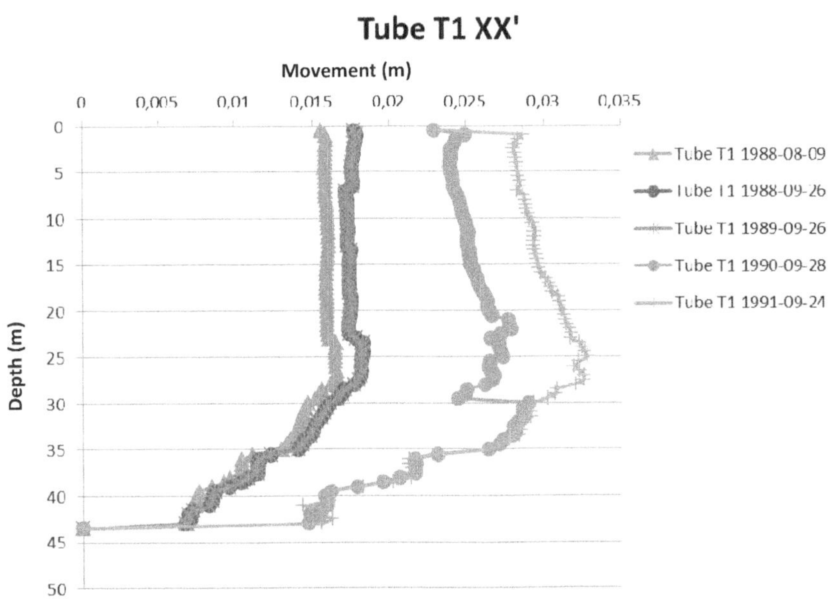

Figure 4.16. *Inclinometer measurement results obtained from the landslide which took place in the Mas d'Avignonet, between 1987 and 1992*

4.1.3.2. *Practical implementation of the survey*

Every measurement contains inherent errors. The quality of inclinometer measurements is susceptible to several external factors:

– outside temperature;

– the grooved tube's quality;

– the probe's temperature;

– the rise's speed, etc.

In order to obtain accurate measures, it is necessary, among other things:

– to let the probe's temperature stabilize at the bottom of the borehole;

– to take the time to wait for the angle measurement's stability;

– to correctly read the device's display;

– not to roll up the electric cable during the rise, but only once the measurements are over, as the sliding contact does not have a uniform resistance;

– to input the A+A– direction (which is the plane of the four wheels) in the direction of the movement being measured.

4.1.3.3. *Interpretation*

The inclinometer helps to determine, in the order of a few tenths of a millimeter, the position of the drill at a given moment. It provides:

– the level, the value of the movement relating to potential failure surfaces (Figure 4.16);

– the thickness of the sliding layers;

– potentially, the tilting, due to the creep of the ground's layers;

– the overall displacement of the surface versus the substratum (2.9 cm from 1988 to 1991 on Figure 4.16).

4.2. Measuring the settlement

To measure the settlement of a backfill, or any soil in general, several mechanisms are available, please use the same char policy.

4.2.1. *Plate settling gauge*

A rigid plate is connected to a rigid rod, which is protected by a flexible tube (Figure 4.17). The plate is fixed to compressible ground. The settlement is measured by the difference, at different time (Figure 4.18) of the rod's emerging length. The test's implementation is difficult but accurate.

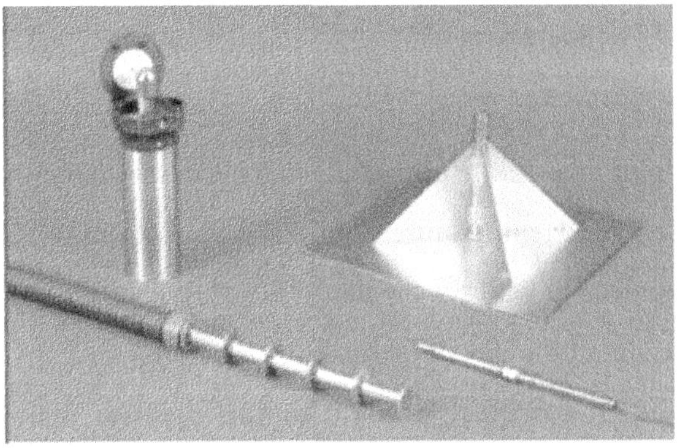

Figure 4.17. *Example of a plate settling gauge*

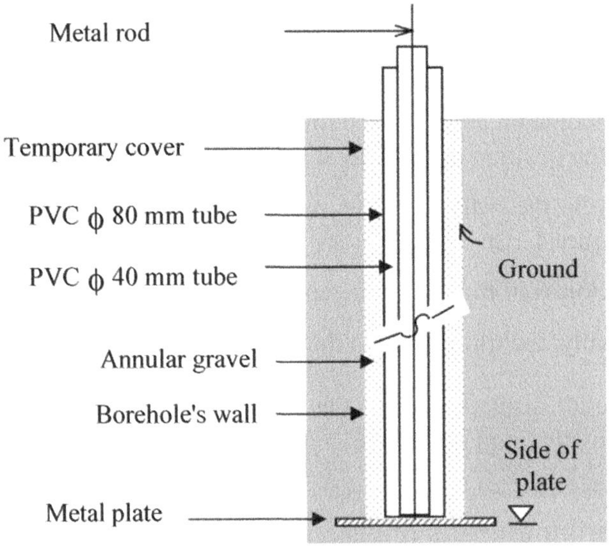

Figure 4.18. *Plate settling gauge's installation*

4.2.2. *Hydraulic settling gauge*

A tank, for which the level has been determined, is positioned under a structure or on the ground where the settling is to be measured. The tank is connected to the measurement panel by two pipes (Rilsan, 4 × 6 mm), one used for liquid (Figure 4.19), the other used for gas. The same pressure (p_0) is applied to the liquid in the pipe from test to test; the water column's height (h_0) is, therefore, always balanced. As the ground settles, this column's transfer is read directly on the graduated measuring tube T, on the measuring table. When the settling's amplitude exceeds the tube's length, the reference pressure (p_0) needs to be changed. An air outlet introduces a constant pressure. Another method can be used which is to adjust the hydraulic level toward a fixed vertical mark. Air is blown into the pipe to adjust the water pressure, bringing it to the original measurement taken during the settling gauge's installation (Figure 4.20). Under these conditions, the settling is measured by formula [4.3].

$$H_t = z_b - z_a = \frac{p_{air}}{\gamma_w} \qquad [4.3]$$

The difference in hydraulic levels determines the settling. Its precision is in the order of millimeters. This type of settling gauge is used in highway projects or backfills.

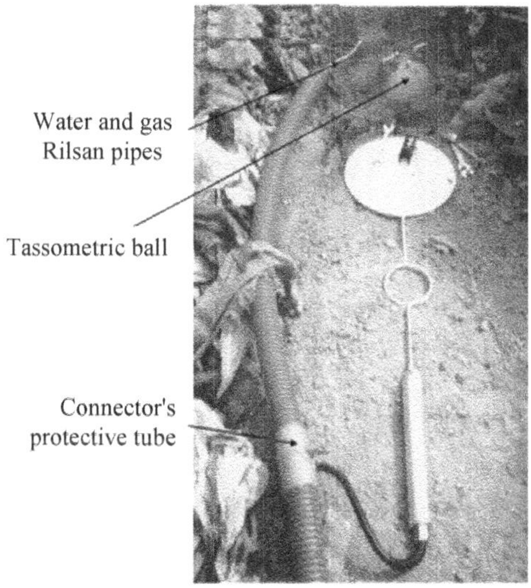

Figure 4.19. *A hydraulic settling gauge*

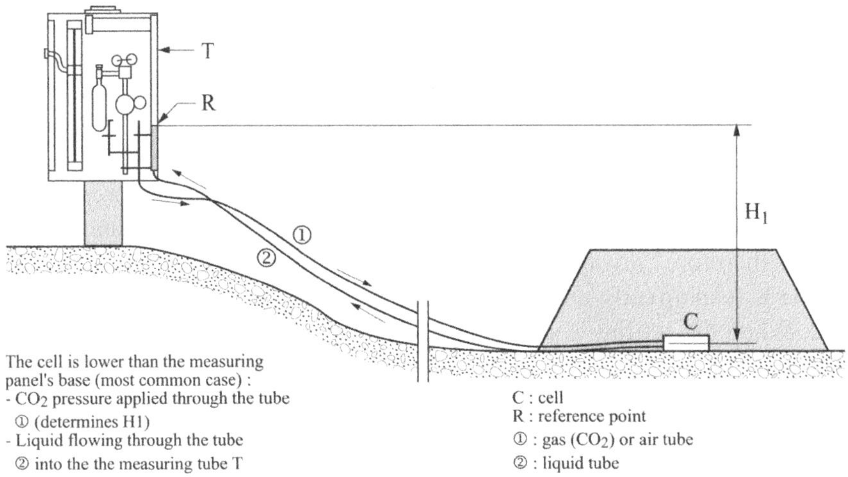

Figure 4.20. *Use of a hydraulic settling gauge*

4.2.3. *Magnetic settling gauge*

Magnetic marks are introduced along a borehole (Figure 4.22), which can be placed either during construction or inside the borehole. The tubing is deformable. A probe (Figure 4.21) is lowered into the tube and emits a sound signal as it passes the magnetic marks.

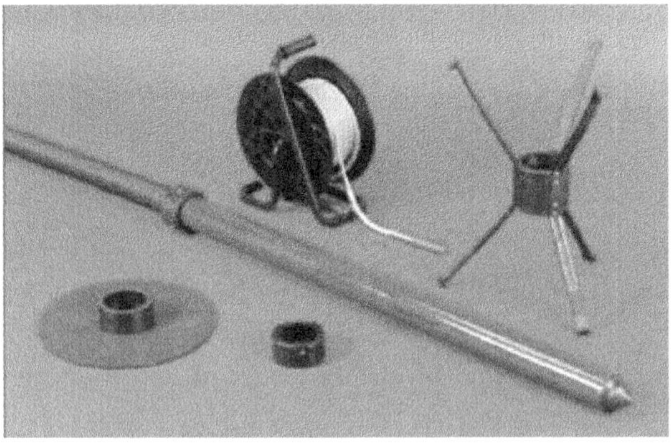

Figure 4.21. *Example of a magnetic settling gauge*

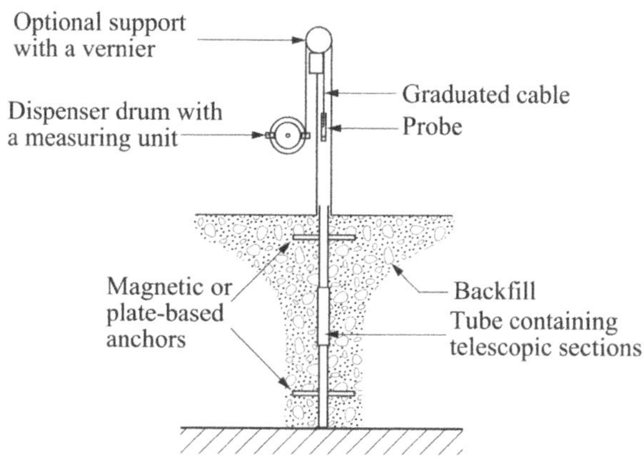

Figure 4.22. *Magnetic settling gauge's installation*

4.3. Force transducers

Force transducers are used to control the tension of ground anchors, prestressing cables, etc. (Figures 4.23 and 4.25). The AFNOR.– NF P 94-153, 1993 standard specifies the necessary measures for nail pull-out tests. A force transducer is a pad consisting of high-strength annular disks, with flexible inner and outer edges. The hollow part of the pad is filled with a liquid (Figure 4.24) allowing for the transduction of hydrostatic pressure into liquid (Figure 4.24). When the force transducer transmits the pressure to the central measurement unit, care must be taken to respect hydrostatic conditions, namely, waiting for the stabilization of the measurement and avoiding circulating the hoses under the sun so as to maintain a constant temperature in the circuit.

Figure 4.23. *Example of a force transducer*

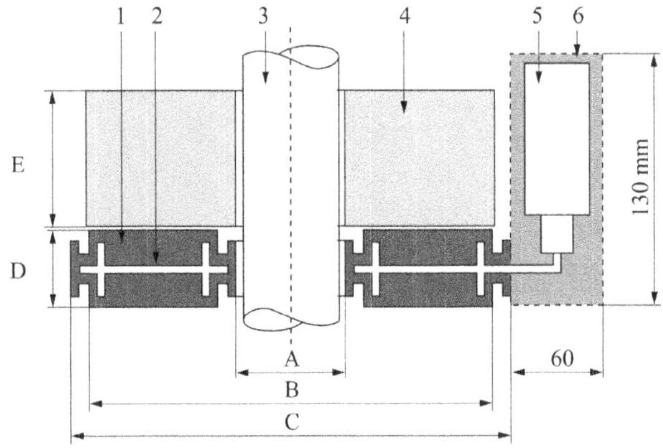

Figure 4.24. *Diagram of a force transducer [PIR 12]*

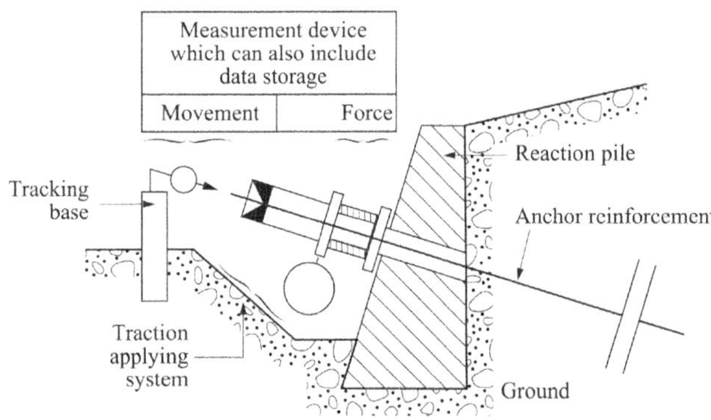

Figure 4.25. *Installation of a force transducer [AFN 93b]*

5

Static Loading Tests

5.1. Plate loading test

Two types of tests can be distinguished:

– low-pressure loading tests, to determine the soil's stiffness;

– high-pressure loading tests, to determine the safety coefficient related to the limit load.

5.1.1. *Low-pressure loading test*

The French standard AFNOR NF P94-117-1 2000 determines the test's operation mode.

5.1.1.1. *Test principle*

The test consists of applying a supposedly even pressure to the surface of the ground by means of a rigid plate. Its settling is then measured by a probe, which helps to determine the soil's vertical elasticity modulus.

The Boussineq formula, corresponding to the settling of an evenly charged and circular soft base, is written as follows:

$$s = 2.\frac{(1-\upsilon^2)}{\mathrm{E}_\upsilon}.p.R \qquad [5.1]$$

where s: deflection or settling at the center of the base;

ν: Poisson's ratio (generally equal to 0.33);

E_v: the soil's vertical elasticity modulus;

p: plate's uniform loading pressure;

R: plate's radius.

In the case of a rigid circular plate of radius *R* and uniformly charged with a pressure *p*, it is proven that:

$$s = \frac{\pi.(1-\upsilon^2)}{2.E_\upsilon}.p.R \qquad [5.2]$$

In the case of a rigid square plate of side *a*, and uniformly charged with a pressure *p*, it is accepted that:

$$s = 1{,}76.\frac{(1-\upsilon^2)}{E_\upsilon}.p.a \qquad [5.3]$$

Other operation modes exist for plate loading tests *m*, which we will describe in more detail further on.

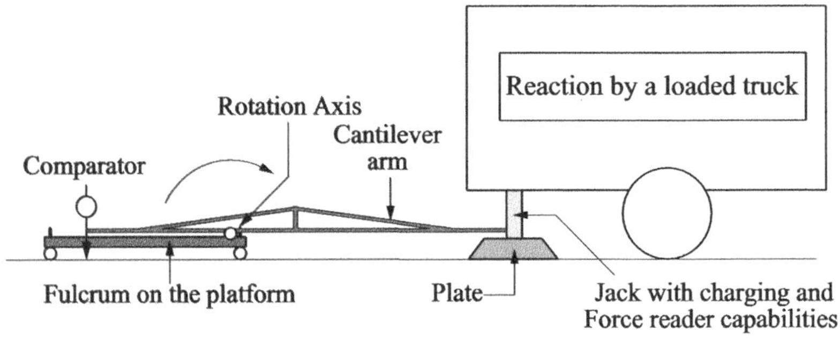

Figure 5.1. *Plate loading test's installation*

5.1.1.2. *Test implementation and necessary precautions*

The circular plate (Figure 5.1) is usually charged by a hydraulic jack. The deflection is carried over into a stable zone in the ground by a cantilever arm (Figure 5.2). A fulcrum is used as a reaction fund, in general a truck chassis or a weighted portico on which the hydraulic jack rests. The cantilever arm is usually between 3 and 5 m in length, and requires a space with sufficient clearance. It supports the comparator on which the plate's settling will be read. The arm's rotation axis support, as well as the comparator's support, needs to be far from the loading plate in order to stay in a stable area of the ground's surface.

Figure 5.2. *Plate loading test*

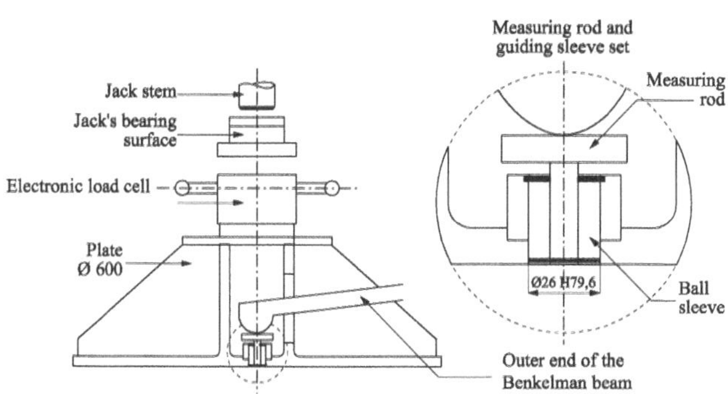

Figure 5.3. *Circular charging plate*

Care must be taken regarding the very limited depth for this test, of around 1.5 times the diameter (Figure 5.4). This ground surveying is primarily used for roads and airfields.

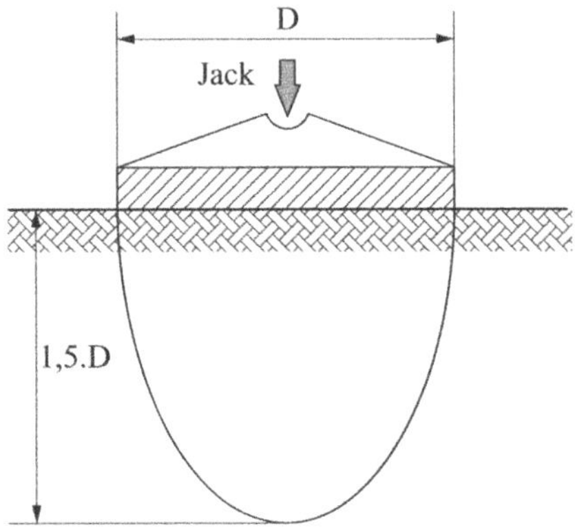

Figure 5.4. *The depth in the area's soil tested with a plate-loading test*

A movement sensor is generally used within a hundredth of a millimeter of the cantilever arm's side. For some tests, carried out without a cantilever arm, three or four movement sensors are used, placed on the edge of the plate on two orthogonal diameters.

Before placing the plate on the terrain, the ground must be stripped to the level projected for the test. Good contact between the ground and the plate is ensured by depositing a layer of fine sand, plaster or cement mortar, the thickness of which should not exceed 1 cm (Figure 5.5). A pressure of 10 kPa is applied during 30 s in order to properly set up all of the device's elements. The pressure is then relieved and the comparators reset.

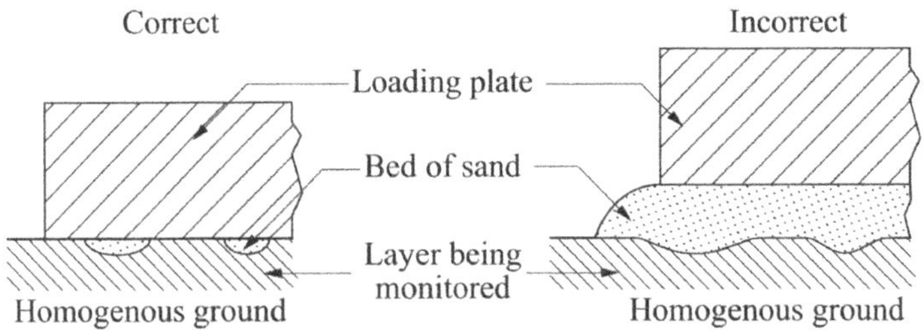

Figure 5.5. *Recommended width for the sand layer*

5.1.1.3. *Plate loading deformability test – the Swiss VSS Test*

Different methods exist for plate loading deformability testing.

The plates' diameter is of 16 or 30 cm, (Swiss norm SN 670 317b) and they are loaded in stages, in a single cycle.

The loading stage is changed when the plate's sinking falls below 0.02 mm per minute.

The ground's modulus is obtained with formula [5.4]:

$$M_E = 2.\mathrm{R}..\frac{\Delta\sigma_i}{\Delta s_i} \cong \frac{E_v}{0{,}75} \qquad [5.4]$$

where M_E: the ground's modulus;

$\Delta\sigma_i$: the loading stage (see Figure 5.6);

Δs_i: the settling stage (see Figure 5.6).

Keeping in mind that M_E is different from the ground's elasticity modulus (E_v). The ratio between the two close to a coefficient of 0.75 [5.4]. Plates with a small diameter can lead to errors relating to the ground's plastification on the plate's edge.

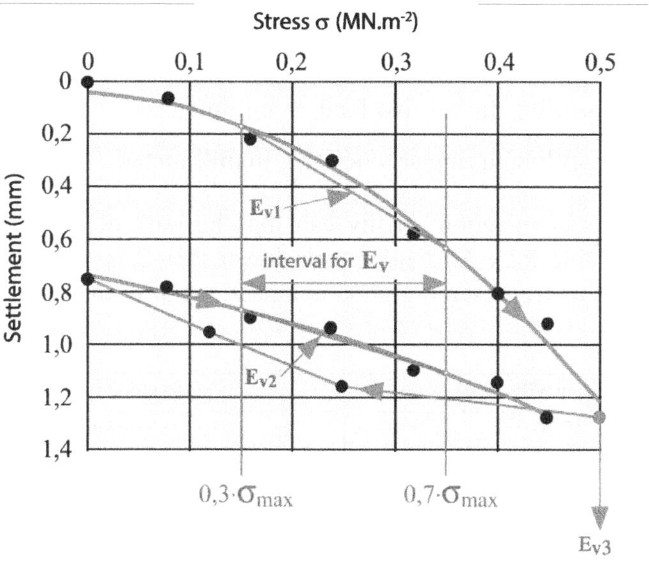

Figure 5.6. *Loading plate test interpretation*

5.1.1.4. *Plate loading deformablity test – The standardized plate test*

This test follows the standard AFNOR NFP94-117-1. The plate has a diameter of 60 cm. The following two loading cycles are carried out:

– the first cycle uses a loading power of 0.25 MPa (7,068 daN) applied during at least 30 s. The pressure is maintained until the deformation is stable (<0.02 mm in 15 s). The total settling is written as e_1;

– the second cycle is pushed up to a load equal to 0.2 MPA (5,654 daN);

– E_{v1} is calculated during the first cycle [5.5]. It represents the soil's deformability in its original state.

The second modulus, E_{v2}, is calculated during the second cycle [5.6]. It represents the deformability's evolution with regard to compaction:

$$E_{v1} = \frac{\pi.(1-v^2)}{2}.\frac{\Delta\sigma_i}{\Delta s_i}.R = \frac{112{,}2}{e_1} \quad [5.5]$$

$$E_{v2} = \frac{\pi.(1-v^2)}{2}.\frac{\Delta\sigma_i}{\Delta s_i}.R = \frac{90}{z_2} \quad [5.6]$$

where E_{v1}: ground's modulus at first load;

E_{v2}: ground's modulus at second load;

e_1: total settling during first load, in millimeters;

z_2: total settling during second load, in millimeters.

The ground's compaction quality can thus be ranked in relation to the modulus E_{v2} (Table 5.1). An E_{v2}/E_{v1} ratio lower than 2 is recommended for common platforms (representing good compaction) or even below 1.2 (great compaction).

E_{v2} in MPa	Roadbed rank
0–20	Unusable
20–50	PF1 class
50–120	PF2 class
120–200	PF3 class
> 200	PF4 class

Table 5.1. *Roadbed ranking according to the E_{v2} modulus*

Because we operate below p_{max}, the influence of plastic deformation is very limited on E_{v2}; this gives a modulus that is very similar to the conditions of the soil's elasticity. A graphical representation of a plate-loading test is shown in Figure 5.7.

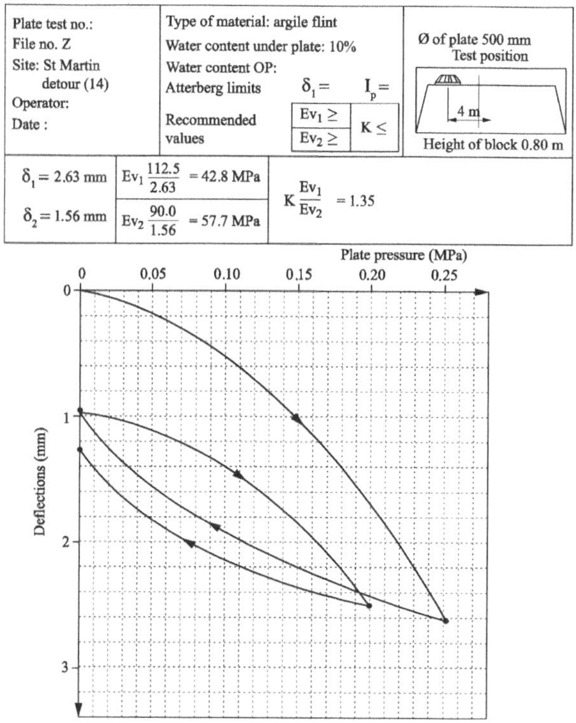

Figure 5.7. *Plate loading test*

5.1.1.5. *Plate loading deformabilty test – The Westergaard test*

This test is subject to the AFNOR NFP94-117-3 standard. The plate has a diameter of 60 or 75 cm. Only one loading cycle is carried out. The test's maximum load is 70 kPa. When the sinking falls below 0.05 mm per minute, the pressure stage is changed.

Unloading and reloading cycles can potentially be included in this test. The plate's thickness must at least 25 mm. It is composed of a stack of four layers with a minimum thickness of 10 mm and declining diameters (75, 60, 45 and 30 cm).

The Westergaard modulus is then calculated with the following formula:

$$K_w = \frac{70}{\Delta s} \qquad [5.7]$$

where K_w: Westergaard modulus in kPa;

Δs: plate's deflection under the 0–70 kPA load.

K is not an elasticity modulus, it integrates plastic deformation.

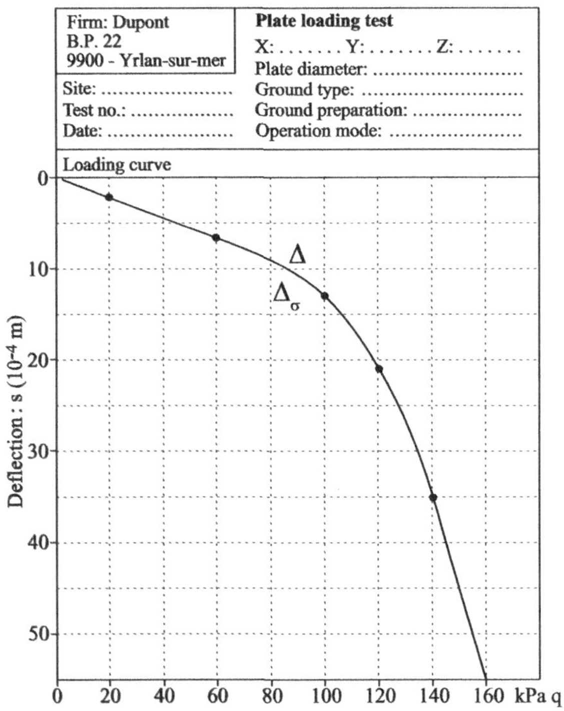

Figure 5.8. *Punching loading test*

5.1.2. *High-pressure loading test*

This test is carried out with a small section bearing plate (100–1,000 cm^2) located in the bottom of a well, at the presumed level of the foundation. The loading is carried out in stages and the corresponding settling measurements are noted and reported.

5.1.2.1. *Ground loading up to punching point*

The section of the plate must be between 100 and 400 cm^2. The test is carried out to punching point, or to the device's limit.

Each pressure stage must be kept constant until the curve representing the deflection in relation to the logarithm of time is linear (Figure 5.9).

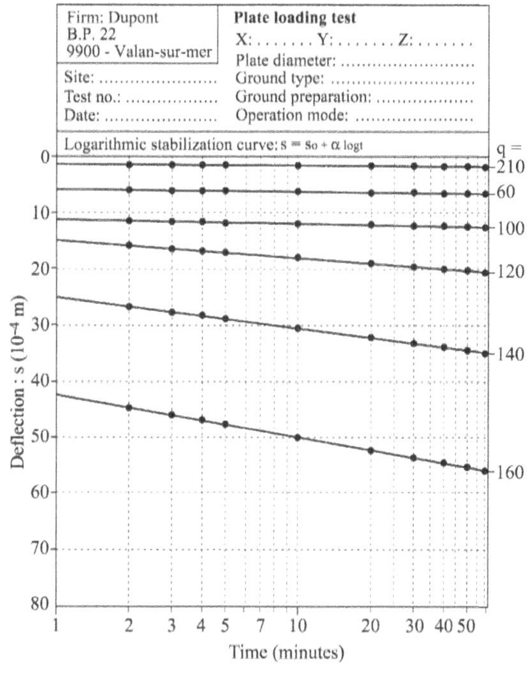

Figure 5.9. *Punching test; evolution of settling over time*

A minimum of four measurement points over a period of 1 h are sufficient.

The test helps to determine the following parameters:

– the slope of the straight line's angular coefficient (α) giving the deflection's evolution in relation to the decimal logarithm of time [5.8], as shown in Figure 5.9;

– a critical pressure, or creep pressure q_f, corresponding to the first point where the slope changes, for the curve representing evolution α in relation to load q (Figure 5.10). Another way of measuring q_f is to take the slope's change on the curve linking the settling to load q, but this method is less precise than the previous method (Figure 5.8);

– a deformation modulus E_c, calculated with the formula [5.9];

– the foundation's allowable stress value Q_a, taken as a fraction of creep pressure q_f at the time of punching [5.10]. This allowable stress value contains an inherent flaw as it depends on the plate's diameter, as does the tested ground itself.

$$s = s_0 + \alpha . Log_{10} t \tag{5.8}$$

$$E_c = 0{,}7 . B . \frac{\Delta\sigma}{\Delta s} \tag{5.9}$$

$$Q_a = 0{,}3 . q_f \tag{5.10}$$

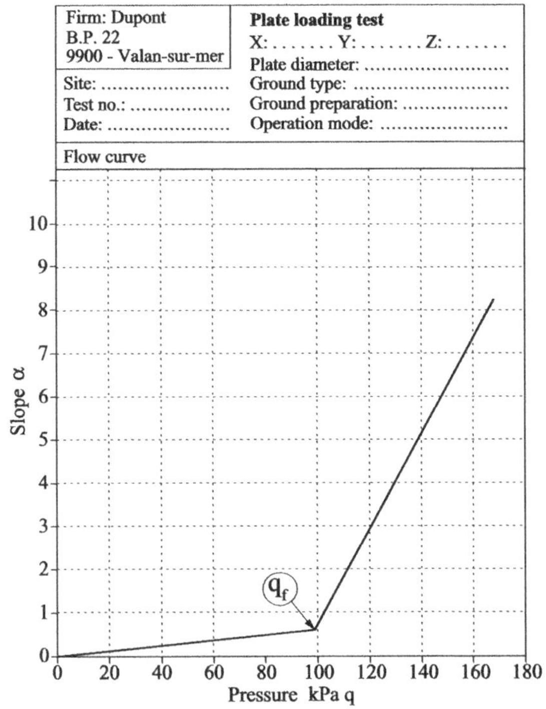

Figure 5.10. *Punching test; slope evolution in relation to load*

where: s_0: deflection's value at the start of each pressure stage;

t: deflection (s)'s corresponding time (0 at the start of each stage);

B: plate's diameter;

$\Delta\sigma$: elastic load increments (see Figure 5.8);

Δs: Δq's corresponding deflection value (see Figure 5.8).

5.1.2.2. *Ground loading up to a settling reference point*

This method was developed in order to anticipate the settling of foundation structures.

It is based on the work by Kogler and Scheidig [KOG 38], which relates the settling (s) of the foundation's base, the settling (s_1) of the plate, subjected to the same surface-load and the foundation's size (B) using formula [5.11]. The test is done on a 1,000 cm^2 plate and the load is applied gradually from 10 kPa to 1.5 times the admissible working rate. The admissible working rate will be either:

– half the load corresponding to the settling of 1 cm;

– or the load which provides the settling measurement s_{1l} of the plate, compatible with the foundation's settling limit (s_l), according to the formula [5.12].

$$s = s_1 . \left(\frac{2.B}{B+0.3}\right)^2 \qquad [5.11]$$

$$s_{1l} = s_l . \left(\frac{B+0.3}{2.B}\right)^2 \qquad [5.12]$$

where B: width of the proposed foundation structure (m).

5.2. Static pile-loading test

5.2.1. *Test principle*

The pile-loading test consists of loading a pile vertically or horizontally up until its load-bearing capacity or its critical load, in order to verify that the effective load is in line with the expected load. Three types of tests exist:

– pile sinking test;

– pull-out test;

– lateral loading test.

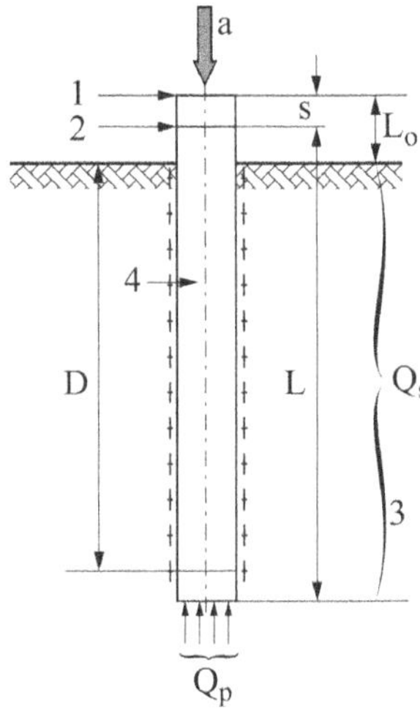

Figure 5.11. *Shaft-loading test principle (NFP94-150 standard)*

These three applications are different and depend upon the design of the structure being built. The most common is the pile-sinking test. It will be described here and must follow the AFNOR NFP94-150, 1990 standard. It consists of:

– applying, on the pile's tip, a static axial force according to a given program;

– measuring the axial movement at the shaft's tip as well as potentially measuring force at different levels along the shaft.

The pile's load, Q, consists of the peak force (Q_p) and the lateral friction force, Q_S.

On a section of the pile of any given length ΔL between the extremities of which the total force on the shaft varies by ΔQ, the lateral friction force, Q_s, can be determined by formula [5.13] with *p* the perimeter of the pile.

$$q_s = \left(\frac{\Delta Q}{p.\Delta L}\right) \quad [5.13]$$

5.2.1.1. *Test class*

There are two ways of using the pile-loading test, either as a preliminary test or as a control test.

The preliminary test helps to:

– check that the pile can be subjected to its planned applications;

– reach the ground's bearing capacity limit, as long as the pile's constituent materials' conventional yield strength limit is not reached;

– optimize the characteristics of the piles that will be used in the structure being built;

– offer more precision to the geotechnical study's findings.

The control test helps to:

– ensure the shaft's quality of performance, its integrity and behavior under the planned applications.

5.2.1.2. *Necessary precautions*

In order to be correctly carried out, the pile sinking test must follow certain conditions:

– the test is expensive and must, therefore, be absolutely necessary;

– the test must be used to test the foundation's ground, not the yield strength of the concrete in the shaft;

– the test can be carried out if its results can be extended to all the shafts being used in the structure;

– the test can be carried out if the use of various geotechnical studies lead to differing results;

– the test can be carried out if its construction mode can influence the result.

5.2.2. *Practical realization of the test*

The practical realization of the test is described in the sections that follow.

5.2.2.1. *Waiting period*

A waiting period must be respected between the pile's construction and the loading test. The values highlighted in Table 5.2 are recommended.

	Compact sand	Saturated silt Saturated fine sand	Clay
Pile implemented on-site	1 month	1 month	1 month
Driven pile	8 days	20 days	1 month

Table 5.2. *Recommended waiting periods before the loading test*

5.2.2.2. *Reaction funds*

The loading of the pile can be done with:

– a weight-based reaction mass (Figure 5.12);

– an anchor-based reaction system (Figure 5.13). This can be achieved by injected tied rods, or by using neighboring piles connected by metal beams.

Weight-based reaction mass can only be used for vertical loads. They are affordable only for loads lower than 3,000 kN and are unusable on soft-surface soils.

The anchor-based reaction systems are more useful, but require a longer installation.

5.2.2.3. *Test preparation*

Before beginning the test, all the information relating to the site and the structure being built must be gathered, as well as the data relating to the pile being used in the test.

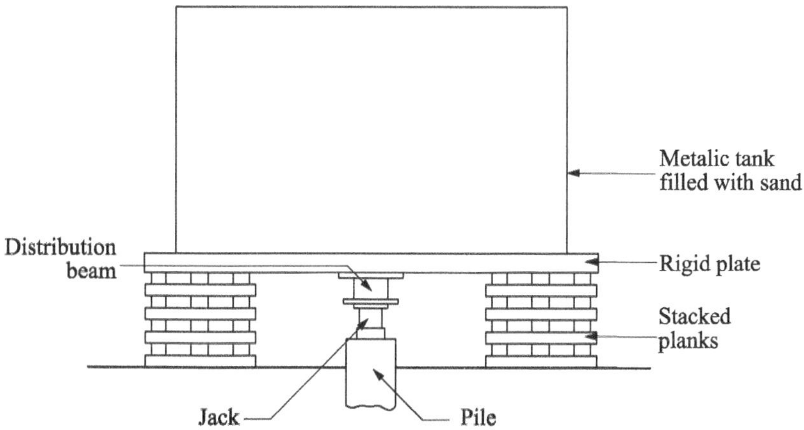

Figure 5.12. *Example of a weight-based reaction mass [CAS 81]*

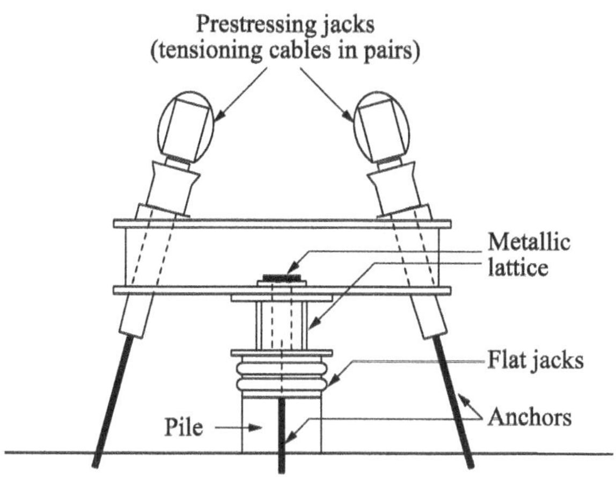

Figure 5.13. *Example of an anchor-based reaction system [CAS 81]*

5.2.2.3.1. Site and structure data

The following must be recorded:

– the mass plan;

– the complete report on the ground surveying that has already been carried out, containing the survey implementation plan, survey log and geotechnical tests;

– the hypothesis used relating to the pile's dimensions (geotechnical sections, mechanical ground characteristics, hydraulic regime, number of piles that need to be built by type and efforts to be transmitted);

– the topographic landmark and its coordinates.

5.2.2.3.2. Data relating to the pile being used

The following must be recorded:

– the implementation plan;

– the mode and technology for implementation;

– the date of the pile's production and potentially, its chronological incidents;

– the characteristics of the pile's nature:

- wood (origin, species, nature of the surface and its potential processing),

- concrete (composition, concreting dates, compression resistance of samples of which the origin, the sampling method and the position are specified; elasticity modulus, reinforcement scheme and nature of the steel),

- metal (grading, elasticity modulus and profile type).

– the piles characteristics relating to its realization:

- boring,

- sinking (impact, vibration, deflection, screwing and the combination of these techniques),

- injection-based boring,

- lost sheath (length and thickness),

- the process of partial or total emptying of tubular metal piles, their filling method, the different materials in the tube.

– the pile's geometrical characteristics are:

- its length, L,

- its cross-sectional area, A,

- its potential frame's cross-sectional area,

- its inclination relative to the vertical,

- the pile's materials' yield point, Q_G,

- its limit state design (LSD) and serviceability limit state (SLS), if known,

- the position and characteristics of reserve piles for the potential use of removable extensometers, or for the measurement of the tip or other sections of the pile's movement,

- the estimated value of the limit load Q_L, gathered from geotechnical data, and of the maximum load, Q_{max} [5.14], to which the test pile will be subjected at bearing capacity,

- the estimated maximum elastic shortening of the pile, in the absence of the terrain's lateral friction,

- for the control test, the maximum force Q is determined in relation to the axial compression force that is applied to the pile being used [5.15].

$$Q_{max} = 1{,}3.\, Q_L \text{ and } Q_{max} \leq 0{,}9.\, Q_G \quad [5.14]$$

$$Q_e \leq 1{,}3.\, Q_{ELS} \quad [5.15]$$

5.2.2.3.3. Pile preparation

This operation consists of:

– making the tip of the pile, which will be attached to the horizontal distribution plate and which must resist the stress induced by the application of the load;

– constructing a window through which the cables needed for the measurement devices will pass;

– marking the position of the movement probes and determining their position in relation to North;

– positioning the removable extensometers (if used) in the reserve piles that are provided for that purpose and fixing them to the contractually predetermined levels.

5.2.2.3.4. Controls and checks

Measurement instruments (gags, electronic sensors and their measurement chain, spring scales and dynamometer rings) must be calibrated every year:

– the removable extensometers must be calibrated before every test;

– the extensometers embedded into the pile must be supplied with their calibration curves.

A verification report must be issued including, for each device or measuring system, the following information:

– the date of their last calibration;

– the device's ID and its place of calibration;

– the temperature during its last verification or calibration;

– the name of the certified verification organization;

– the validity's expiration date.

Once the axial load's, the extensometer's (if any) and the measurement device's reset values have been noted, the control of the installation and the devices' smooth running is done by loading the pile with the recommended program (Figure 5.14).

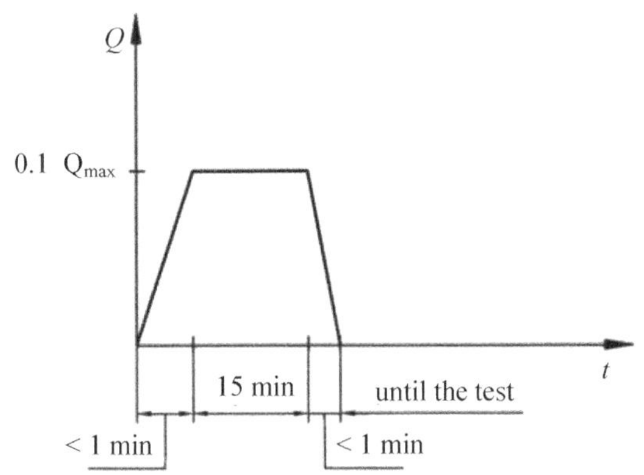

Key

Q Compression's axial force

t Time

Figure 5.14. *Loading and reloading program in preparation for the test [AFN 90c]*

A force of 0.1 Q_{max} is applied for 15 min. The force and movement measurements are taken at the following times t = 1, 2, 3, 5, 7, 10, 15 min. The extensometers' measurements (if any extensometers are being used) are taken at time t = 10 min.

After unloading, the measurements are taken with the same frequency in order to detect any abnormality or eccentricity of the force being applied in relation to the bearing surface's center of gravity.

The equipment will remain in place until the beginning of the test so as to maintain the settings used during the preparation phase.

5.2.2.4. *Measurements*

The primary measurements taken are the load at the pile's tip, and the sinking.

5.2.2.4.1. Measuring the load at the pile's tip

Manometers connected to the hydraulic jack's circuit measure the load at the pile's tip effortlessly. The following conditions must be respected for the measurement to be correct:

– development of a ball-jointed link between the jack and the reaction pile (Figure 5.15). A wooden plate held between two steel sheets is a simple way of achieving this;

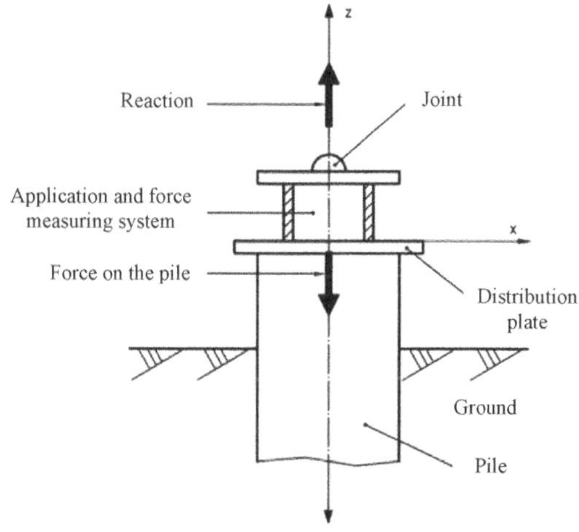

Figure 5.15. *Measurement device for the force at the tip of the pile [AFN 90c]*

– the laying of the piston's locking nut in relation to the jack's cylinder, in order to avoid any waste during the loading of the pile, in the case of a sudden leak in the supply circuit;

– the jack must be able to slide without any friction while it is being loaded;

– the axial force's measurement uncertainty must be lower than 5% of the measured value and 1% of the nominal load Q_{max}.

5.2.2.4.2. Measuring sinking at the pile's tip

The sinking is measured easily enough with comparators (Figure 5.16) fixed to stationary beams, themselves sitting a good distance away on small piles or concrete bases. The measurement principle can be represented graphically (Figure 5.17). The fluctuation of the baselines, due to the following, must be avoided as much as possible:

– wind (the tip of the pile could be covered in order to avoid this);

– temperature (in the case of a metal base);

– water (swelling or expanding of the base, if made of wood);

– the settling of the ground (the settling of the base's support in relation to the pile);

– the operator's inexperience (visitors leaning on the measuring base, etc.).

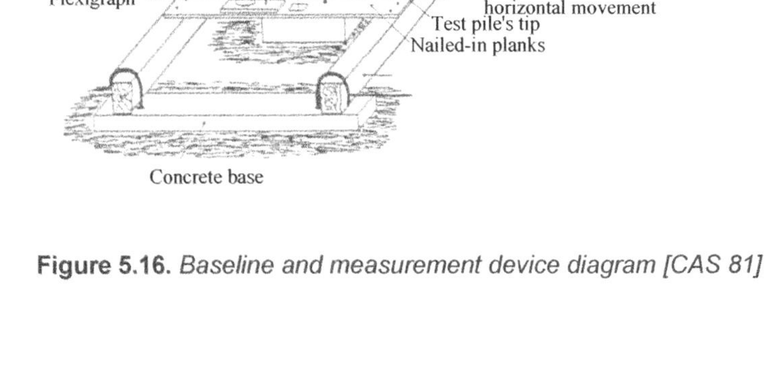

Figure 5.16. *Baseline and measurement device diagram [CAS 81]*

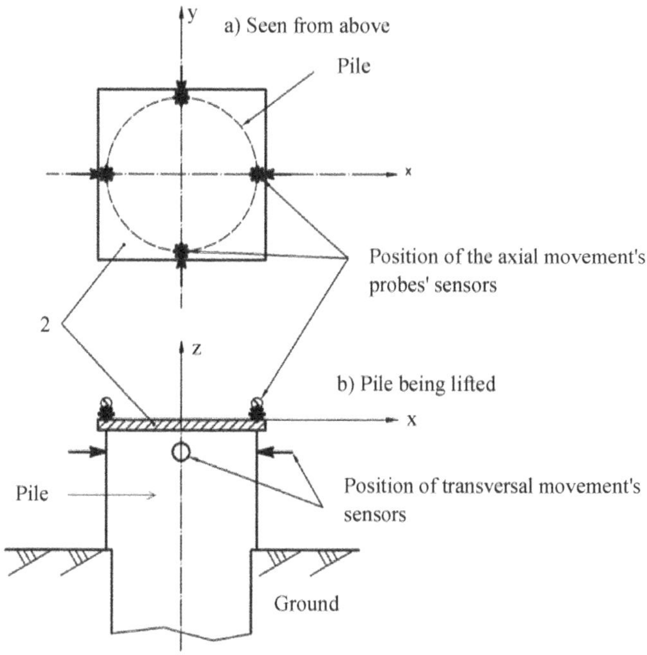

Figure 5.17. *The principle of the measures for movement of the tip of the pile [AFN 90c]*

The following is, therefore, recommended:

– installing the measuring base at least 3 m from the pile;

– covering the pile and the fixed base;

– limiting the circulation of people to the bare minimum;

– establishing a security perimeter in order to deny access to non-experts;

– tying all of the measurements to a fixed point situated about 30 m from the test.

The movement's measurement range must be at least 1.2 times the size of the expected maximum movement. The measurements must be taken away from any temperature variations. The vibrations on-site must be kept at a reasonable level throughout the test.

The test's accuracy must be above 0.1 mm and of 2/1,000 of the given value. The movement must be measured in at least three points of which the barycenter coincides with the pile's axis. The unit for the measurement of time is the second.

5.2.3. *Loading cycles*

Two types of loading can be distinguished:

– progressive loading with no intermediary unloading;

– progressive loading with intermediary unloading (standard, Chellis method in the USA, Soletanche method, operating mode of the Ponts et Chaussées, etc.).

5.2.3.1. *The ASTM standard*

The maximum target test load E is double the pile's supposed working load. The load is applied in increments of 25, 50, 75, 100, 125, 150, 175 and 200% of the nominal charge Q_n.

The loading stage is changed after having obtained a settling speed lower than 0.25 mm per hour. Each stage must be maintained for at least 2 h. The final load is maintained for 24 h.

The unloading is subject to the same stages.

The cyclical loading is optional. In the case where cycles are requested, they are carried out at 50, 100 and 150% of the pile's nominal charge.

5.2.3.2. *Chellis method*

The plastic settling of the pile is to be determined. This is done by means of 2 or 3 loading stages and a total of 3 or 4 intermediary unloading stages.

The following is thus determined (Figure 5.18):

– the settling, for each load, in relation to time;

– the plastic settling, established as residual settling after total unloading;

– the comparison between the measured elastic settling of the pile (δ_m) and the theoretical elastic settling (δ_i) if the ground were a perfect elastic solid.

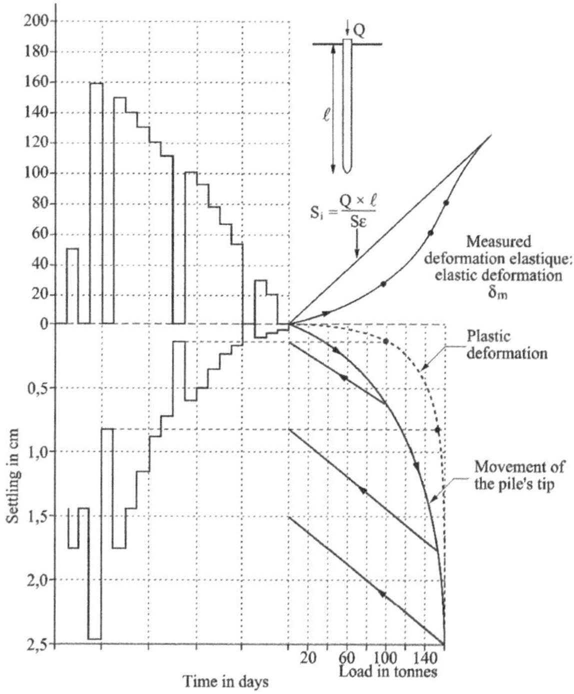

Figure 5.18. *Loading test according to the Chellis method (USA)*

This test's interpretation is entirely empirical. There are rules, which limit movement:

– the the pile nominal load is the half of the load that give a total settllement of 0.025 cm for every 10 kN (Chicago Building Code);

– the pile's nominal load is taken as half the load for which the total settling's increase is higher than 0.75 cm for every additional 10 kN (Bureau of Bridge, Ohio);

– the pile's nominal load is the one for which the total settling's increase is higher than 0.125 cm for every additional 10 kN, or for which the plastic deformation increase is higher than 0.075 cm for every additional 10 kN.

There are rules that limit the settling speed:

– the nominal load is equal to half the test load at which the total settling does not exceed 0.6 cm every 48 h (Louisiana Department of Highways);

– the nominal load is equal to half the test load at which none of the following phenomena occur:

- the settling or elastic rise exceeds 0.6 cm every 24 h, or,

- the permanent sinking, after the removal of all loads, exceeds 1.2 cm. The maximum test load must have been maintained for at least 24 h (Boston Building Code).

5.2.3.3. *Loading method according to the standard [AFN 90c]*

The load is done in two cycles (Figure 5.19):

– during the first cycle, the load will be applied from 0 to 0.5 Q_n (nominal load) by 0.1 Q_n stages, each stage being maintained for 60 min. During the unloading, each stage will last 5 min;

– during the second cycle, the load will be applied from 0 to 0.5 Q_n by 0.1 Q_n stages, each stage will be maintained for 30 min. The load will then be applied from 0.5 Q_n to Q_n, by 0.1 Q_n stages, each stage will be maintained for 60 min. The unloading stages will last 5 min, with unloading increments of 0.2 Q_n.

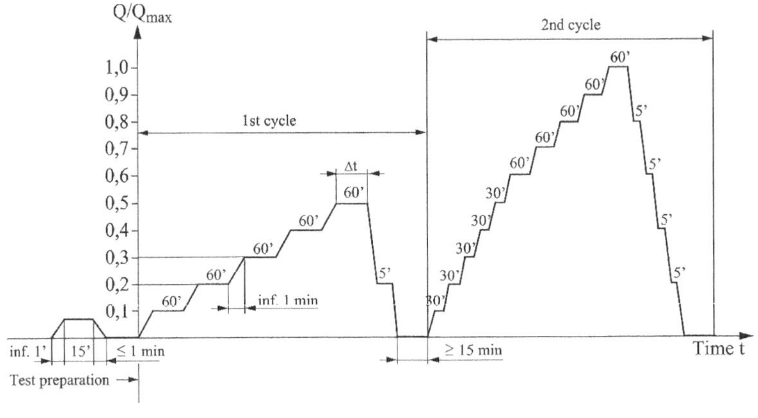

Figure 5.19. *Loading test following NFP 94-150 standard*

The tip of the pile's force and movement measurements are taken from the start of the expected load's application and at the following times: 1, 2, 3, 4, 5, 10, 15, 20, 30, 45, 60 min. During the unloading cycle, the measurements are taken after 1, 2, 3, 4 and 5 min. After the complete unloading of the pile, the measurements are taken at 10 and 15 min.

If extensometers are being used, at least two measurements must be taken:

– the first measurement must start at t = 5 min;

– the second measurement must have finished by t = 60 min;

– the duration of the extensometers' measurements, which are subject to the number and type of extensometers, must not exceed 5 min.

5.2.3.4. *Loading method for control tests [AFN 90c]*

This test is carried out on one of the building structure's piles in order to test it before it is put into use. The test load must not exceed Q_e nor, under any circumstances, the pile's constituent materials' conventional yield strength limit:

– after the test preparation phase, the loading is done in stages of 0.2 Q_e up until Q_e is reached. Each stage is maintained at a constant load for 1 h. The unloading is done in 0.2 Q_e stages, each maintained for 5 min;

– the measurements are taken at the same times as for the punching test.

5.2.3.5. *Soletanche method [CAM 61]*

The Soletanche method arises from the work of Cambefort and Chadeisson [CAM 61], who suggest using smaller load increases as the punching point draws nearer (Table 5.3).

Stage load as parts of Qn										
E=1,5.Q_n	0,1	0,25	0,5	0,75	1	1,25	1,4	1,5		
E=2.Q_n	0,1	0,25	0,5	0,75	1	1,25	1,5	1,75	1,9	2

Table 5.3. *Values for each of the Soletanche method's loading stages*

Three cycle-based tests are used:

– the first test calls for frequent load resets (this allows for a certain reduction in lateral friction). This provides one measurement by settling excess;

– the second test uses a restricted number of cycles (complying with NFP 94-150);

– the third test is carried out quickly in order to simulate overloading. The Soletanche method recommends maintaining each stage until a straight line is obtained on the chart s (settling) in relation to the decimal logarithm of time according to formula [5.16]:

$$s = s_0 + \alpha . Log_{10} t \qquad [5.16]$$

Formula [5.16] is similar to the formula used for plate-loading tests. If the experiment's points do not align, a curve must then be pursued using the formula [5.17]:

$$s = s_0 + \alpha . Log_{10}\left(1 + \frac{1}{m}\sqrt{t}\right) \qquad [5.17]$$

The stages' minimum duration must be 20 min for low loads and 200 min for maximum loads. The Soletanche method supplies the following:

- the sinking curve in relation to time [5.16], [5.17],

- the sinking curve in relation to loads (extrapolation to 100 or 1,000 h based on the previous curves),

- the limit load (or creep load) determined by the chart's sharp elbow, relating the load Q to the formulas' α slopes [5.16], [5.17],

- the pile's nominal load, determined with a safety coefficient F from 1.25 to 1.67 in relation to the creep load Q_f:

$$Q_n = \frac{Q_f}{F} \qquad [5.18]$$

5.2.3.6. *Other methods*

Cycles measuring the elastic rise per unloading can be included.

Around 10 quick unloading and reloading cycles can be included at the end of each stage in order to cause plastic deformations.

5.2.4. *Pile-test interpretation*

The NFP94-150 standard indicates the way of interpreting the sink test.

5.2.4.1. *Movement an at every stage*

This movement is measured between 30 min and 1 h according to the following formula [5.19]:

$$a_n = (s_{60} - s_{30}) \qquad [5.19]$$

where n: number of stages;

s_{60}: movement after 60 min;

s_{30}: movement after 30 min.

These movements are measured at α the first loading.

5.2.4.2. *Creep curve (α_n) between the middle and end of each stage*

The section (α_n) of the movement's curve, in relation to the decimal logarithm of time between t = 30 and t = 60 min, is calculated at each stage (Q_n), according to formula [5.20].

$$\alpha_n = \frac{(s_{60} - s_{30})}{\mathrm{Log}_{10}(2)} \qquad [5.20]$$

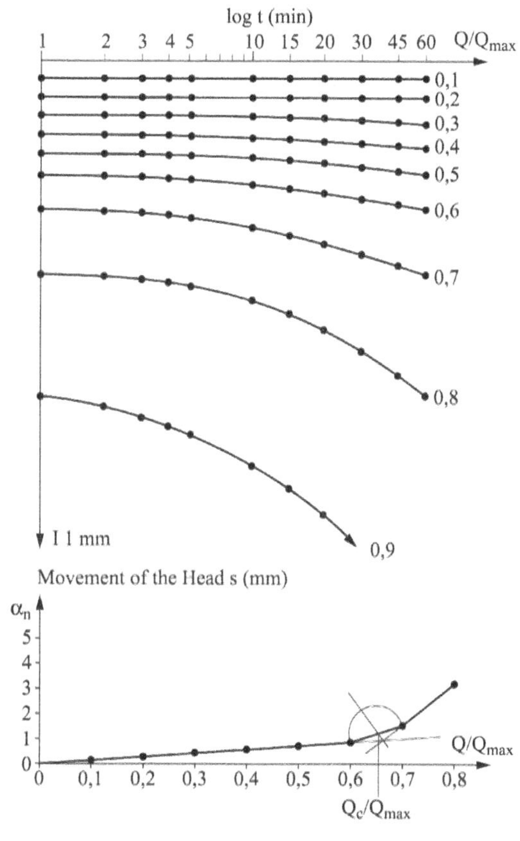

Figure 5.20. *Creep curve and determination of the creep's critical load Q_c (NFP 94-150)*

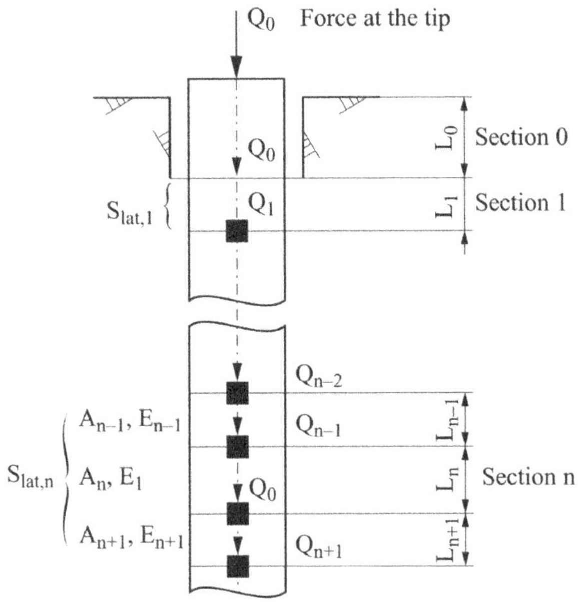

Figure 5.21. *The definition of a pile's different sections (NFP 94-150)*

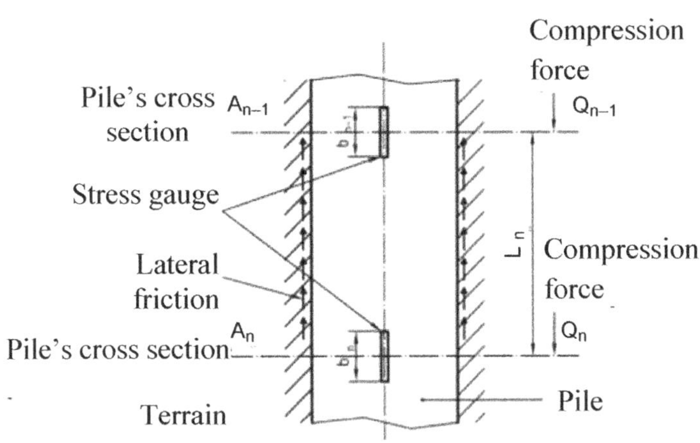

Figure 5.22. *Example of measurements taken with non-recoverable extensometers (NFP 94-150)*

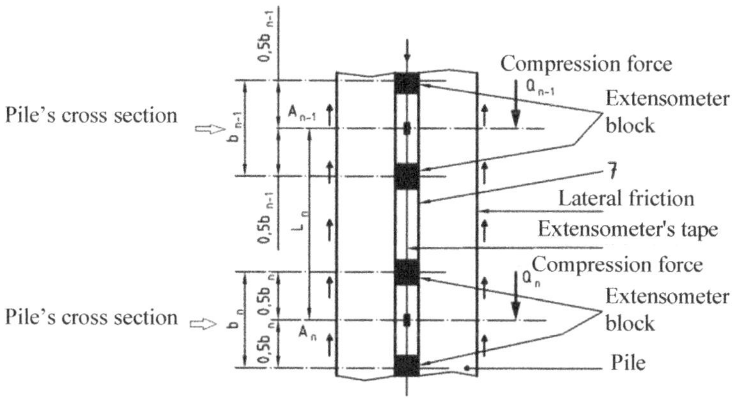

Figure 5.23. *Example of measurements taken with non-recoverable extensometers (NFP 94-150)*

5.2.4.3. *Rupture load Q_{le}*

The conventional rupture load Q_{le} is the force that, when applied to the pile, produces a movement at its tip equal to the larger of the following two values:

– 0,1 B (0,1 B_e when the pile's cross-sectional area is neither circular, square nor rectangular);

– 20 mm.

5.2.4.4. *Critical creep load Q_c*

The critical creep load Q_c is determined by a construction carried out on a curve depicting the creep's evolution in relation to time (Figure 5.20). The linear sections at the beginning and end of this curve are extended until they intersect. The bisector of the angle thus formed cuts the curve at the Q_c/Q_{max} point (see Figure 5.20).

5.2.4.5. *Force distribution along the pile's shaft*

The use of extensimeter measurements provides force distribution along the pile's shaft.

For both removable and non-recoverable extensometers attached to a pile, their measurement base's deformation (of length bn) is conventionally the same as that of a pile's section of identical length.

The deformation's e_n measurement, taken with extensometers (Figures 5.21, 5.22, 5.23), helps to determine the graphical representation of force along the pile (Figure 5.24).

As the pile's constituent materials' behavior is supposedly elastic and linear, the axial force in the pile is calculated at the middle of the measurement base of the extensometer used, at a given load stage, using formula [5.21], where E_n is the pile's elasticity modulus and A_n is the pile's straight section.

In the case of a composite pile, the elasticity modulus is determined with formula [5.22], where E_i and S_i designate, respectively, Young's modulus and the area of the pile's constituent materials in the cross-section A.

$$Q_n = e_n . E_n . A_n \qquad [5.21]$$

$$E_e = \sum_{i=1}^{m} S_i . E_i \Big/ \sum_{i=1}^{m} S_i \qquad [5.22]$$

In the case of a pile made of reinforced concrete, in which the percentage of steel is lower or equal to 0.5%, the E modulus will be the same as for concrete.

5.2.4.6. *Determining the average lateral friction unit*

Based on the axial forces (Q_n and Q_{n+1}) measured at the extremities of a section of the pile, of rank n and lateral surface $S_{lat,n}$ (Figures 5.22 and 5.23), the lateral friction unit for the section is determined by formula [5.23]:

$$q_{sn} = |Q_n - Q_{n+1}| \Big/ S_{lat,n} \qquad [5.23]$$

5.2.4.7. *Lateral friction mobilization in relation to relative pile-ground movement*

The ground is conventionally expected to remain motionless during a pile-loading test.

The movement of the pile's cross-section is calculated in the middle of the section where the average lateral friction unit has been measured. It is the tip pile settlement, minus the elastic deformations between the tip of the pile and the considered level.

The mobilization curve graphically represents the lateral friction unit (q_{sn}) in relation to movement (s_{ln}) (Figure 5.25).

$$s_{sn} = s - \frac{(Q_0.L_0)}{E_{m,0}.A_{m,0}} - \sum_{i=1}^{n-1}\left(\frac{Q_{m,i-1}+Q_{m,i}}{A_{m,i-1}+A_{m,i}}\right).\left(\frac{2.L_{m,i}}{E_{m,i-1}+E_{m,i}}\right) - \left(\frac{Q_{m,n-1}+Q_{m,n}}{A_{m,n-1}+A_{m,n}}\right).\left(\frac{L_{m,n}}{E_{m,n-1}+E_{m,n}}\right) \qquad [5.24]$$

$$A_{m,i} = 0{,}5.(A_{i-1} + A_i) \qquad [5.25]$$

$$Q_{m,i} = 0{,}5.(Q_{i-1} + Q_i) \qquad [5.26]$$

$$L_{m,i} = 0{,}5.(L_{i-1} + L_i) \qquad [5.27]$$

$$E_{m,i} = 0{,}5.(E_{i-1} + E_i) \qquad [5.28]$$

5.3. In conclusion: pile-loading tests

Several implications can be drawn from the previous information:

– pile-loading tests are expensive;

– they are difficult to perform, requiring safety precautions, stable baselines, correct load centering and reaction piles that will not influence the measurements;

– the correct waiting times must be imperatively observed before each loading test;

– the use of the standard operating mode [AFN 90c] is recommended, as it is the only test that uses a standardized methodology, which allows for different tests to be validly compared with each other;

– no method exists that can determine the lateral friction force and peak force's respective parts at every moment of the loading test.

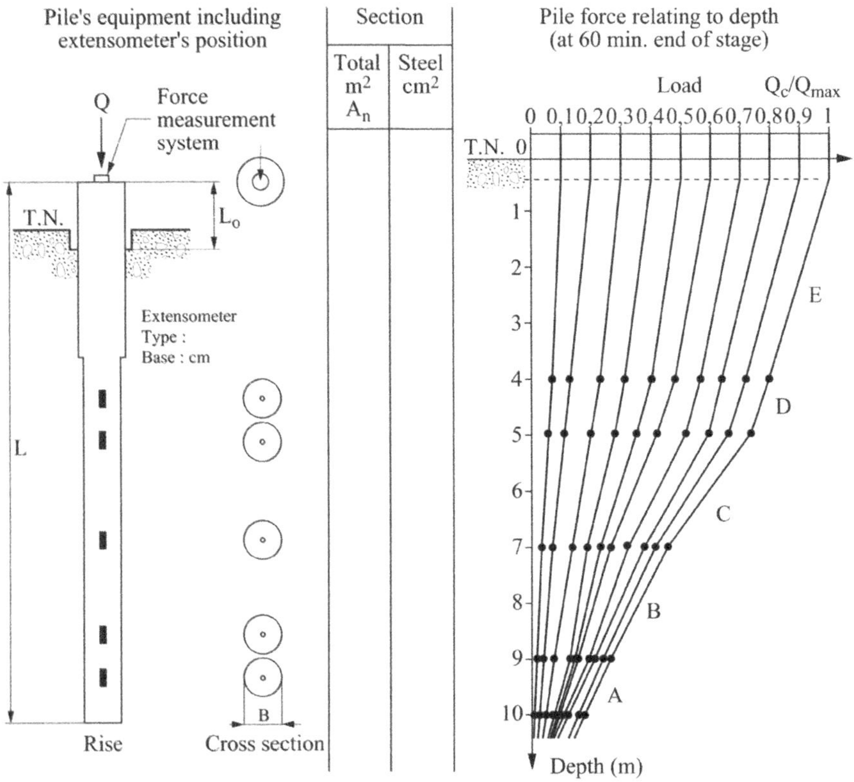

Figure 5.24. *Distribution of force along the pile in relation to depth (NFP 94-150)*

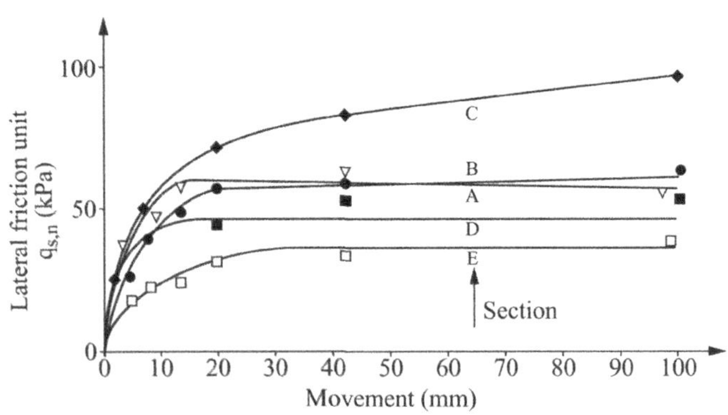

Figure 5.25. *Lateral friction unit mobilization in relation to pile-ground movement (NFP 94-150)*

6

Tests by Flat Dilatometer (DMT)

The Marchetti flat dilatometer is a normalized test [AST 07, EUR 07, TEC 01] that provides an estimate of the settlement of foundations and coefficient K_0 of the soil.

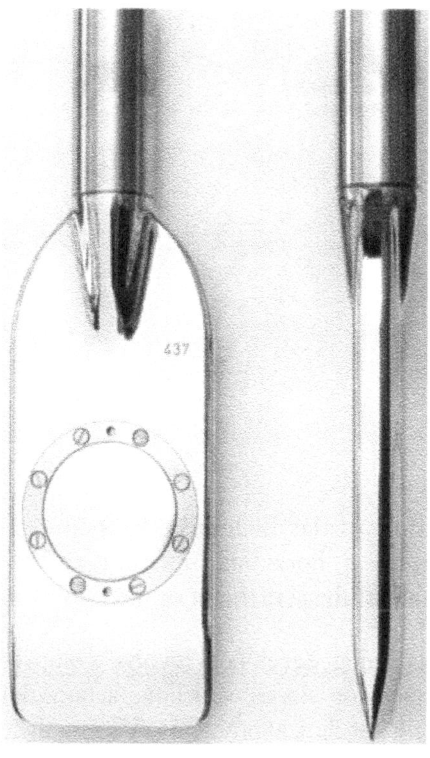

Figure 6.1. *Marchetti dilatometer [TEC 01]*

6.1. Principle of the test

A Marchetti dilatometer test involves inserting a flat blade (Figures 6.1 and 6.2) located at the end of a series of rods. Once the test depth is reached, a circular steel membrane located on one side of the blade is horizontally bent into the soil. The pressure is measured at the two levels of the bend in the blade.

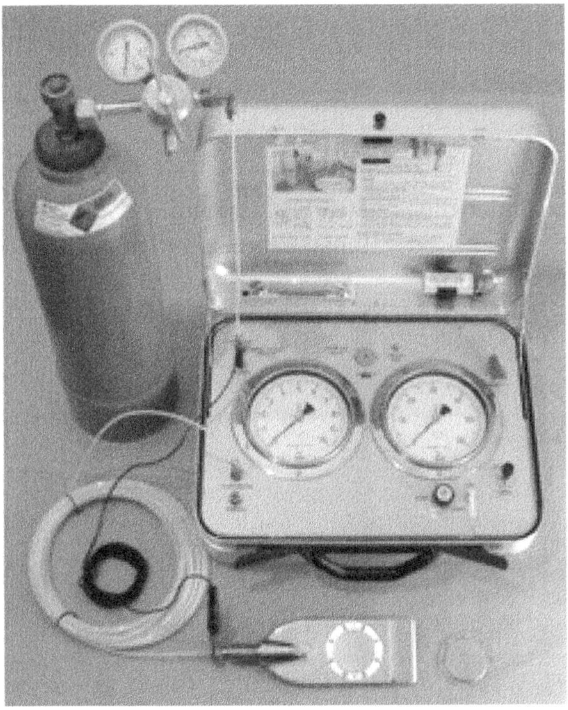

Figure 6.2. *The complete Marchetti dilatometer apparatus [TEC 01]*

The main application of flat dilatomer Marchetti test (DMT) is to rapidly and precisely measure a horizontal modulus of the soil deformation, resulting in an estimate of the settlement of the foundations.

Other applications being developed involve compacting control systems, detecting sliding surfaces on slopes, soil liquefaction, deformation of laterally loaded piles and other geotechnical problems by using the correlations.

6.2. Modus operandi

The general layout of the dilatometer test is shown in Figure 6.3. The test occurs in the following manner:

– at the office, and first at the test campaign, we measure the correction membrane pressures A and B required to, respectively, begin the deformation of the membrane, and reach a movement of 1.1 mm of the membrane. The measurement for A must be located between 5 and 30 kPa and the measurement for B must be located between 5 and 80 kPa;

– on the field, the string of rods is inserted so as to position the dilatometer at the desired depth in the soil;

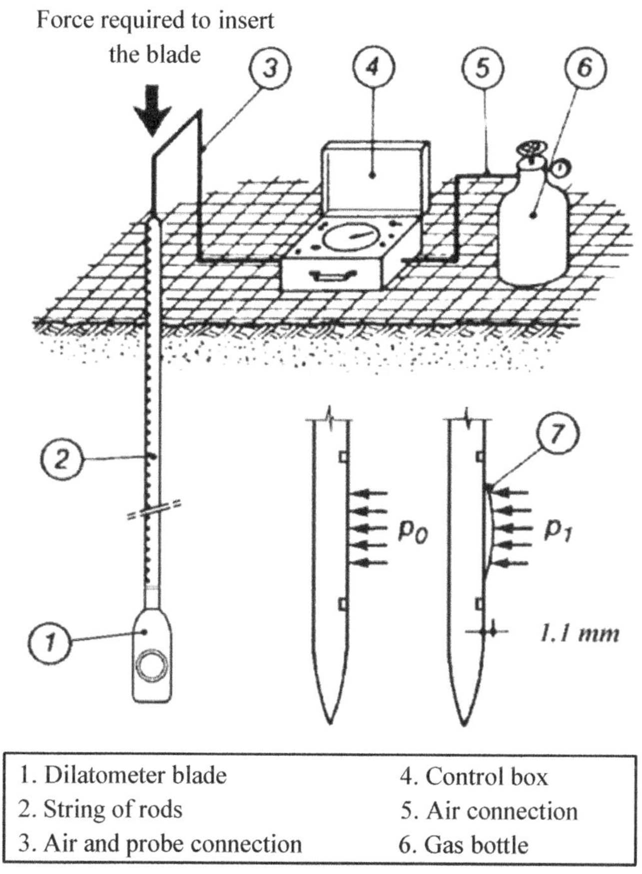

Figure 6.3. *Diagram of the complete Marchetti dilatometer apparatus [TEC 01]*

– the operator fills the membrane and measures the pressure A required to move the membrane ("lift-off"). The measurement must be made in 15 s at the least and 1 min at the most;

– the operator measures the pressure B required to move the center of the membrane by 1.1 mm. The measurement must be made in 20 s at the least and 1 min at the most;

– a third reading, of pressure C ("closing pressure"), can also be taken, less than 1 min after taking reading B, to completely deflate to the membrane and return to the initial state of deformation;

– the blade is then pushed into the soil by one increment of depth (10 cm at least, usually 20 cm) for the next test;

The scheme measurement of the membrane deformation is shown in Figures 6.4 and 6.5.

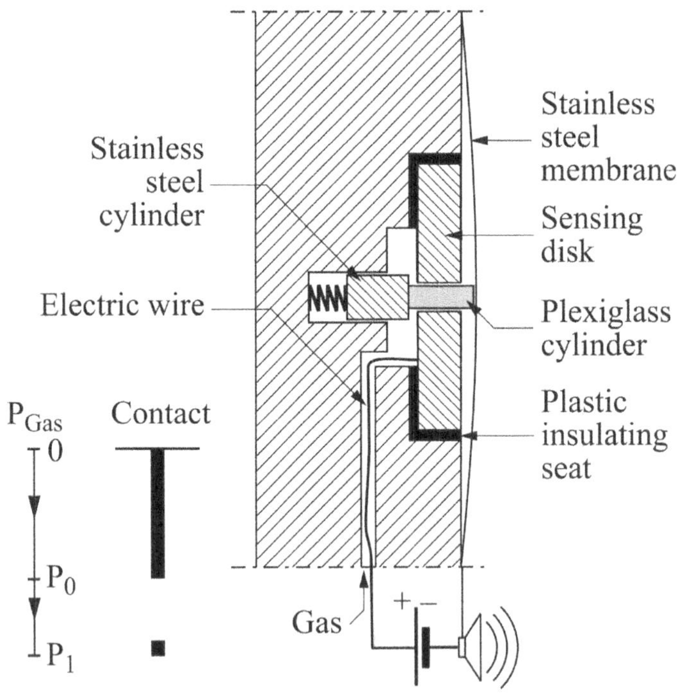

Figure 6.4. *The scheme measurement of the membrane deformation [TEC 01]*

Figure 6.5. *Disassembled membrane [TEC 01]*

6.3. Interpretation

Interpretation involves correcting pressures A and B using membrane correction values of ΔA, ΔB which are determined by free air calibration.

With this device, we can measure precisely the values obtained directly (I_D, K_D, E_D, u_0). However, the other parameters come from correlation (K_0, OCR, c_u, Φ ,C_{vh}, k_h, γ, M) and must be used carefully.

The measurement of the venting pressure z_M must not be used for recent devices, and does not appear in the formulas for p_0 and p_1 and u_0.				[6.1]
Symbol	Description	Relation		Condition
p_0	Corrected first reading	$p_0 = 1{,}05.(A + \Delta A) - 0{,}05.(B - \Delta B)$	[6.2]	See [6.1]
p_1	Corrected second reading	$p_0 = (B - \Delta B)$	[6.3]	See [6.1]
I_D	Material index	$I_D = {(p_1 - p_0)}/{(p_0 - u_0)}$	[6.4]	
K_D	Horizontal$_{stress}$ index	$K_D = {(p_0 - u_0)}/{\sigma'_{v0}}$	[6.5]	
E_D	Dilatometer modulus	$E_D = 34{,}7.(p_1 - p_0)$	[6.6]	

$K_{0,MDT}$	Coefficient of Earth pressure in situ	$K_{0,DMT} = \left(K_D/_{1,5}\right)^{0,47} - 0,6$	[6.7]	If I_D<1,2
OCR	Overconsolidation ratio	$OCR_{DMT} = (0,5.K_D)^{1,56}$	[6.8]	If I_D<1,2
c_u	Undrained shear strength	$c_{u,DMT} = 0,22.\sigma'_{v0}(0,5.K_D)^{1,25}$	[6.9]	If I_D<1,2
Φ	Friction angle	$\Phi_{DMT} = 28 + 14,6.Log(K_D) - 2,1.[Log(K_D)]^2$	[6.10]	If I_D>1,8
C_{vh}	Consolidation coefficient	$C_{vh} = 7/t_{flex}$ ($C_v$in cm²:s)	[6.11]	t_{flex}, measured on the curve Log(t)-A
k_h	Permeability coefficient	$k_h = C_{vh}.\gamma_w/M_h$	[6.12]	$M_h \approx K_0.M_{DMT}$
γ	Specific weight of the soil	Marked correlation (Figure 6.6)		
M	Vertical drained constrained modulus	$M_{DMT} = R_M.E_D$	[6.13]	
		$R_M = 0.14 + 2.36.Log(K_D)$	[6.14]	if $I_D \leq 0,6$
		$R_M = 0.5 + 2.Log(K_D)$	[6.15]	if $I_D \geq 3$
		$R_M = R_{M,0} + (2.5 - R_{M,0}).Log(K_D)$	[6.16]	if $0,6 < I_D < 3$
		$R_{M,0} = 0.14 + 0,15.(I_D - 0,6)$	[6.17]	
		$R_M = 0.32 + 2.18.Log(K_D)$	[6.18]	if $K_D > 10$
		$R_M = 0.85$	[6.19]	if $R_M > 0,85$
u_0	Initial interstitial pressure	$u_0 = p_2 = C + \Delta A$	[6.20]	In draining soils. See [6.1]

Table 6.1. *Main correlation relations for the Marchetti flat dilatometer [TIC 01]*

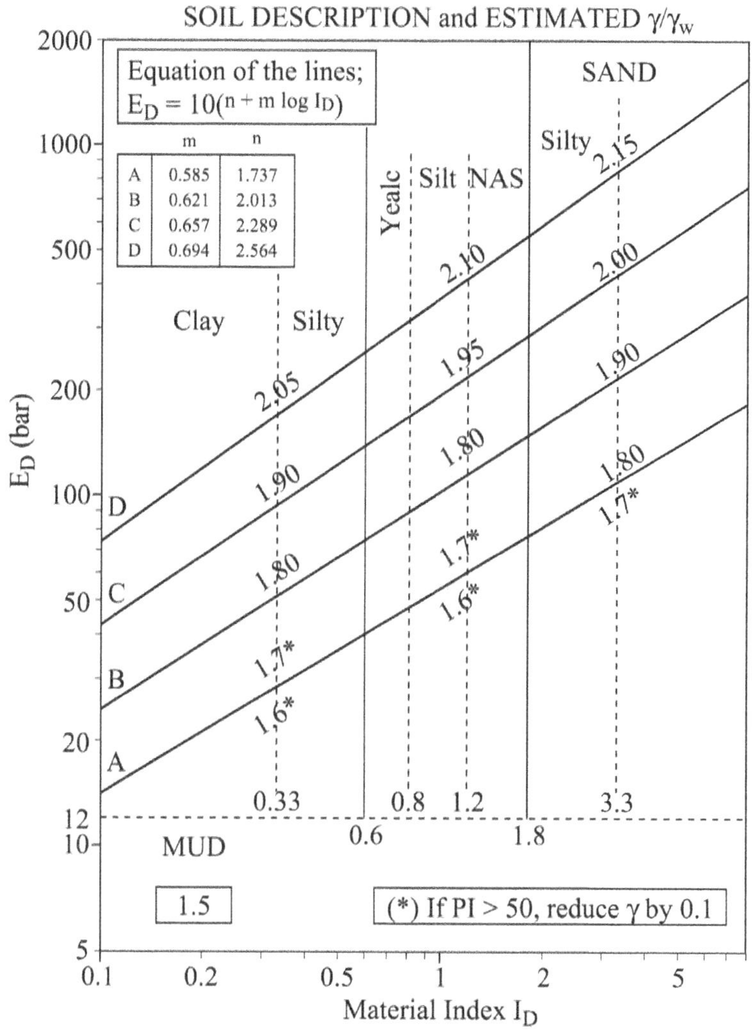

Figure 6.6. *The relation between the elastic modulus of the soil ED and the density index ID [TEC 01]*

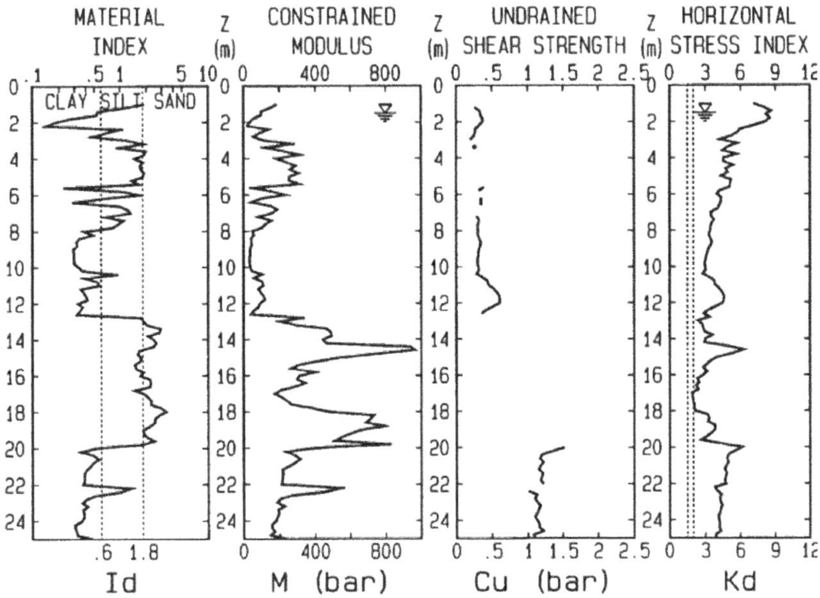

Figure 6.7. *An example of results obtained using the Marchetti flat dilatometer [TEC 01]*

7

Penetrometer Test (CPT, CPTU, SPT, DCPT) and Variants

A cone is inserted into the ground (Figure 7.1) tied to a string of rods. This result is achieved by pipe jacking or pile driving. There are several types of penetrometer:

– the static penetrometer (SP), which is inserted using a system of jacks;

– the dynamic penetrometer, which is inserted by pile driving;

Figure 7.1. *Example of a penetrometer probe*

– the dynamic-static penetrometer, which combines the advantages of the two previous processes, while avoiding their disadvantages;

– the piezocone, which measures the cone effort and the interstitial pressure on advancement.

The penetrometer has been the subject of several dedicated works [SAN 72, CAS 78].

7.1. Static penetrometer (or cone penetrometer test, CPT)

7.1.1. *Principle of the test*

The static penetration test is normalized [AFN 96d]. It involves inserting into the ground a reduced-scale pile with a diameter of 30–100 mm by continuous pipe jacking, at a slow and constant rate (2 cm per second +/ – 0.5 cm per second). The cone resistance of the penetrometer allows us to estimate the load-bearing capacity of the foundations.

We directly measure:

– the cone effort Q_c;

– the total insertion effort Q_t;

– on some devices, we can directly measure the lateral friction Q_s on a sleeve.

We have shown (Figure 7.2) a scheme of the penetrometer probe. The normalized dimensions of the static penetrometer are shown in Table 7.1.

Designing the seal at the level of the joining of the penetrometer and the friction sleeve must be done carefully (Figure 7.3).

A practical example of a device used is shown in Figures 7.4 and 7.5.

The principle of the measurement is shown in Figure 7.6. A typical result is shown in Figure 7.7.

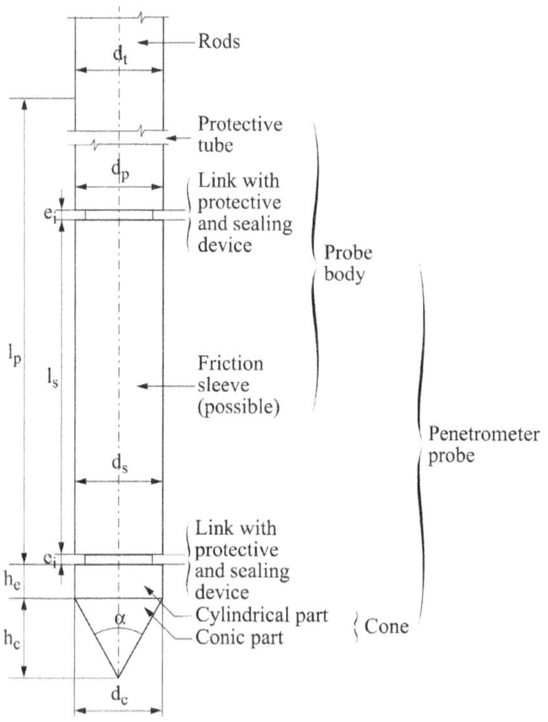

Figure 7.2. *Diagram of the penetrometer probe*

Designation			Symbol	Unit	Value	Tolerance of usage
Probe hobbing system		Hobbing speed	v	mm/s	20	± 5
Probe	Probe body	Length	l_p	m	$\geq 1,0$	–
		Diameter	d_p	mm	$d_p \leq d_c$	–
	Cone	Diameter	d_c	mm	35,7	± 0,3
		Height of the conic element	h_c	mm	31,0	– 2 + 0,3
		Height of the cylindrical element	h_e	mm	$7 \leq h_e \leq 10$	–
	Friction sleeve	Diameter	d_s	mm	$d_c \leq d_s \leq 36,5$	–
		Length	l_s	mm	133,7	0 +1
	Link with sealing and protective device		e_i	mm	≤ 5	–
			A_i	mm^2	< 10	–

Table 7.1. *The normalized dimensions of the static penetrometer*

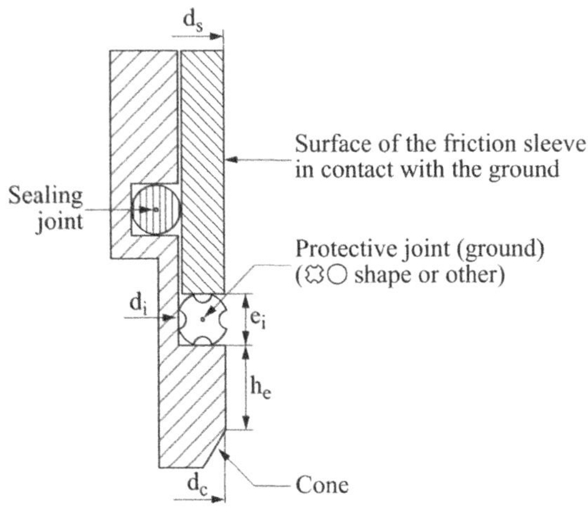

Figure 7.3. *Scheme of the penetrometer probe [AFN 96d]*

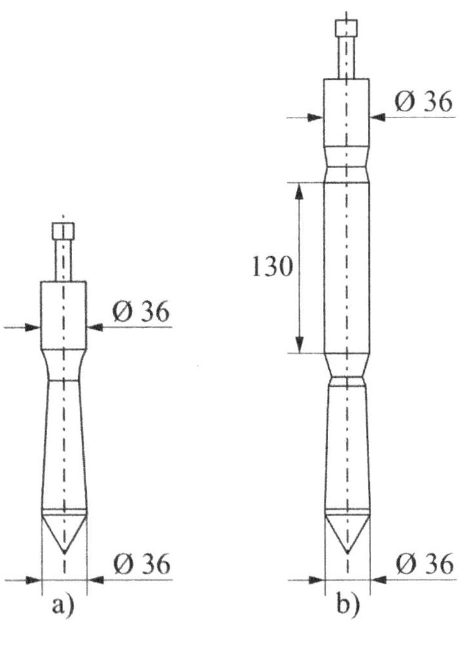

Figure 7.4. *Two types of penetrometer probes from the Gouda penetrometer (1970): a) simple probe and b) probe with a sleeve for the measurement of lateral friction*

Figure 7.5. *Example of a static penetrometer, the Pagani penetrometer (from eurofor.com)*

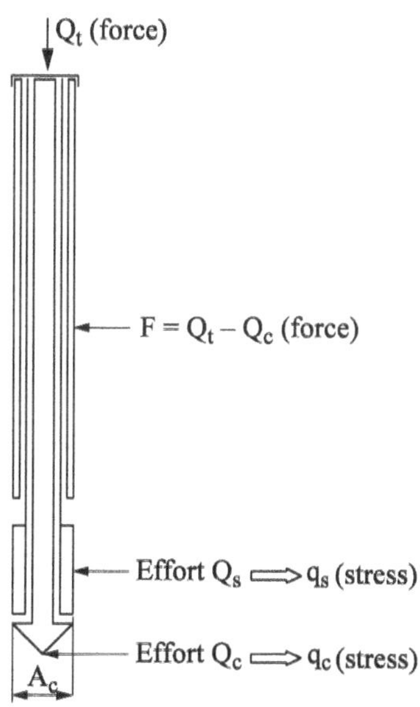

Figure 7.6. *The principle of penetrometer measurements*

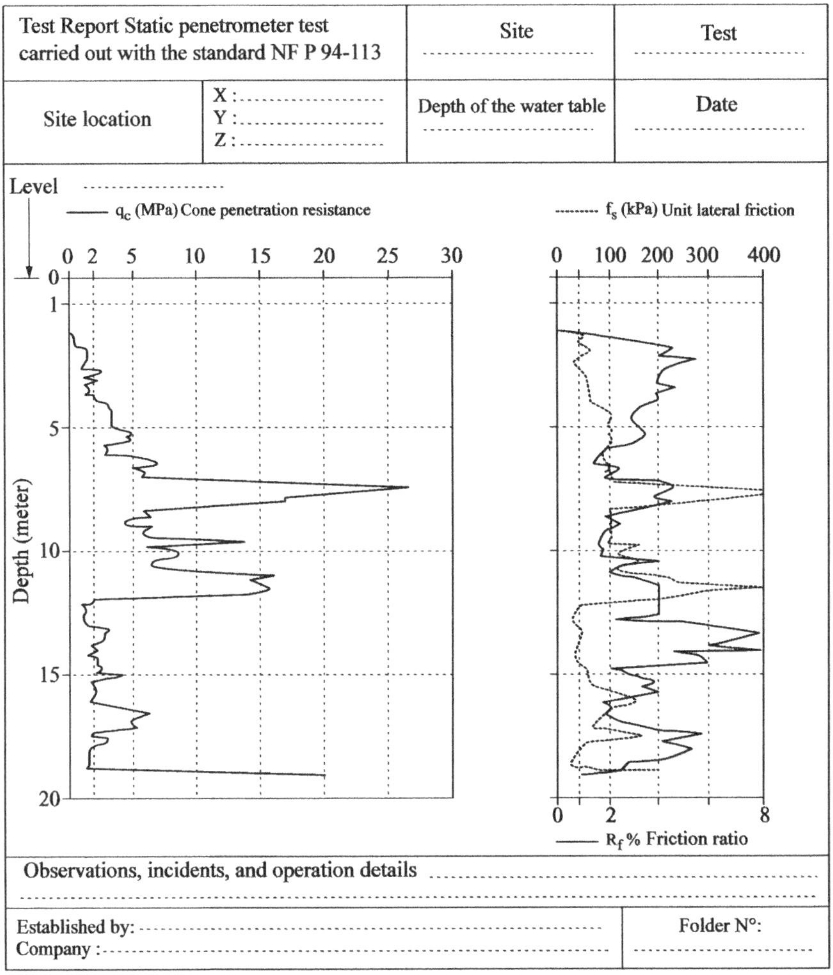

Figure 7.7. *Example of a static penetrogram [AFN 96d]*

7.1.2. *Measurement methods*

The cone penetration resistance is measured every 10 cm, at the most.

7.1.2.1. *Discontibuous survey*

The rod string is pushed down to the depth of measurement by controlling the total insertion effort. The external tube (Figure 7.6) is acted on over a length of 10 cm at the most. The cone is then inserted alone over a length of 4 cm and the cone effort is controlled.

7.1.2.2. *Contibuous survey*

The penetrometer is pushed in without measuring the total effort, only noting the peak effort, which is then measured continuously.

7.1.3. *Interpretation*

The penetrometer allows us to measure several parameters.

7.1.3.1. *Resistance to the penetration of the cone, or peak resistance*

This resistance, written q_c, is obtained by dividing the cone effort Q_c by the surface area A_c of the largest straight section of the probe, according to relation [7.1]; q_c is expressed in MPa:

$$q_c = {Q_c}/{A_c} \quad [7.1]$$

7.1.3.2. *Total insertion effort*

This is written as Q_t and expressed in kN. It is measured on the external tube of Figure 7.6.

7.1.3.3. *Lateral friction effort on the whole shaft*

This is written as Q_{st} and expressed in kN. It is obtained using the following difference:

$$Q_{st} = Q_t - Q_c \quad [7.2]$$

7.1.3.4. *Unit lateral friction*

Unit lateral friction is the ratio between the friction effort measured on the special sleeve behind the probe Q_s and the surface of the sleeve:

$$f_s = {Q_s}/{A_s} \quad [7.3]$$

7.1.3.5. *Friction ratio*

The friction ratio is the ratio between the unit lateral friction f_s and the penetration resistance of the cone q_c measured at the same depth (and not at the same time). R_f is expressed as a percentage:

$$R_f = {}^{f_s}/_{q_c} \quad [7.4]$$

7.1.3.6. *Friction index*

The friction index is the ratio between the penetration resistance of the cone q_c and the unit lateral friction f_s measured at the same depth (and not at the same time):

$$I_f = {}^{q_c}/_{f_s} \quad [7.5]$$

The evolution of these results is represented as a function of the depth on a standard graph (Figure 7.7).

7.1.4. *Use of the static penetrometer to calculate foundations*

The penetrometer allows us to approximately determine the geological section of the terrain, the level of the foundations and their bearing capacity.

7.1.4.1. *Geotechnical section of the terrain*

Feedback from a number of tests carried out allows us to estimate the nature of the soil as a function of the cone effort and the total effort.

An extract from document Fond 72, shows in Table 7.2 the relationship between total cone resistance and the nature of the soil.

Table 7.3, an extract from the Sanglerat works, shows the relationship between the mean unit friction and the nature of the soil.

In Table 7.4, an extract from the Sanglerat works, there can be found the relationship between both cone resistance *and* mean unit friction with the nature of the soil.

At its summit, the cone has an angle of 60° and its base, A_c, has a surface area of 10 cm^2.

The friction sleeve has a lateral surface area of 150 cm^2.

	q_c	Nature	Cone resistance (Q_c)	Total Effort Q_t	Observations
Inconsistent fine soils	3 MPa	Silt	q_c does not increase much with the depth (Z)	Q_t varies little	Need for precise measurements
		Peat		Q_t increases very slightly with Z	
		Soft clay		Q_t increases with Z	Example: use of an electric probe
Consistent fine soils	1 MPa 3 MPa	Silt	q_c varies little with Z	Q_t increases slightly with Z	In the presence of gravel or rocks, Rp can present a "saw-tooth"-shaped curve
		Clay	q_c increases slightly with Z	Q_t increases linearly with Z	
Grainy soils	3 MPa 20 MPa	Loose sand	q_c increases slightly with Z	Q_t varies slightly with Z	q_c and Q_t can present variations caused by the heterogeneity or variations in the compactness of the sable, and by the presence of gravel
		Compact sand	q_c is practically constant in function of Z	Q_t remains practically constant	
Carbonate soils	1 MPa 6MPa	Limestone	q_c increases with Z	Q_t varies little with Z	q_c and Q_t are strongly influence by the presence of moduli and in the soft limestone
		Marl	q_c increases with Z	Q_t increases with Z	Q_t is sensitive to an increase in compactness (parabolic shape)

Table 7.2. *Qualitative identification of the nature of soils from cone resistance and total effort (Fond 72, 1997)*

Type of soil	Schermann	Begemann	Fugro	Sanglerat
Gravel sand	$0 < R_f < 0{,}5$	–	–	–
Sand	$0{,}5 < R_f < 2$	$1{,}25 < R_f < 1{,}6$	$0{,}5 < R_f < 1{,}5$	$R_f < 2$
Silty sand	$2 < R_f < 2{,}5$	$1{,}6 < R_f < 2{,}5$	$1{,}5 < R_f < 2$	$1 < R_f < 2{,}75$
Silt	$2{,}3 < R_f < 3{,}5$	$2{,}5 < R_f < 3{,}6$	$2 < R_f < 2{,}5$	$2{,}75 < R_f < 3{,}5$
Silty clay	$3 < R_f < 4{,}5$	$3{,}6 < R_f < 4$	$2{,}5 < R_f < 3$	$3{,}5 < R_f < 7$
Clay	$4 < R_f$	$4 < R_f < 7$	$3 < R_f < 6$	$3 < R_f < 8$
Peat	–	$7 < R_f$	$6 < R_f$	–

Table 7.3. *Classification of soils as a function of the friction ratio R_f [SAN 72]*

Type of soil	Cone resistance (100 kPa)	Friction ratio (%)
clayey gravel	$10 < q_c$	$R_f < 5$
Sandy gravel	$20 < q_c$	$R_f < 2$
Clay sand	$3 < q_c < 60$	$0{,}5 < R_f < 8$
Silty sand	$8 < q_c < 40$	$1{,}1 < R_f < 5$
Sand	$15 < q_c$	$R_f < 2$
clayey silt	$10 < q_c$ $q_c < 10$	$R_f < 3$ $3 < R_f < 9$
Silt	$7 < q_c < 20$	$1{,}5 < R_f < 4$
Sandy silt	$4 < q_c < 30$	$1 < R_f < 7$
Soft clay	$q_c < 8$	$4 < R_f$

Table 7.4. *Lithological classification of soils [SAN 72]*

7.1.4.2. *Estimation of the load-bearing capacity and packing of foundations*

This estimation is based on an empirical correlation between penetrometer measurements and the terms needed for the calculation of foundations [AFN 88, FAS 93, EUR 04].

7.1.5. *Types of static penetrometer*

In practice, there is a whole range of different penetrometers, some of which are described here.

7.1.5.1. *Begemann penetrometer (1953)*

This is presented in Figure 7.8:

– the mobile cone is equipped with a sleeve;

– the section of the probe has an area of 10 cm^2 (diameter of 36 mm), with an angle at the summit of 60°;

– the string of rods has a diameter of 36 mm.

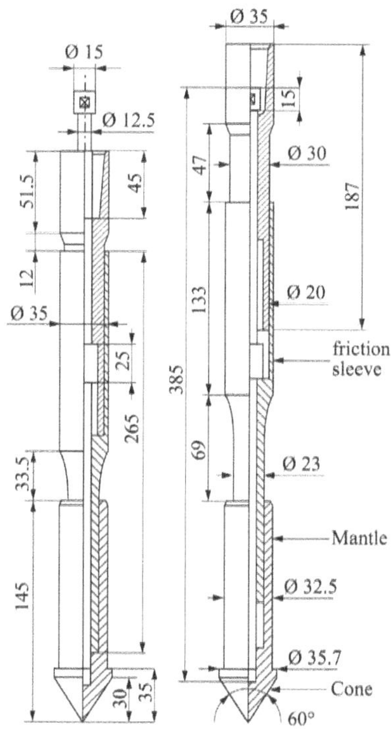

Figure 7.8. *The Begemann penetrometer (1953)*

7.1.5.2. *Andina penetrometer (1967)*

This is presented in Figures 7.9 and 7.10. The reaction is provided by the weight of a truck carrying the penetrometer:

– the cone is fixed or mobile;

– the probe has a large cross-section with an area of 50 cm^2 (diameter 80 mm) with an angle at the summit of 90°;

– the sleeve for the measurement of lateral friction has a diameter of 80 mm with a length of 25 cm;

– the string of rods has a diameter of 80 mm.

When the soil is very compact, the small probe is used (Figure 7.9, on the right), which has a cross-section with an area of 12 cm^2 (and a diameter of 39 mm) and angle at the summit of 90°.

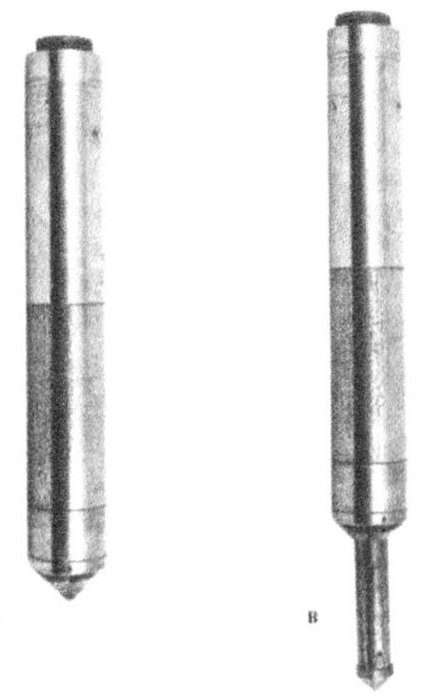

Figure 7.9. *The Andina penetrometer (1967), the two mobile cone systems*

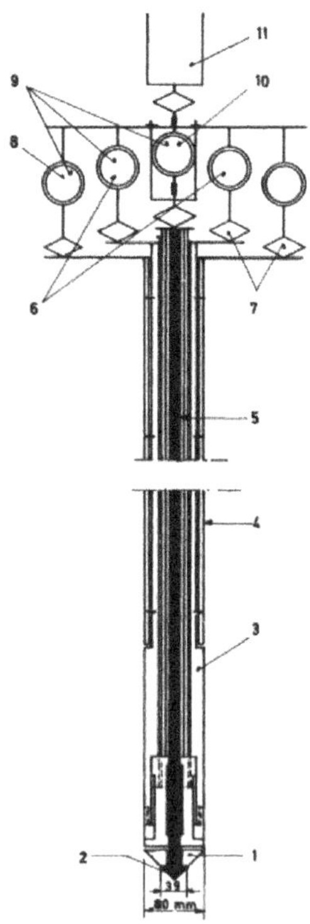

Figure 7.10. *Diagram of the Andina penetrometer (1967), the measurement system used*

With this penetrometer, when immobile the following can be simultaneously measured:

– the peak resistance over 50 cm^2;

– the lateral friction on the sleeve;

– the total insertion effort.

When an effort of 90 MPa is reached on the main cone, this is followed by a dynamic penetration.

7.1.5.3. *Gouda penetrometer (1970)*

This possesses the following characteristics (Figure 7.4):

– the mobile or fixed cone can be of the simple type (Figure 7.4 a) or the sleeved type (Figure 7.4 b);

– the cone has a cross-section with an area of 10cm^2 (and a diameter of 36 mm), with an angle at the summit of 60°;

– the string of rods has a diameter of 36 mm;

– the string of rods is inserted using a system of jacks.

7.1.5.4. *Parez-type penetrometer (1953)*

The device consists of a probe whose cone is attached to a hydraulic jack placed at the tip of the string of rods (Figure 7.11):

– the probe has a diameter of 45, 75, 100 or 200 mm;

– the rods have a diameter that is identical to that of the probe.

The oil pressure is read continuously on a manometer throughout the duration of the insertion.

7.1.5.5. *Fondasol-type penetrometer*

Due to the presence of different force sensors (Figure 7.12), this device is able to continuously measure the cone effort and the unit lateral friction effort.

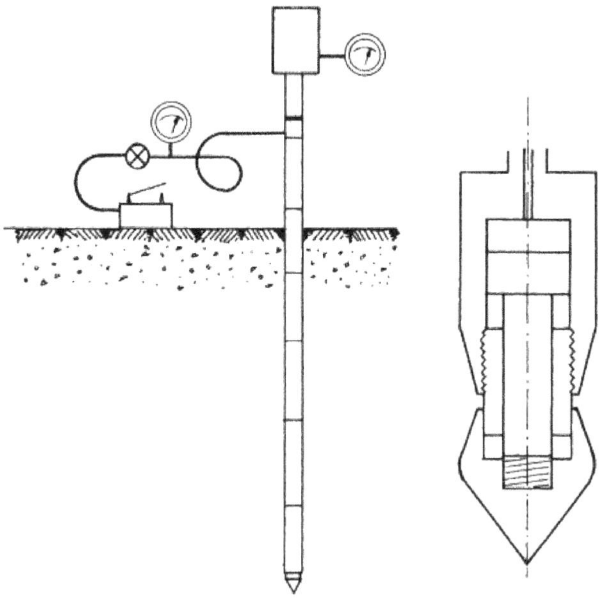

Figure 7.11. *The Parez penetrometer (1953)*

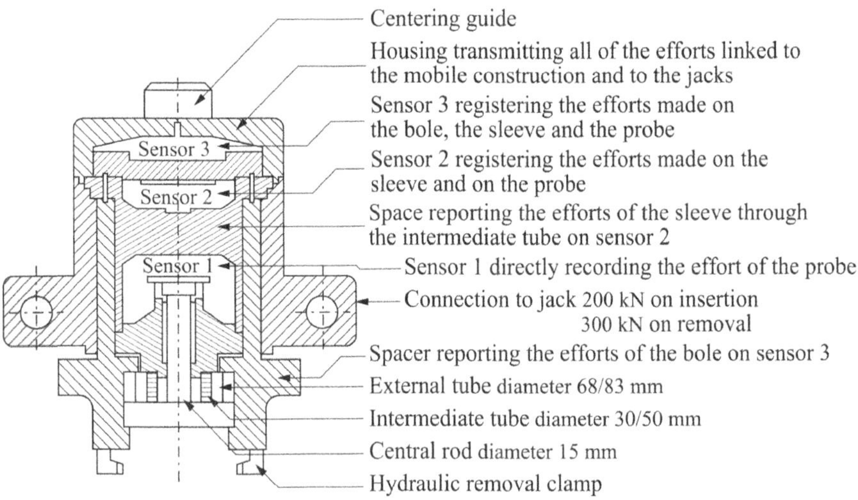

Figure 7.12. *The whole electronic measurement block for the Fondasol-type penetrometer*

7.2. The piezocone (CPTU)

This device is covered by standard (NF P 94-119; ASTM D5778-12). The device is represented in the diagram in Figure 7.15.

7.2.1. *Principle of the test*

Experiments have shown that the use of static penetrometers in fine soils leads to an overestimation of the resistance to insertion of the cone q_c. This is caused by the interstitial overpressure that occurs when the cone moves into the ground. The principle behind piezocone tests is to carry out a static penetration test by simultaneously measuring the interstitial pressure on advancement. An example of a static piezocone is presented in Figure 7.13.

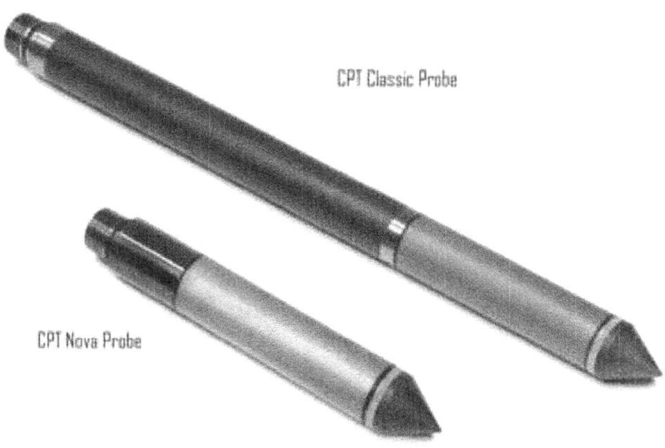

Figure 7.13. *Example of a piezocone*

At the depth chosen for the test, the jacking is stopped, and the interstitial pressure is measured as a function of time so as to observe its evolution. Measurement of the interstitial pressure u_T is stopped when the difference between it and the initial interstitial pressure u_0 before the test has decreased by a half compared to the pressure during the stopping of the jacking u_h:

$$u_t - u_0 = 0{,}5.(u_h - u_0) \qquad [7.6]$$

The time required for this dissipation is written t_{50}.

The piezocone test allows us to measure the cone effort with an electric sensor, and also to record the interstitial pressure upon contact with the cone, or immediately after (Figure 7.16). The appearance of an interstitial overpressure upon advancement is then indicative of the presence of a layer of soil that is more or less impermeable. Inversely, the absence of the interstitial overpressure indicates the presence of granular material (sands and gravel).

7.2.2. *Modus operandi*

7.2.2.1. *Equipment and test conditions*

The dimensional characteristics of the piezocone are shown in Table 7.5. In this table, the cone has an angle at its summit of 60°, with a straight section with an area of 10 cm^2. The lateral friction sleeve has a surface area of 150 cm^2.

The principle behind the construction of the piezocone is identical to that of the static penetrometer for the shared technical parts.

The filter must resist abrasion and clogging, while maintaining a level of permeability greater than that of the layers that is goes through. In media saturated with saltwater, if the filter is made of metal, it must have the same constitution as the cone. A new, perfectly saturated filter is used before each probing.

The variation in volume of the interstitial pressure sensor must not be more than 0.2 mm^3 for the entirety of the measurement domain. The total volume of the liquid contained in the channels, the measurement chamber, the pores and the filter must be less or equal to 600 mm^3.

The sensor and the measurement chain must be compensated for with regard to variations in temperature from –10 to +40°C.

At a given depth, the measurement time and the storage time must be less than or equal to 5.10^{-3}s. The measurement system must allow for variations of the interstitial pressure $u_t – u_0$ to be recorded as a function of the time as soon as the jacking stops.

Designation					Symbol	Unit	Value	Usage tolerance
Probe jacking system	Jacking system				V	mm/s	20	± 5
Probe	Probe body	Length			l_p	m	> 1,0	
		Diameter			d_p	mm	$d_p \leq d_c$	
	Cone	Diameter			d_c	mm	35,7	− 0,3 + 0,3
		Conic part	Height		h_c	mm	31,0	− 2,0 + 0,3
		Cylindrical part	Total height		h_e	mm	≤ 10	
			Height of the lower cylinder		e_c	mm	> 2	
			Height of the Upper cylinder		e_s	mm	≥ 1	
			Filter casing	Height	e_j	mm	≤ 5	
				Lower diameter of the filter casing	d_j	mm	$d_j < d_c$	
			Detachable filter	Height	e_f	mm	< 5	
				External diameter	d_f	mm	$d_c < d_f < d_s$	
	Friction sleeve		External diameter		d_s	mm	$d_f < d_s < 36{,}5$	
			Length		l_s	mm	133,7	− 0 + 1
	Connection with sealing and protection device		Height		e_i	mm	≤ 5	
			Surface		A_i	mm2	≤ 10	

Table 7.5. *The normalized dimensions of the piezocone*

7.2.2.2. *Before the test*

The location of the piezoconic test is identified on a situation map. The tests are separated on the plan by at least 2 m. The manometers are calibrated every 6 months; the sensor scales and the force washers are calibrated every 3,000 m of jacking and every 6 months at least.

The protective joints of the probe are examined to remove any soil particles.

The probe is completely saturated in a container of deaerated water. A new filter is installed on the probe. The cone is enveloped in a thin protective membrane so as to not lose the saturation of the filter before the test, and it is left in place until it is in the starter hole.

A starter hole is created up to 0.30 m under the level of the water table.

The depth of the surface of the water table is measured in relation to the natural terrain and within a precise referential system.

7.2.2.3. *During the test*

The axis of the string of rods must be made to coincide with the axis of the jacking effort.

The rods are pushed in vertically with a maximal incline of 2% in relation to the vertical. A stop is marked after a penetration of 0.5–1 m below the level of the water table to stabilize the temperature of the cone.

The structure of the reaction system is checked to ensure it is stable.

The speed of penetration must remain constant over a minimal length of 0.80 m during the insertion of the probe with a tolerance of +/– 5 mm/s/.

After a penetration of 3 m into the saturated soil in place, jacking is stopped to determine the variation of interstitial pressure as a function of time (Figure 7.14).

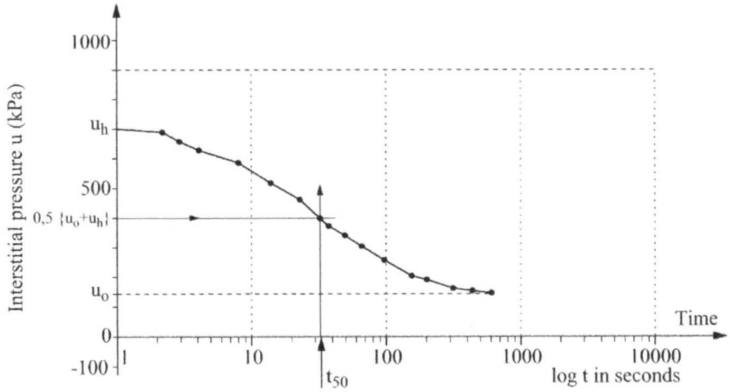

Figure 7.14. *Dissipation of the interstitial pressure of the piezocone*

A continuous measurement of all of the data is recommended. The difference between the depth of two measurements must never be greater than 5 cm.

The following must be noted:

– depth;

– the apparent effort exerted on the cone Q_c;

– interstitial pressure u.

These can also be noted:

– the total penetration effort Q_t;

– the lateral friction effort Q_s;

– the incline of the probe;

– changes in the interstitial pressure as a function of time after stopping the jacking.

Uncertainty with regard to the length of penetration must not exceed 1/100. The maximal uncertainty with regard to the other measurements must be the smallest of the following values:

– 5% of the value measured;

– 1% of the maximal value of the measurement domain.

7.2.2.4. *Test report*

An example of a test report is shown in Figure 7.17. The test report must contain the following pieces of information:

– the reference to standard NF P 94-119;

– the name of the company that carried out the test;

– identification of the test, the date, reference to the layout plan and eventually the planimetric coordinate system;

– the level of the start of the piezocone test in relation to the surface of the ground, and the altimetric level of the ground in relation to a given reference;

– the depth of the water table in the starter hole and the variation in interstitial pressure u_0 as a function of the depth;

– the different results presented in the form of a graph, with an arithmetic scale, as well as the results of the tests for the dissipation of the interstitial pressure with a semi-logarithmic scale;

– the observations linked to carrying out the test as well as the incidents and operative details not covered by the standard.

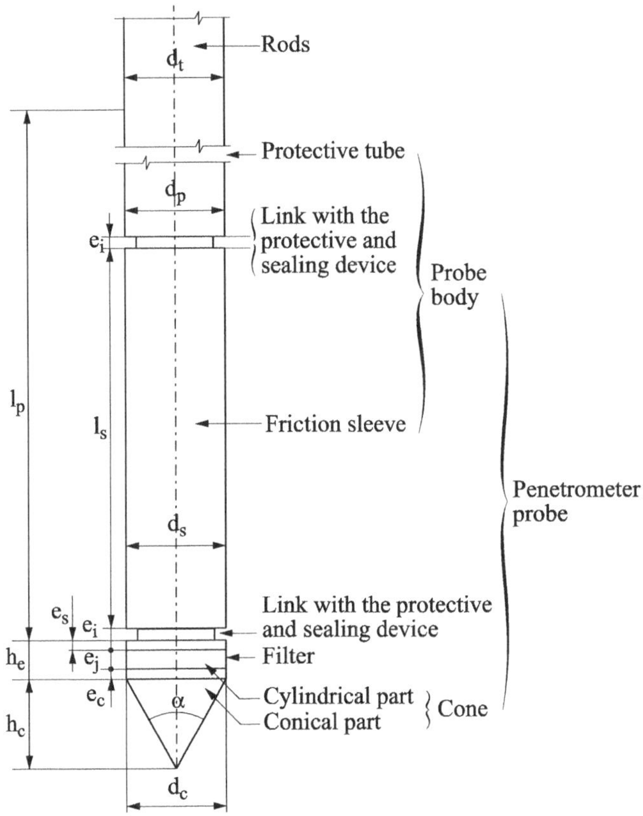

Figure 7.15. *Diagram of the principle of the piezocone*

7.2.3. *Interpretation*

The design of the piezocone is the same as that of the static penetrometer, and this can be seen in the variables measured, the values of the cone's apparent resistance to penetration q_c [7.1], the lateral friction effort Q_{st} [7.2], the unit lateral friction f_s [7.3], the friction ratio R_f [7.4] and the friction index I_f [7.5]. In addition to these measurements, the device can measure the interstitial penetration pressure u and allow access to two new quantities as shown below.

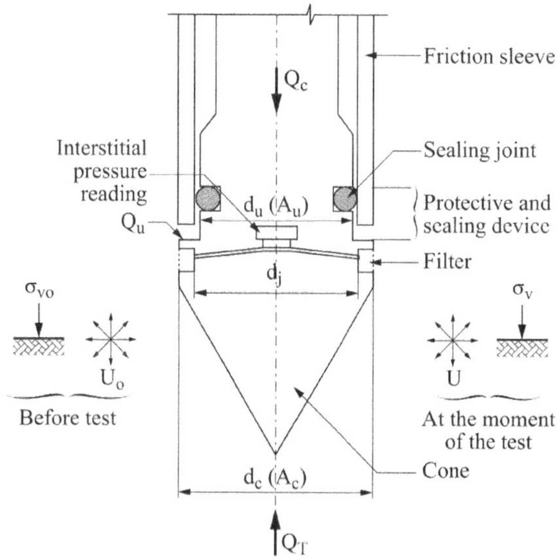

Figure 7.16. *Diagram of the extremity of the piezocone*

7.2.3.1. *Total cone resistance at the base of the cone q_T*

This resistance accounts for the action Q_u of the interstitial penetration pressure u on the cone. This resistance q_T is given by [7.9] with a the ratio of the internal transverse sections A_u (above the cone) and the external ones A_c (cylindrical part of the cone) following [7.10]:

$$q_T = {}^{Q_T}\!/_{A_c} = {}^{(Q_c + Q_u)}\!/_{A_c} \qquad [7.7]$$

$$q_T = {}^{Q_c}\!/_{A_c} + {}^{(A_c - A_u).u}\!/_{A_c} \qquad [7.8]$$

$$q_T = q_c + u.(1 - a) \qquad [7.9]$$

$$a = {}^{A_u}\!/_{A_c} \qquad [7.10]$$

7.2.3.2. *Interstitial pressure coefficient β_q*

This coefficient is calculated using relation [7.11], with σ_{v0} the total vertical constraint of the ground in place at the level of the test (before test).

In the absence of measured data, the specific weight of the soil is taken as equal to 18 kN/m^3:

$$\beta_q = {(u - u_0)}\Big/{(q_T - \sigma_{v0})} \tag{7.11}$$

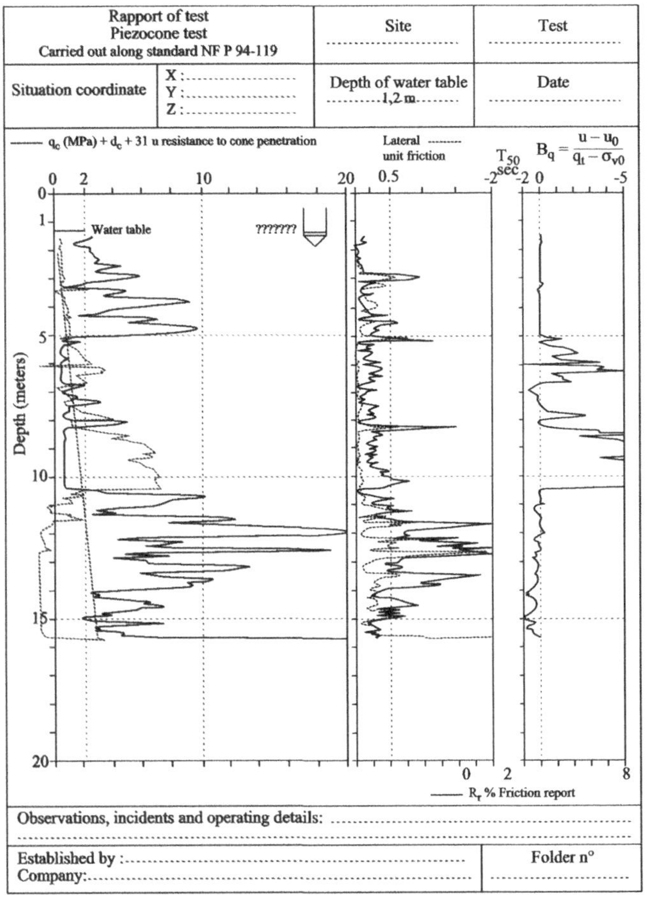

Figure 7.17. *An example of a piezocone test result (Standard NFP94-119, 1995)*

7.2.4. *Exploitation of the test*

Jones and Rust [JON 82] constructed an abacus linking the value of the interstitial overpressure to the nature of the ground penetrated (Figure 7.18).

Robertson [ROB 82] classified the nature of the soil into 12 classes depending on the results of the piezocone (Figure 7.19).

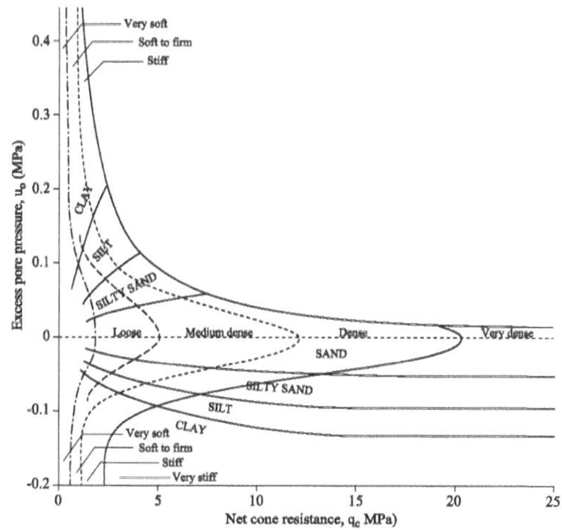

Figure 7.18. *The soil identification abacus according to Jones and Rust [JON 82]*

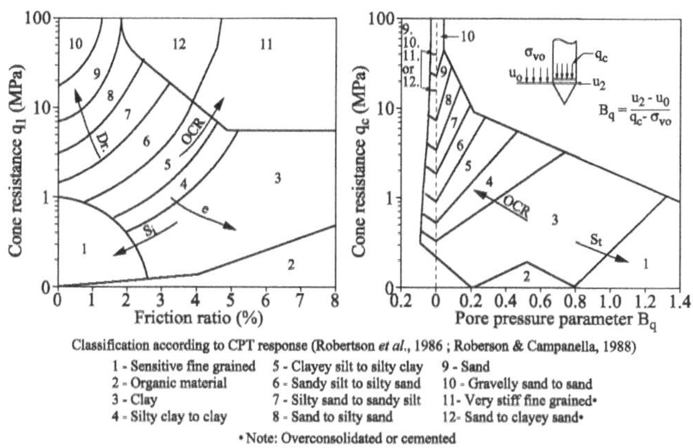

Figure 7.19. *The abacus for the identification of soil layers by piezocone [ROB 89]*

7.3. Standard penetration test (SPT)

This test has been developed extensively in the USA [AST 11], South America, Germany and the United Kingdom. In France, it is addressed in standard [AFN 91c].

7.3.1. *Principle of the test*

The principle of this test involves pushing a core drill into the ground by pile driving (Figures 7.20 to 7.22) to collect a sample of the soil that will be tested in the laboratory, and to get an estimate of the bearing capacity of the soil using pile-driving formulas.

Figure 7.20. *Example of an SPT core drill tube (photo Internet)*

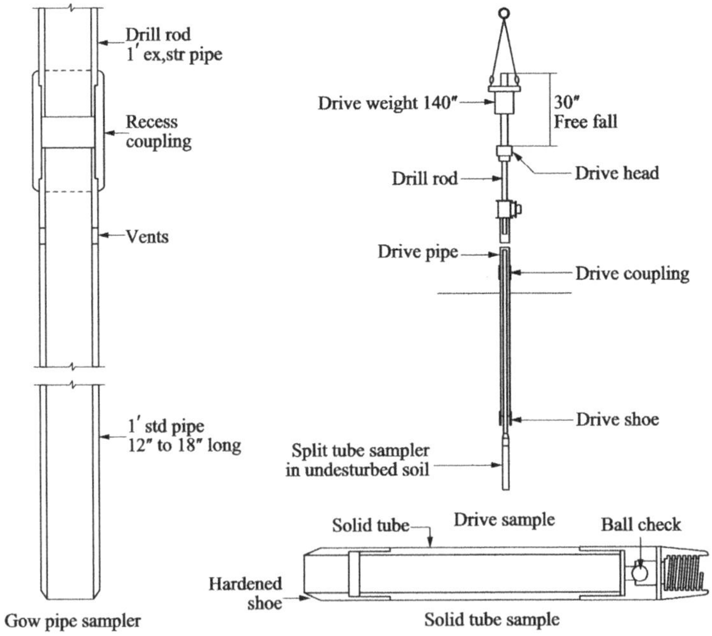

Figure 7.21. *Diagram of the principle of SPT tests*

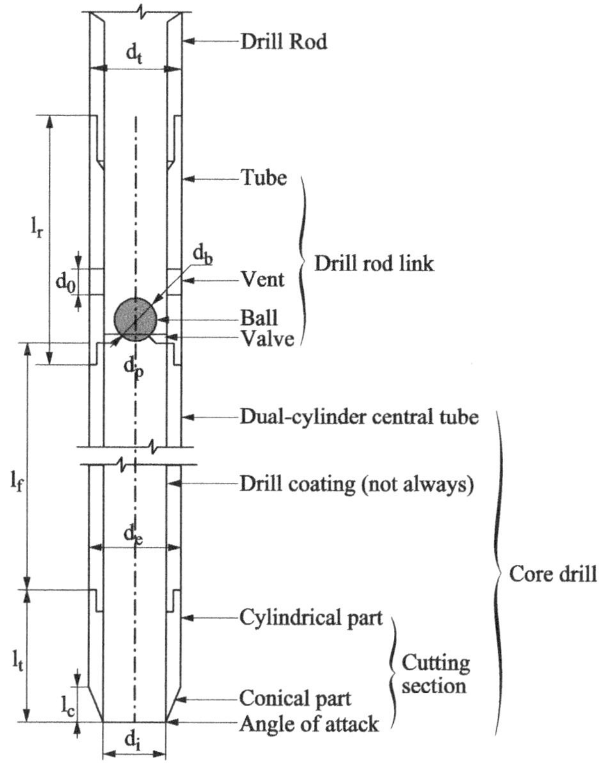

Figure 7.22. *SPT core drill tube*

7.3.2. *Modus operandi*

The dimensions of the devices are given in Table 7.6.

7.3.2.1. *Controls to be made before the test*

Before the probe is put in place, the following must be measured:

– the masses and dimensions of the various constitutive elements;

– the height of the fall of the pile driver hammer;

– the speed of the percussion of the hammer;

– the incident percussion energy.

The penetration test must be located on a site map.

Designation			Symb.	Unit	Value	Tolerance of usage
Hammer		Mass	M	kg	64	+ 0 – 0.7
		Drop height	H	m	0.75	± 0.02
		Beating frequency	–	hits /min	15 to 30	–
		Launch length over width ratio D_M of the hammer	–	–	1 to 2	–
Anvil	Diameter		–	cm	< 0.5 × D_M D_M > 10	–
	Mass of anvil alone		–	kg	10 to15	–
	Mass of anvil and guide connected		–	kg	≤ 25	–
Beating rods		Linear density	–	kg/m	4.0	– 0.5 + 1.0
		External diameter	d_r	mm	42.5	± 3.5
Core drill	Cutting section	External diameter	d_e	mm	51	± 1
		Internal diameter	d_i	mm	35	± 1
		Total length	l_t	mm	76	± 1
		Length of conical part	l_c	mm	19	± 1
		Attack radius	r	mm	16	± 1
	Central tube	Length	l_f	mm	≥ 450	–
Rod core drill link		Link length	l_e	mm	150	± 2
		Vent: diameter	d_o	mm	≥ 13	–
		Ball: diameter	d_b	mm	25	± 1
		Valve: diameter	d_p	mm	22	± 1
Borehole		Diameter	–	mm	≤ 130	–

Table 7.6. *Normalized SPT dimensions*

Every month, the rod head must be tested by being rolled around 360° on a plane.

After each probing, the following must be monitored:

– the straightness of the rods, using a reference ruler;

– the state of the core drill and its cutting section;

– the counting system for the number of blows depending on the extent of insertion.

7.3.2.2. *Preparing the borehole*

The boring is carried out using a pneumatic or a mechanical drill.

The hole must be carefully cleaned to the level of the test depth. A borehole casing or borehole mud can be used if necessary. The auger-style casings must not be used for tests located below the level of the water table.

When boring is carried out below the level of the water table, the level of the water or the boring fluid must be kept constant during the removal of the boring tools or during tests.

The boring tools must be removed slowly to avoid suction effects, which decompress the soil. The casing must not be pushed in below the test start level.

7.3.2.3. *Insertion of the core drill*

A core drill with normalized dimensions is lowered into the center of the hole (Figure 7.23). Imperial measurements have been converted into the SI equivalents (Table 7.6). The external diameter is 51 mm (2 inches), the internal diameter is 35 mm (1.375 inches) and the length is greater than 45 cm and can go up to 76 cm (30 inches). The core drill is made up of two cylindrical half-shells that allow for the recovery of a more or less disturbed soil sample (Figure 7.24).

Figure 7.23. *The drill and hammer used for SPT [AGG 01]*

Figure 7.24. *Disassembled core tube for the removal of the sample [ROG 06]*

This core drill is driven in using a hammer of 64 kg (1400 +/– 2 lbs), which falls from a height of 75 cm (30 inches) (Figure 7.25).

At the bottom of the borehole, the core drill is pushed in by 15 cm (6 inches) into the ground (starter depth), and the corresponding number of blows N_0 is noted. If a penetration of 15 cm cannot be achieved in 50 blows, the corresponding depth marks the end of the test.

Next, the number of blows N_1 and N_2 required to push the core drill a further 15 cm each time is counted. The value $N = N_1 + N_2$ is called resistance to penetration.

The rope that pulls the core drill back up, and then drops it down again onto the anvil must perform 1.75 rotations around the holding axle or the drive shaft (Figure 7.25).

This is stopped when:

– 50 blows are required to push the core drill by an increment of 150 mm;

– 100 blows are required to push the core drill in 305 mm;

– 10 successive blows have no effect on penetration.

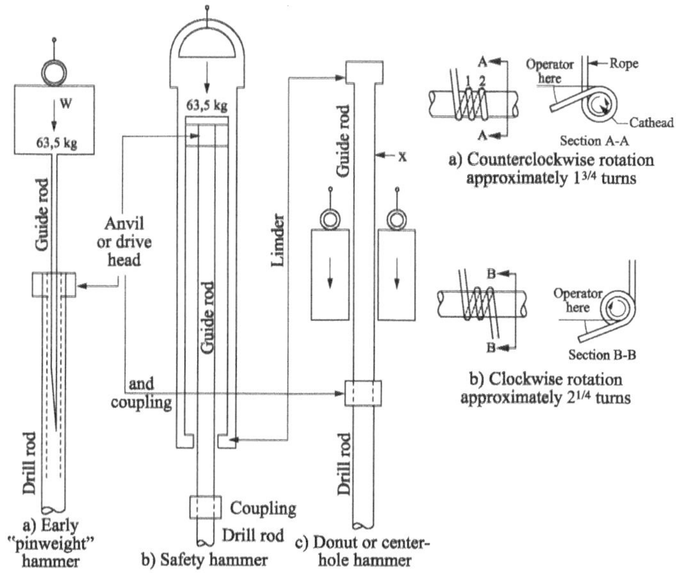

Figure 7.25. *The different hammers used for SPT and the principle behind the maintenance of the rope*

7.3.3. *Interpretation of SPT*

The tests are carried out at regular intervals of depth, between 0.50 m at the least and 1.5 m at the most. Preferably, an interval of 1 m is chosen.

An example of an SPT penetration graph is shown in Figure 7.27.

Recent studies have shown that the theoretical penetration energy [7.12] is greater than the real energy measured from the impact velocity of the hammer on the anvil.

$$E_{in} = M.g.h = 63{,}5.9{,}807.0{,}762 = 474{,}5\text{J} \qquad [7.12]$$

If we take a base of 100 for the theoretical energy, the experimental energy is brought to between 30 and 80 [KOV 82] and 70–100 [RIG 83]. This is due to the techniques used to maintain and free the hammer, friction of the maintenance rope, etc. In these conditions, the number of blows measured N comes down to the number of blows that would be needed with a machine delivering 60% of the theoretical energy of formula [7.12], using relation [7.13]:

$$N_{60} = N.C_N.C_E.C_B.C_R.C_S \quad [7.13]$$

$$C_N = \sqrt{p_a / \sigma'_{v0}} \quad [7.14]$$

Coefficient C_N is an adjustment coefficient for the pressure of soils at rest, whose value [7.14] is calculated from p_a, the reference pressure (100 kPa) and σ'_{vo} the effective vertical constraint.

The other coefficients are given in Table 7.7 and take into account the percussion method (C_E), the bore diameter (C_B), the length of the rope (C_R) and the sampling method (C_S).

Factor	Variable equipment	Term	Correction
Overburden pressure		C_N	$(P_a/s'_{vo})^{0.5}$ but $C_N \leq 2$
Energy ratio	Donut hammer	C_E	0.5-1.0
	Safety hammer		0.7-1.2
	Automatic hammer		0.8-1.5
Borehole diameter	65-115 mm	C_B	1.0
	150 mm		1.05
	200 mm		1.15
Rode length	3-4 m	C_R	0.75
	4-6 m		0.85
	6-10 m		0.95
	10-30 m		1.0
	> 30 m		< 1.0
Sampling method	Standard sampler		1.0
	Sampler without liners	C_S	1.1-1.3

Table 7.7. *Correction coefficients of the SPT [ROB 97]*

The interpretation of this test is given by correlations between the corrected number of blows N_{60} and the nature of the soil, its consistency or its mechanical characteristics.

This test can be used for the dimensioning of superficial foundations (Figure 7.26) due to the relation between the corrected number of blows N and the bearing capacity coefficients N_γ, N_q and Φ [AFN 88].

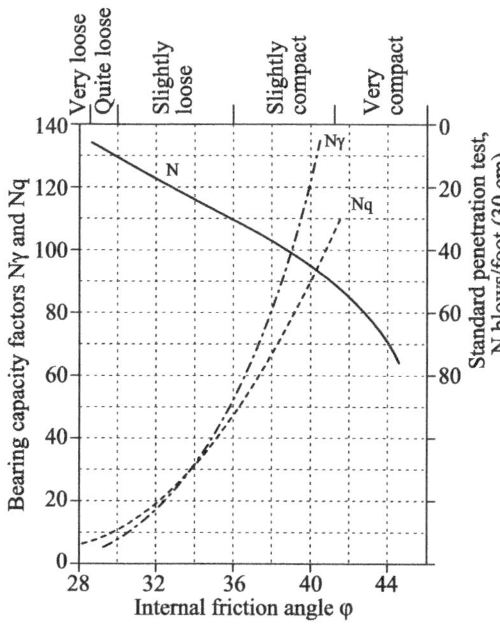

Figure 7.26. *Using the SPT for the dimensioning of foundations [AFN 88]*

7.3.3.1. *Soil with friction*

The first interpretation of the SPT for non-cohesive soils was made by Terzaghi and Peck [TER 48], followed by Peck *et al.* [PEC 53]. In the case of sands located above the water table, the number of blows N_{60} can be linked to the density of the soil using the correlations (Table 7.8). Correlations have since been suggested (Table 7.9) in the case of granular soil at depths of around 6 m, normally consolidated, as well as relation [7.15] by Wolf [WOL 89], relation [7.16] proposed by Kulhawy and Mayne [KUL 90] and relation [7.17] by Hatanaka and Uchita [HAT 96]:

$$\phi' = 27 + 0{,}3.N_{30}. - 0{,}00054.N_{60}^2. \qquad [7.15]$$

$$\phi' = tan^{-1}\left\{{N_{60}}/{[12{,}2 + 20{,}3.(\sigma_v'/p_a)]}\right\}^{0{,}34} \qquad [7.16]$$

$$\phi' = \sqrt{20.C_N.N_{60}} + 20. \qquad [7.17]$$

Relative Density		Consistency		
Sand and Gravels	Blows/Ft (N_{SPT})	Silts and Clay	Strength (tsf)	Blows/Ft (N_{SPT})
Very loose	0-4	Very soft	0-1/4	0-2
Loose	4-10	Soft	1/4-1/2	2-4
Medium	10-30	Firm	1/2-1	4-8
Dense	30-50	Stiff	1-2	8-16
Very dense	Over 50	Very stiff	2-4	16-32
		Hard	Over 4	Over 32

Table 7.8. *Correlation providing the consistency of sands and gravels for SPT [PEC 53]*

7.3.3.2. *Soil with cohesion*

Terzaghi and Peck [PEC 48] also proposed correlations between the number of SPT blows N_{60} and the undrained strength of the clay (Table 7.10). Similarly, relation [7.18] by Kulhawy and Mayne [KUL 90] as well as relation [7.19] by Hara *et al.* [HAR 74] provide an estimate of the undrained consistency.

Description*		Very loose	Loose	Medium	Dense	Very dense
Relative density D_r	0	0.15	0.35	0.65	0.85	
SPT N'_{70}	fine	1-2	3-6	7-15	16-30	?
	medium	2-3	4-7	8-20	21-40	> 40
	coarse	3-6	5-9	10-25	26-45	> 45
Ø:	fine	26-28	28-30	30-34	33-38	
	medium	27-28	30-32	32-36	36-42	< 50
	coarse	28-30	30-34	33-40	40-50	
γ_{wet}, pcf		70-100†	90-115	110-130	110-140	130-150
(kN, m^3)		(11-16)	(14-18)	(17-20)	(17-22)	(20-23)

† Excavated soil or material dumped from a truck will weigh 11 or 14 kN/m^3 and must be quite dense to weigh must over 21 kN/m^3 . No existing soil has a D_r = 0.00 nor a value 1.00 –common ranges are from 0.3 to 0.7.

Table 7.9. *Correlation giving the friction angle in SPT*

No of blows. N	Consistency	q_u ton/sq ft
Below 2	Very soft	Below 0.25
2-4	Soft	0.25-0.50
4-8	Medium	0.50-1.00
8-15	Stiff	1.00-2.00
15-30	Very stiff	2.00-4.00
Over 30	Hard	Over 4.00

Table 7.10. *Correlation providing the level consistency in SPT [TER 48]*

An example of a penetration graph is shown in Figure 7.27.

$$c_u/p_a = 0{,}06.N_{60} \quad [7.18]$$

$$c_u/p_a = 0{,}29.N_{60}^{0{,}72} \quad [7.19]$$

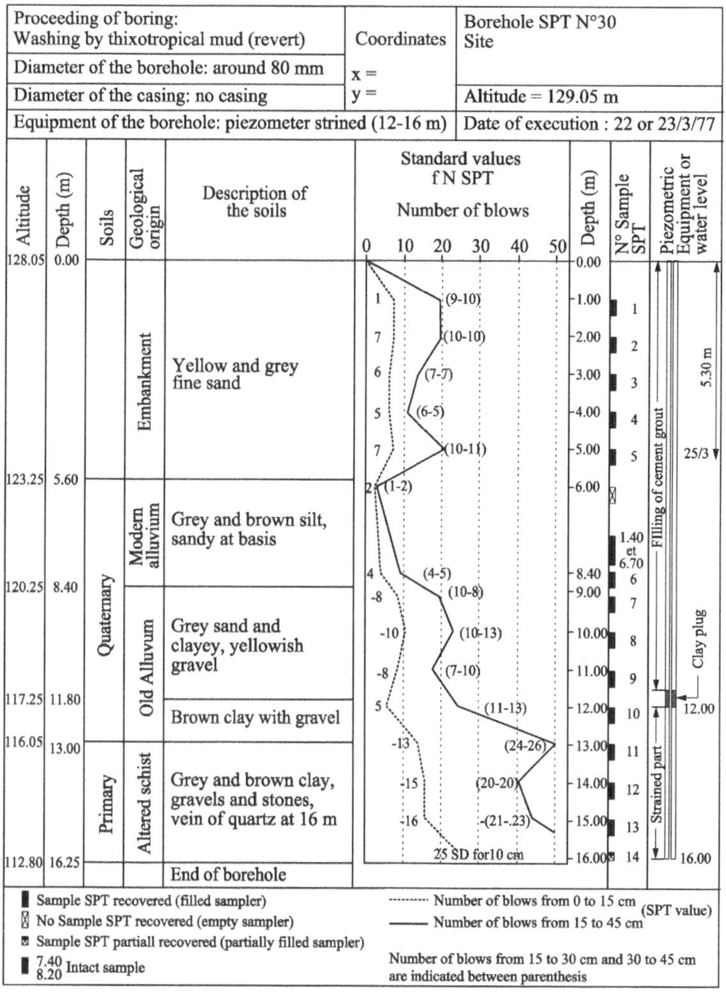

Figure 7.27. *Example of an SPT test report*

7.4. The dynamic penetrometer (DCPT)

Examples of manual, light, dynamic penetrometers are given in Figures 7.28 and 7.29. The test is normalized, whether it is the test with drilling mud, written PDA in France [AFN 90a], or the test without drilling mud, written PDB in France [AFN 90b].

Figure 7.28. *Example of a self-driven Sedidrill penetrometer GTR 790 (Doc. Sedidrill®)*

Figure 7.29. *Example of a light dynamic penetrometer, the Panda® penetrometer*

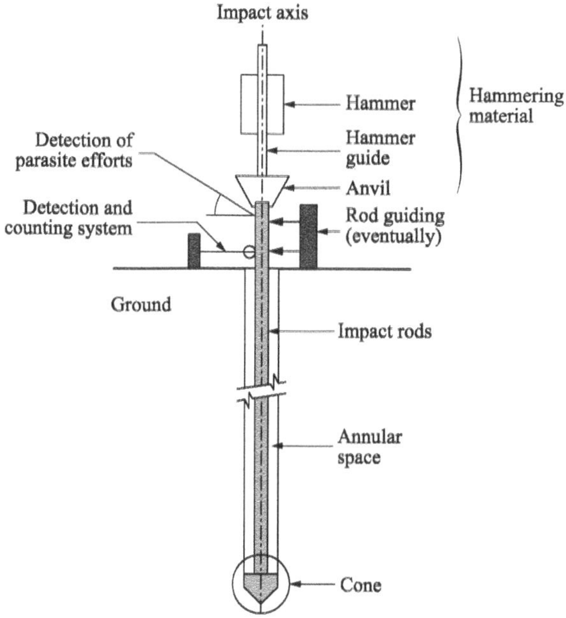

Figure 7.30. *Diagram of the dynamic penetrometer*

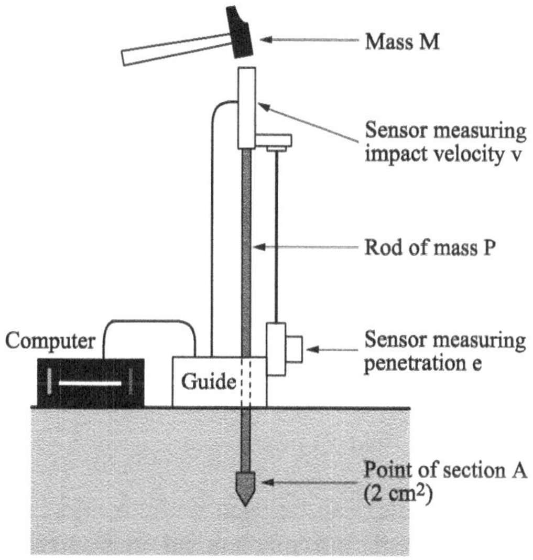

Figure 7.31. *Diagram of the Panda light dynamic penetrometer*

7.4.1. *Principle of the test*

This test consists of hammering a string of rods, armed with a cone at its toe, into the ground (Figure 7.31). For the hits, a hammer of mass M is used, falling from a given and fixed height of H. The hammer hits an anvil tied to a string of rods, causing penetration. N_d is the number of blows necessary to push the cone in for a given height h (most often equal to 20, 25, 30 or 75 cm). This measurement allows us to determine the value of the resistance to dynamic penetration q_d, which can be correlated with the resistance of the cone q_c.

7.4.2. *Modus operandi*

7.4.2.1. *Case of the motorized dynamic penetrometer*

Under the effect of the hammer falling at a rate of 15–30 times per minute, the string of rods is pushed into the ground in a continuous manner. Any stop of more than 5 min must be marked on the test report.

The number of blows needed to push the cone in by 10 cm is written as a function of the total length of the rods introduced into the ground.

The mass of the hammer must be adapted during the hitting and chosen among four masses of 32, 64, 96 and 120 kg so that the number of blows needed for a 10 cm penetration between 2 and 30 is, excluding local heterogeneities (blocks, hollow sections, weak sections, etc.).

The end of the test is marked by one of the following conditions being met:

– the previously determined depth is reached;

– penetration from 30 hammer blows is less than or equal to 10 cm with a mass of 128 kg;

– the rebound is greater than 5 cm.

The dynamic penetrometer can be used with an injection of drilling mud, called DPA. Its geometrical and mechanical characteristics are given in

Table 7.11. In the case of DPA, a surface pump injects the drilling mud into the string of rods. This mud returns just behind the cone of the dynamic penetrometer.

The dynamic penetrometer can be used without drilling mud, the DPB. Its geometrical and mechanical characteristics are given in Table 7.12.

	Designation		Symbol	Unit	Value	Tolerance of usage
Hammer	Mass		m_1 m_2 m_3 m_4	kg kg kg kg	32 64 96 128	0 % – 1%
	Drop height		H	m	0,75	± 0,2
	Launch - Length over width ratio D_M of the hammer		–	–	1 to 2	–
	Beating frequency		–	Hits/ min	15 to 30	–
Anvil	Diameter		–	cm	< 0,5 m > 10	–
	Mass	Anvil alone	–	kg	10 to 15	–
		Anvil and joined guide	–	kg	≤ 25	–
Rods	Length		–	m	1 to 2	± 0,1%
	Mass		–	kg/m	4,0	– 0,5 + 1,0
	External diameter		d_t	mm	42,5	± 3,5
Probe	Area of the straight section		A	cm^2	30	–
	External diameter		d	mm	61,8	+ 0,4 – 1,8
	Length of the conical part		l_2	mm	30,9	+ 0,2 – 2,0
	Length of the cylindrical part		l_1	mm	61,8	+ 0,4 – 5
Probe-holder	Mass		–	kg	≤ 4	–
	External diameter		d_1	mm	42,5	± 3,5
	Injection orifices	Diameter	–	mm	5	– 0 + 0,2
		Distance to the probe	–	mm	< 20	–

Table 7.11. *Geometrical characteristics of the type A dynamic penetrometer with injection of drilling mud: DPA*

<table>
<tr><th></th><th colspan="2">Designation</th><th>Symbol</th><th>Unit</th><th>Value</th><th>Tolerance of usage</th></tr>
<tr><td rowspan="4">Hammer</td><td colspan="2">Mass</td><td>m</td><td>kg</td><td>64</td><td>0
– 0,7</td></tr>
<tr><td colspan="2">Drop height</td><td>H</td><td>m</td><td>0.75</td><td>± 0.02</td></tr>
<tr><td colspan="2">Launch - Length over width ratio D_M of the hammer</td><td>–</td><td>–</td><td>1 to 2</td><td>–</td></tr>
<tr><td colspan="2">Beating frequency</td><td>–</td><td>hits/ min</td><td>15 to 30</td><td>–</td></tr>
<tr><td rowspan="3">Anvil</td><td colspan="2">Diameter</td><td>–</td><td>cm</td><td>< 0.5 D_m
> 10</td><td>–</td></tr>
<tr><td rowspan="2">Mass</td><td>Anvil alone</td><td>–</td><td>kg</td><td>10 to 15</td><td>–</td></tr>
<tr><td>Anvil and joined guide</td><td>–</td><td>kg</td><td>≤ 25</td><td>–</td></tr>
<tr><td rowspan="3">Rods</td><td colspan="2">Length</td><td>–</td><td>m</td><td>1 to 2</td><td>± 0,1%</td></tr>
<tr><td colspan="2">Mass</td><td>–</td><td>kg/m</td><td>≤ 8.5</td><td>–</td></tr>
<tr><td colspan="2">External diameter</td><td>d_t</td><td>mm</td><td>34</td><td>± 2.5</td></tr>
<tr><td rowspan="5">Probe</td><td colspan="2">Area of the straight section</td><td>A</td><td>cm^2</td><td>20</td><td>–</td></tr>
<tr><td colspan="2">External diameter</td><td>d</td><td>mm</td><td>50.5</td><td>± 0,5</td></tr>
<tr><td colspan="2">Length of the cylindrical part</td><td>l_1</td><td>mm</td><td>50.5</td><td>+ 0.5
– 5</td></tr>
<tr><td colspan="2">Length of the conical part</td><td>l_2</td><td>mm</td><td>25</td><td>– 0.5
+ 2.0</td></tr>
<tr><td colspan="2">Length of the truncated part is the probe can be recuperated</td><td>l_3</td><td>mm</td><td>≤ 50.5</td><td>–</td></tr>
</table>

Table 7.12. *Geometrical characteristics of the type B dynamic penetrometer without mud injection forage: DPB*

The total loss of injection mud is also a criterion for the stopping of the test, which can, however, be continued in non-standard conditions.

The cone has an angle at its summit of 90°.

The cone can have a mobile cone (Figure 7.32), meaning that it can move in relation to the string of rods, it can be fixed to the string of rod (Figure 7.33), which then requires an excess width of 20 mm +/– 2 mm in relation to the string of rods, or it can have a lost cone (Figure 7.34) like the Panda (Figure 7.29), which is a light penetrometer whose penetrative energy is provided by a separate hammer.

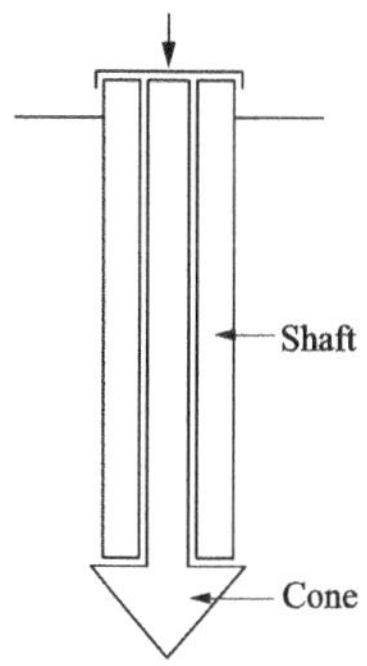

Figure 7.32. *Mobile point dynamic penetrometer*

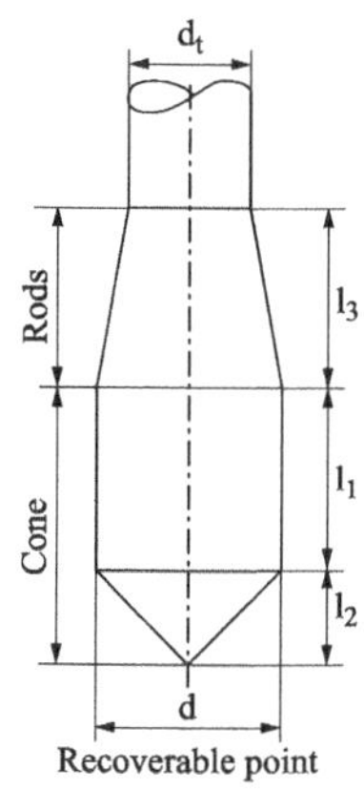

Figure 7.33. *Fixed cone dynamic penetrometer*

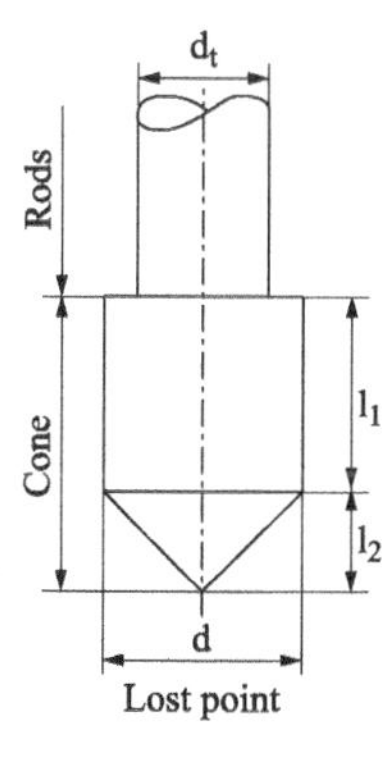

Figure 7.34. *Lost cone dynamic penetrometer*

The characteristics of the main penetrometers used in France are shown in Table 7.13. In this table, the power of penetration is the ratio between the hammering energy compared to the surface of the point or the cone.

Origin	Mass kg	Height of fall m	Penetration power kW/m2	Diameter of the probe mm	Diameter of the rod mm
ANDINA	750	0.16	981	39	35
ANN	5.2	1	53	35	18
BERG	60	0.50	from 106 to 200	60	32
BEVAC	30	0.20	58	36	22
BIG	from 25 to 75	from 0.25 to 0.75	from 22 to 200	60	32
BOTTE	from 50 to 100	0.50	from 166 to 333	43.7	32
DURMEYER	130.75	1	300	75	31.5
ETF	150	0.50	228	65	45
FONDASOL	100	0.30	104	60	60
PANDA	2.35	Variable	from 0 to 50	22 and 16	14
PILCON	63	0.75	167	60	42
SATTEC	320	0.33	367	60	60
SEDITHEC	10	0.50	49 to 98	24.2 and 35.6	22
SERMES	30.60 or 90	0.40	from 30 to 92	70	40
SISCO	50	0.50	166	43.7	32
SOBESOL	60	from 0.50 to 1.50	from 56 to 170	55	42
SOTEC	8	0.80	60	35	25
VERITAS	15	1	76	50	34

Table 7.13. *The different types of dynamic penetrometer used in France*

An example of a penetrometric survey is given in Figure 7.35.

7.4.2.2. *The Panda penetrometer*

The Panda is a standardized [AFN 12a] variable energy dynamic penetrometer. The test consists of manually beating a string of rods into the ground with a hammer (Figure 7.31). At each impact, the Panda© equipment records the speed of the impact, which it then converts into beating energy, using an internal calculator. The system also records the extent of the penetration of the string of rods using a belt system. The penetrograms are immediately available for consultation on the site.

The Panda system benefits from several advantages:

– it is light and can be easily transported to the relevant sites;

– the recording of data is automatic, all of the recordings are safely stored;

– the beating energy can be modulated by decreasing it in soft soils and increasing it in hard soils, so as to maintain as near constant rate of penetration;

– the system records a point of measurement at each hit of the hammer, which corresponds to a few cm of penetration. This results in a nearly continuous penetrogram, able to easily detect interfaces of layers of soil;

– the device is particularly adapted for the control of soil compaction.

However, the Panda has a certain number of disadvantages:

– it is a light penetrometer with low levels of penetration energy;

– its depth of investigation is 2–3 m at the most;

– the friction of the soil on the string of rods at more than 1 m causes an overestimation of the dynamic resistance.

An example of a penetrometric survey is given in Figure 7.35.

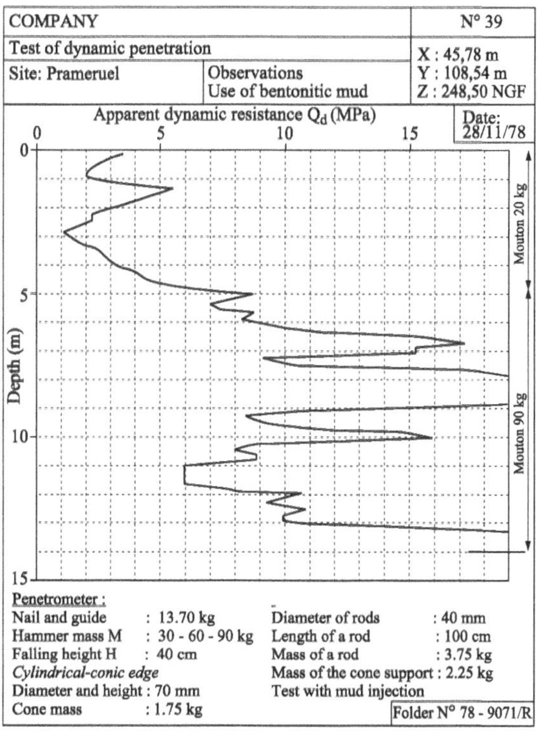

Figure 7.35. *Graph of the dynamic penetrometer penetration*

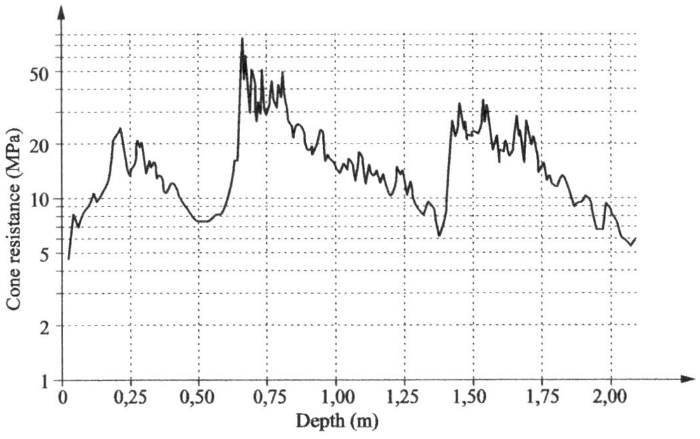

Figure 7.36. *Example of Panda penetrometer penetration (Sol-Solution document)*

7.4.3. *Interpretation*

We can deduce from this test, the dynamic resistance q_d, calculated using the Dutch formula [7.20], with M the mass of the hammer, M' the mass hit (sum of the masses of the string of rods, the anvil and the cone), A_c the surface of the greatest straight section of the cone, g acceleration caused by gravity, e the mean penetration interval [7.21] and N_{dh} the number of blows required to produce a penetration equal to h:

$$q_d = \frac{M^2.g.H}{(M+M\prime).A_c.e} \quad [7.20]$$

$$e = \frac{h}{N_{dh}} \quad [7.21]$$

7.4.4. *Use of dynamic penetration in the calculation of the foundation*

Using the penetration graph, it is possible to:

– detect the different horizons of the site;

– extrapolate this to identification via survey;

– estimate the subgrade bearing stratum of driven piles or steel piles;

– calculate the bearing capacity of deep foundations;

– calculate the bearing capacity of superficial foundations.

7.4.4.1. *bearing capacity of driven piles*

This nominal load-bearing capacity is calculated from the dynamic resistance q_d using relation [7.22], with Q_N the nominal load of the pile, S the section of the pile and F the safety coefficient, comprised between 6 and 12:

$$Q_N = \frac{q_d . S}{F} \tag{7.22}$$

7.4.4.2. *bearing capacity of superficial foundations*

Sanglerat [SAN 72] states that for granular soils, and those located above the water table, relation [7.23] can be applied, in the case where the depth of embedding is greater than the width of the footing:

$$\sigma_{adm} = \frac{q_d}{20} \tag{7.23}$$

Below the water table, and for loose soils, a considerable overestimate appears for the value of q_d, caused by the pressurization of the interstitial water under the cone of the penetrometer.

7.4.5. Comparison between the results of the static and dynamic penetrometers

The comparison between the resistance q_c measured with the static penetrometer, and the resistance q_d measured with the dynamic penetrometer for different soils is shown in Figures 7.37 and 7.38. We can note:

– the negative influence of the water table on q_d (Figure 7.37), which leads to an overestimation of the resistance;

– the constant overestimation of q_d compared to q_c (Figure 7.38).

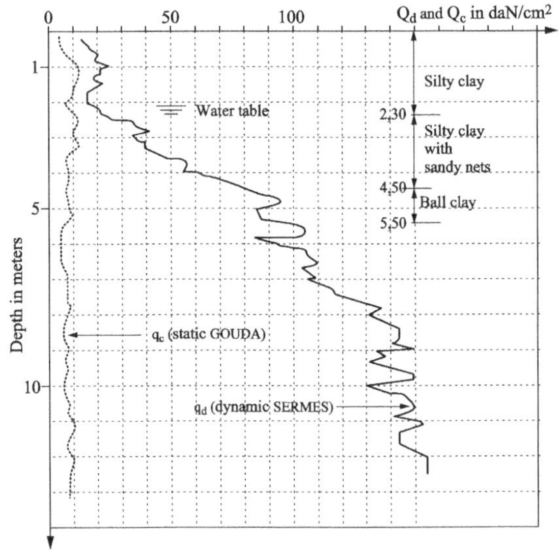

Figure 7.37. *Comparison between the tests of static penetration (Delf cone) and the tests of dynamic penetration in a dense medium*

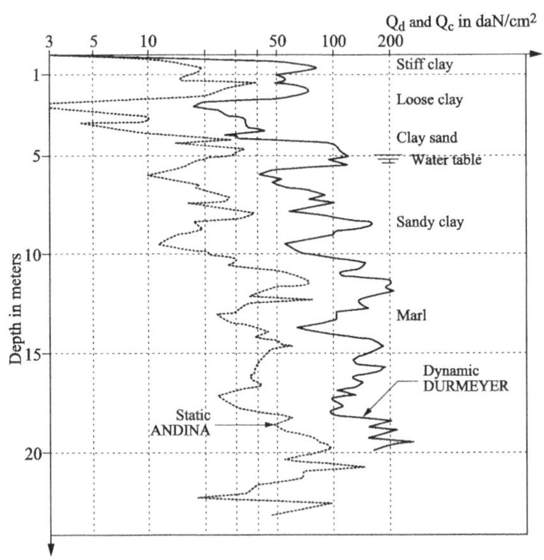

Figure 7.38. *Comparison between the Andina static penetration test and the Durmeyer dynamic penetration test*

8

Direct Shear Tests *In Situ*

In a number of projects of civil engineering, it is necessary to know the shearing characteristics of the soil in place (cohesion c and angle of friction Φ). These can be used to:

– calculate the bearing capacities of the foundations;

– calculate the stabilities of the slope (embankment, banks and cuttings);

– calculate the stabilities of diggings (slurry wall, sheet piling, etc.);

– calculate support structures, etc.

It is not always possible to recover intact samples and analyze them in the laboratory. Some soils cannot be directly sampled, due to their granularity (gravels and soils containing large elements), their water content (saturated sand under the water table) or their degree of fracturing (altered rock). Moreover, some soils are sensitive to the manner in which they are transported and stored (vibration during transport and variation of water content). It is, therefore, necessary to carry out tests *in situ* to determine the shear characteristics that are essential for the project.

8.1. Direct shear box test

A prismatic soil sample is taken (Figure 8.1), linked to the soil mass to be studied. The sample is surrounded by a metal cage to immobilize it horizontally.

The sample is loaded vertically, either through the reaction of a jack on the roof of a gallery (Figure 8.1), or using a mass placed on its summit.

The soil is interacted with using a horizontal shearing through a jack (Figure 8.1) or a screw equipped with a deformable ring.

The soil breaks along a precise shearing plan, at the base of the prismatic sample and in contact with the soil mass.

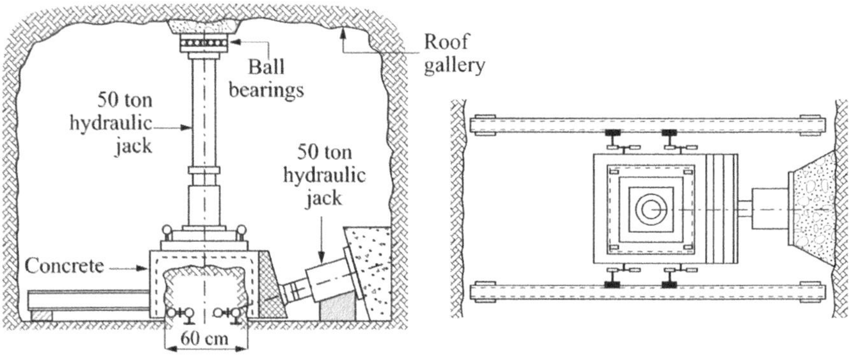

Figure 8.1. *Shear test with a concrete mass in a gallery*

The curves of the shear graphs are then drawn as a function of movement (Figure 8.2). The maxima are linked, which allow us to calculate the cohesion c and the internal friction angle Φ of the soil at peak resistance and during the post-peak plateau.

It can be noted that the shear curves present a peak for low stress normal to the shear plane or for overconsolidated soils, and do not present a peak for high normal stress, or for normally consolidated soils.

While the principle of the test is simple, carrying it out is difficult to do, as it requires three intact samples that are not fissured in the shear plane. Moreover, the test is hard to put in place, and it is very sensitive to activity in the site around it.

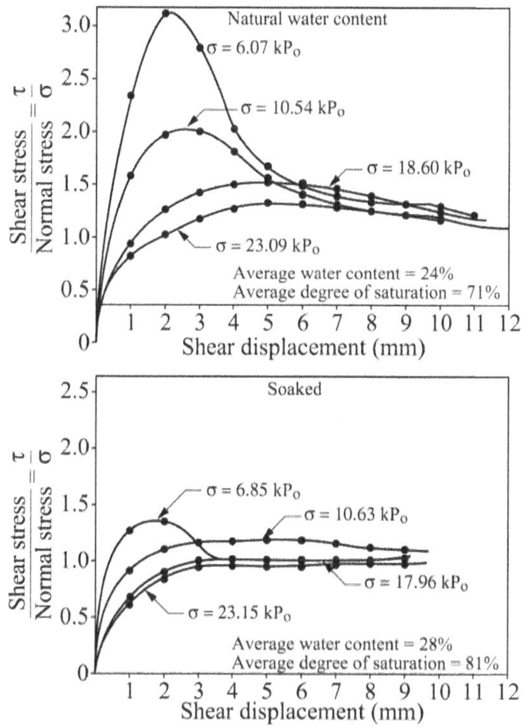

Figure 8.2. *Evolution of shearing during the direct box shear test in a soil*

8.2. The vane test (VST)

Vane tests are covered by standards in the USA [AST 78], the UK (BS 1377, 1990), Australia (AS F2.2, 1977), Germany (DIN 406, 1980), India (IS 4434, 1978) and France [AFN 91b].

8.2.1. *Principle of the test*

Two vanes, perpendicular to each other, are pushed into the ground (Figure 8.3). Once the device is in place, a torque load C is applied on the vanes using a lever or a gear system (Figure 8.4). The value of the torque C is linked to the undrained cohesion of the soils c_u. A diagram of the setup of the vane test with torsion bar is presented in Figure 8.5. An overall view of the connection to the string of rods is presented in Figure 8.6.

Figure 8.3. *The vanes of the Gouda vane tester (Internet document)*

Figure 8.4. *Torque gauge loaded on the top of the string of rods*

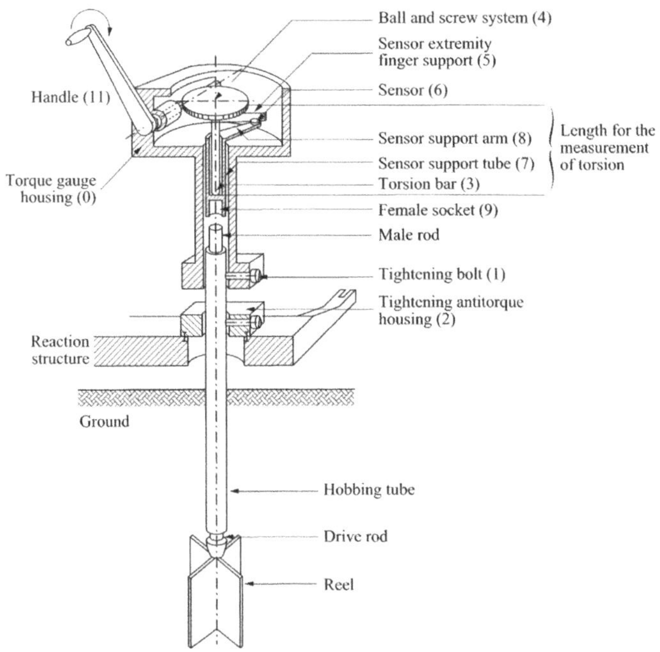

Figure 8.5. *Principle of the torsion bar vane tester [AFN 91]*

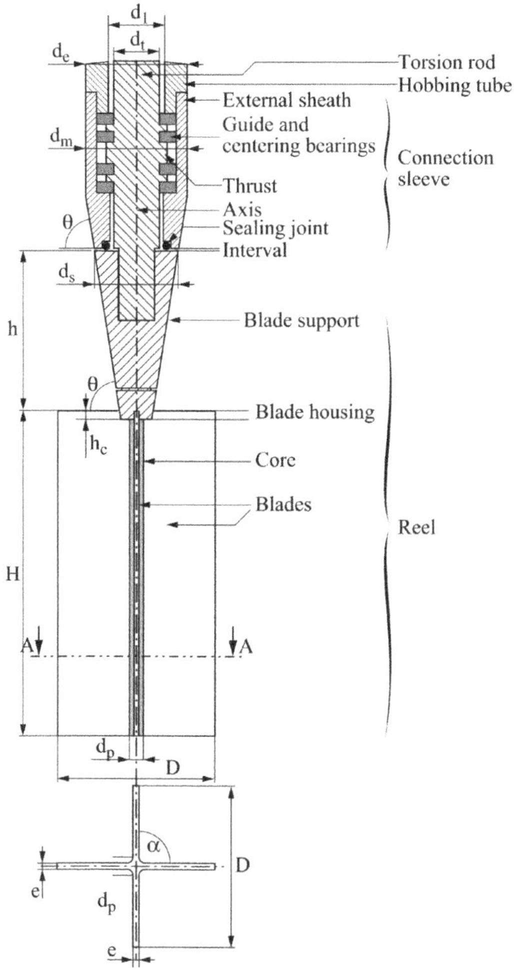

Figure 8.6. *Diagram of the reel and the connection to the string of rods [AFN 91]*

8.2.2. *Practical realization of the test*

The diameters of the vane are 60, 70 and 100 mm. The 100 mm diameter vane is kept for measurements in very soft soils (undrained consistency of less than or equal to 20 kPa). The 60 mm diameter reel is used for consistent clays (undrained consistency of around 100 kPa). The ratio of the height of the vanes to diameter must be equal to 2 at least.

Designation			Refer.	Unit	Value	Tolerance
Reel	Blades	Width	D	mm	70	+ 0 – 1
		Angle between blades	α	degree	90	± 5
		Thickness	e	mm	2	– 0,1 + 0,5
		Height	H	mm	140	– 0 – 1
	Core	Diameter	d_p	mm	≤ 7	–
	Blades core truncated support	Small diameter	d_c	mm	16	+ 0 – 0,2
		Larger diameter	d_s	mm	38	+ 0 – 0,2
		Blades link	h_c	mm	≥ 4	–
		Height	h	mm	70	± 0,2
Hobbing tubes		External diameter	d_e	mm	≥ 38	–
		Internal diameter	d_i	mm	> 1,5 d_t	–
		Length	–	cm	100	± 0,1
Torsion rod		Diameter	d_t	mm	≤ 22	–
		Length	–	cm	100	± 0,1
Torque gauge		Max moment	–	N.m	≥ 120	–
Connection sleeve of hobbing tube to reel		Angle	θ	degree	81	+ 1 0
		External diameter	d_m	mm	≥ d_e	–

Table 8.1. *Dimensional characteristics of the field vane tester*

The test is carried out in the following manner:

– the device is brought to the level of the soil being tested, by direct jacking or by being placed in a previously drilled bore and then driven into the ground;

– the minimal penetration into the ground must be such that the upper part of the reel is located more than 50 cm below the hobbing start level, or at least 5 drill diameters from the bottom of the bore;

– the spacing between two consecutive tests must not be less than 50 cm;

– penetration is carried out without rotation at a maximal speed of 2 cm/s;

– installation of the measurement head and execution of the test must both be done immediately after introduction of the reel into the ground;

– the speed of rotation at the head of the string of rods must be between 6 and 18° per minute (sexagesimal degrees);

– the measurements are made every 10 s and – once the maximum torque is reached – they are followed so as to get six readings beyond this maximum;

– this is followed by a rapid rotation of the transmission rods so that the reel turns around 10 times in the ground. The test is followed-up, with six readings, separated from each other by 10 s, carried out after the speed of rotation used during the test has been resumed;

– the frequency of the tests is recommended with 1 m spacing vertically in the ground;

– with each change of terrain, a calibration of the friction of the rods in the grounds must be done, before connection to the reel.

The results are given in the form of diagrams, which provide the undrained cohesion C_u and the residual cohesion C_r as functions of the depth (Figure 8.10).

We assume that the undrained consistency is strong on the surface, and decreases rapidly in a dried soil area. C_u is constant in an overconsolidated soil area, and increases with depth in normally consolidated soil zones (see Figure 8.10).

Vane test dimensional characteristics are shown in Table 8.1.

8.2.3. *Interpretation: determining the undrained cohesion*

The vanes of the vane tester outline a cylindrical area, with:

– the external surface of the cylinder having a diameter of D and a height of h, and therefore an area of π.D.h;

– the two upper and lower circular sheared faces having a diameter D;

– the lateral surface of the cone supporting the vanes.

We assume that the soil only reacts through its its undrained cohesion. The resisting torque provided by the ground is, therefore:

– along the horizontal inferior face, we consider a circular ring of thickness dr, on which a shearing force of c_u is applied. The moment aimed at this circular ring is [8.1], which is integrated between 0 and D/2 for the inferior surface [8.2], which finally results in [8.4];

– along the lateral surface, the shear strength c_u is exerted on the surface of the cylinder of radius D/2 [8.5];

– along the upper horizontal disk, taken as the difference between the sections of the external enveloping disk and the internal disk supporting the rotation axis [8.6];

– along the frustum of the cone supporting the vanes, with m a coefficient by convention, equal to 0.6 [8.7];

– the total resisting torque is, therefore, the sum of these four values. We can deduce from this coefficient K such that [8.8]:

$$dM_1 = 2.\pi.r.^2\, c_u dr \qquad [8.1]$$

$$M_1 = \int_0^{D/2} 2.\pi.r^2.c_u dr \qquad [8.2]$$

$$M_1 = 2.\pi.c_u.\left[\frac{r^3}{3}\right]_0^{D/2} \qquad [8.3]$$

$$M_1 = \pi.c_u..\frac{D^3}{12} \qquad [8.4]$$

$$M_2 = \pi.\frac{D^2}{2}.h.c_u. \qquad [8.5]$$

$$M_3 = \pi.\frac{c_u}{12}.(D^3 - d_c^3) \qquad [8.6]$$

$$M_4 = \pi.\frac{c_u}{12}.(d_s^3 - d_c^3).m.cos\theta \qquad [8.7]$$

$$K = \frac{(M_1+M_2+M_3+M_4)}{c_u} = 1310cm^3 \qquad [8.8]$$

An experiment moment is measured M_{mes} and from this, we can deduce the undrained consistency using relation [8.9] and the residual consistency, using relation [8.10]. The experimental corrective coefficient k depends on the plasticity index I_p. It was proposed by Pilot to take into account the influence of plasticity on the results of the test (Figure 8.9).

$$c_u = k.\frac{M_{mes}}{K} \quad [8.9]$$

$$c_r = k.\frac{M_{res}}{K} \quad [8.10]$$

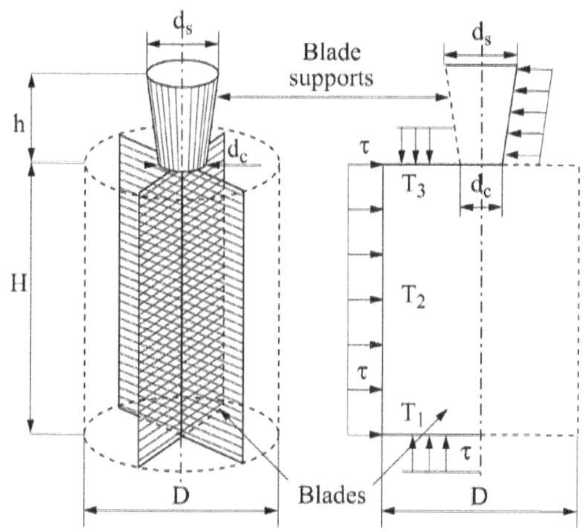

Figure 8.7. *Shear surfaces of the reel [AFN 91]*

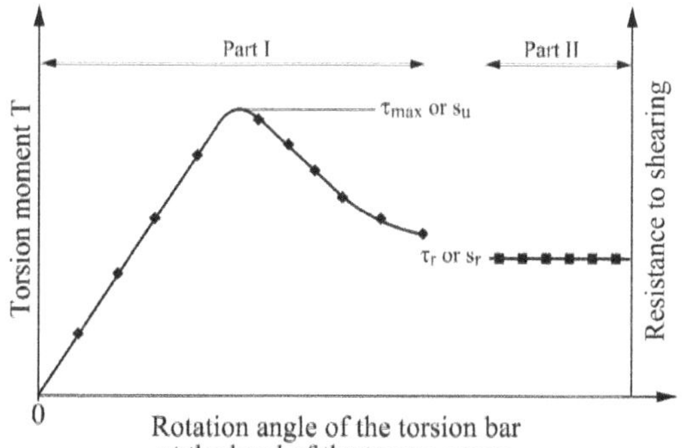

Figure 8.8. *Determining τ_{max} [AFN 91]*

In relation [8.9], the value of the measured moment M_{mes} is used for peak resistance values, or the maximum resistance moment (Figure 8.8), and the relation for the plateau after the peak is used for the plateau resistance moment. We can then deduce the peak consistency c_u and the residual plateau consistency c_r. In these two cases, the undrained angle of friction is equal to zero ($\Phi_u = 0$).

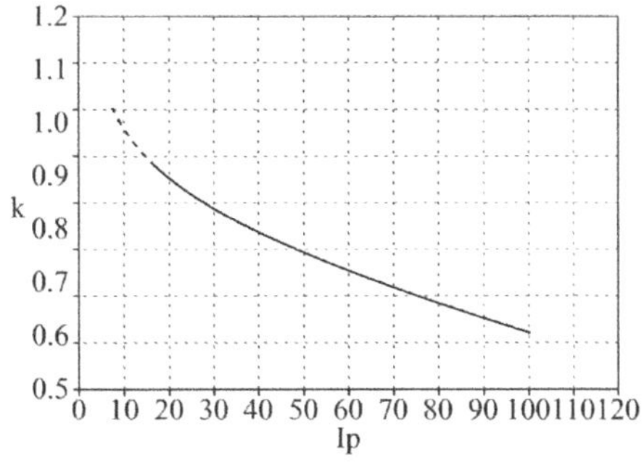

Figure 8.9. *Consistency correction coefficient measured by the vane tester (Pilot)*

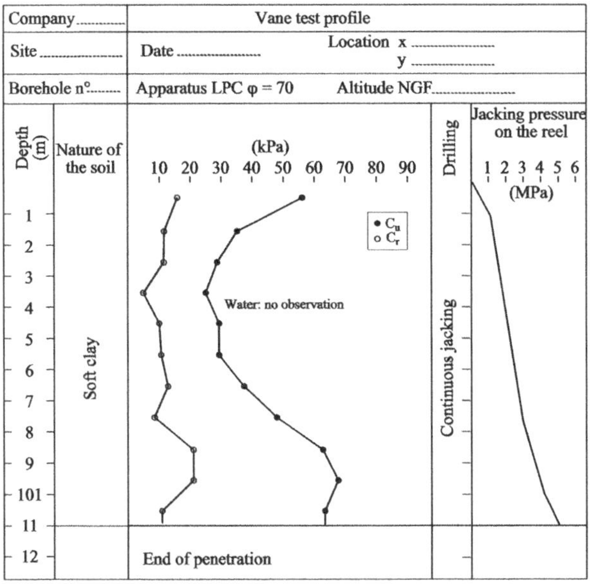

Figure 8.10. *Example of a vane test profile*

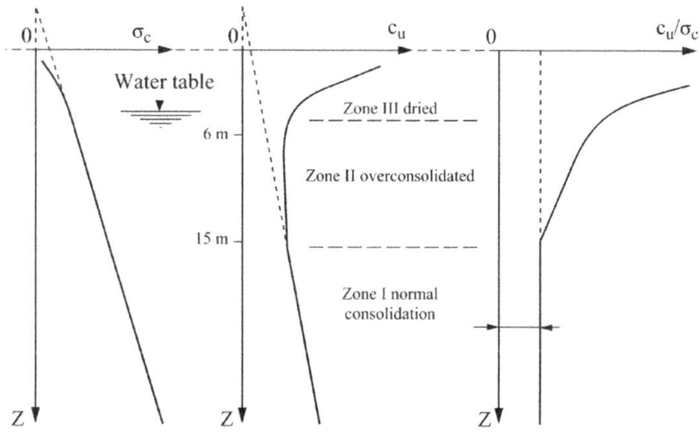

Figure 8.11. *Variations of the consolidation pressure σ_c, undrained consistency c_u obtained through he vane test and the ratio c_u/σ_c as a function of depth (Skempton)*

8.2.4. *Exploitation complement of the vane test*

8.2.4.1. *Relationship between the vane test results and the laboratory values*

From the many tests carried out with the Kovacs vane tester and with triaxial testers in the soft clay of an alluvial plane near Grangemouth in England, Skempton was able to link the value of the ratio of c_u to σ_c (the pressure of overconsolidation) with the liquidity limit of the clay w_L. This relation is presented in the form of a practically linear curve (Figure 8.12).

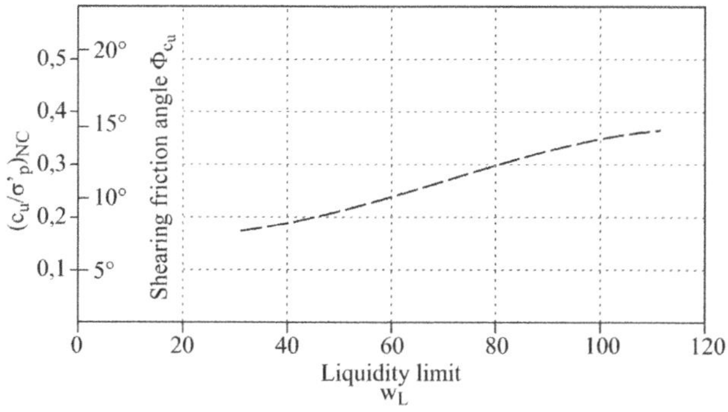

Figure 8.12. *Relation between the liquidity limit and $(C_u / \sigma_c)_{NC}$ normally consolidated clays (Skempton)*

Holtz and Kovacs [KOV 81] improved on the Skempton approach, linking the ratio of C_u to σ_p with the plasticity index I_p. In this analysis, the results from Skempton, Bjerrum, Leonards and Kenney are grouped around an upward curve. Osterman's results, however, are completely different (Figure 8.13).

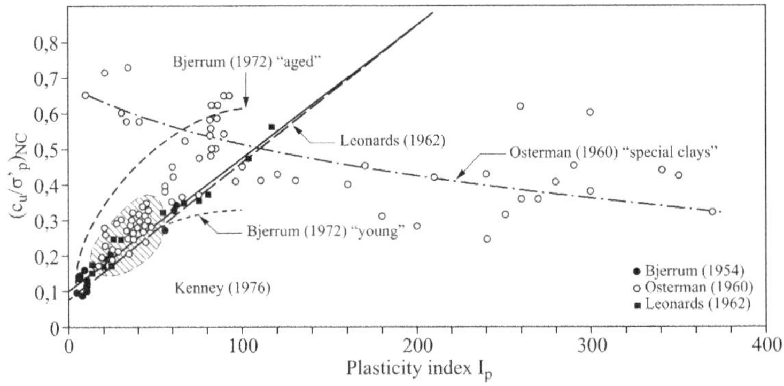

Figure 8.13. *Relation between the ratio (c_u / σ'_c) and the plasticity index for normally consolidated clays [HOL 81]*

Bjerrum and Simons [BJE 60] analyzed the relationship between the ratio of C_u to σ_p and the liquidity subscript. This ratio tends to an assymptote to 0.1 for a liquidity index larger than 1. Bjerrum [BJE 72] analyzed the ratio between the undrained cohesion measured by the vane test *in situ* and at the laboratory as a function of the plasticity index (Figure 8.14). The two values of cohesion are not the same, unlike what might have been presumed. There is some overevaluation of c_u when measured in the laboratory.

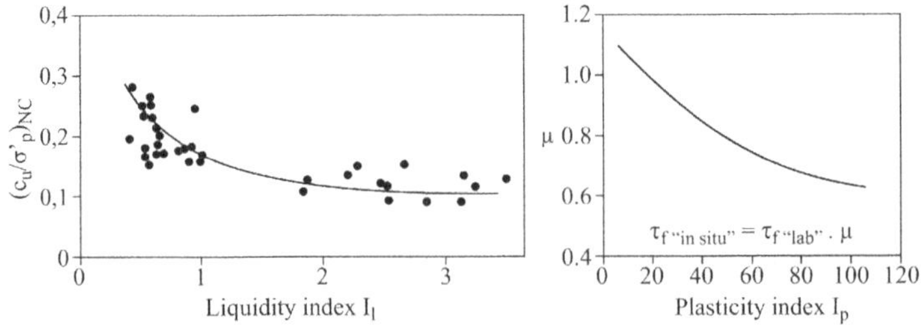

Figure 8.14. *Relationship between ratio (c_u / σ'_c) and the liquidity index a), between c_u measured "in situ" and in the laboratory and the plasticity index [BJE 60] b), for Norwegian clays [BJE 72]*

8.2.4.2. *Relationship between the results using the vane test and the results of other "in situ" tests: the penetrometer and the pressuremeter*

Generally, we consider a relationship between the undrained consistency c_u measured by vane test and with the limit pressure of the pressuremeter, in the form of relation [8.11]:

$$\frac{p_{LM}}{c_u} = 5{,}5 \text{ to } 6 \quad [8.11]$$

We also consider a relationship between the peak resistance measured with the static penetrometer q_c and the undrained consistency measured by vane test, in the form of relation [8.12]:

$$15 < \frac{q_c}{c_u} < 20 \quad [8.12]$$

8.3. The Philiponnat *phicomètre*

The *phicomètre* is covered by a French standard [AFN 97].

8.3.1. *Principle of the test*

It is a direct shearing test carried out into a borehole, based on a different principle from that of the vane test (Figure 8.15).

A probe is lowered into a borehole. The probe is made of a cylinder grooved horizontally (Figure 8.16) and split longitudinally into six parts (c – Figure 8.17). The application of a fluid pressure (using water) on one membrane (20 – Figure 8.17) allows the shells to be applied on the walls of the borehole and the grooves to be entered into the ground (23 – Figure 8.17). In this way, the radial stress σ_r is controlled; this will be the normal stress to the sheared soil cylinder (26 – Figure 8.17).

The whole of the string of rods is pulled longitudinally using a hollow jack (7 – Figure 8.17) that supports itself on blocks (8 – Figure 8.17).

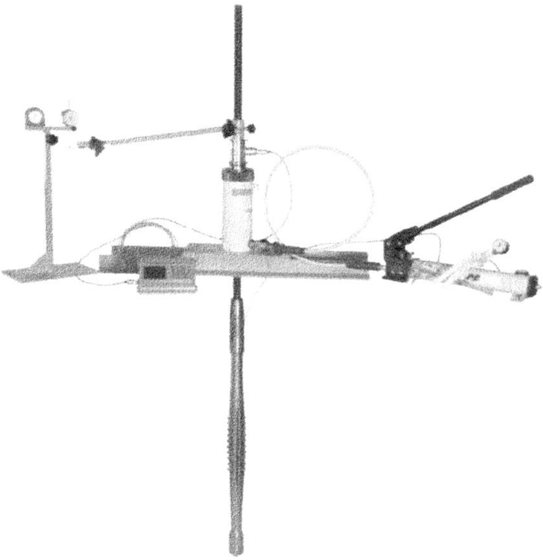

Figure 8.15. *The Philiponnat phicomètre*

Figure 8.16. *The phicomètre probe*

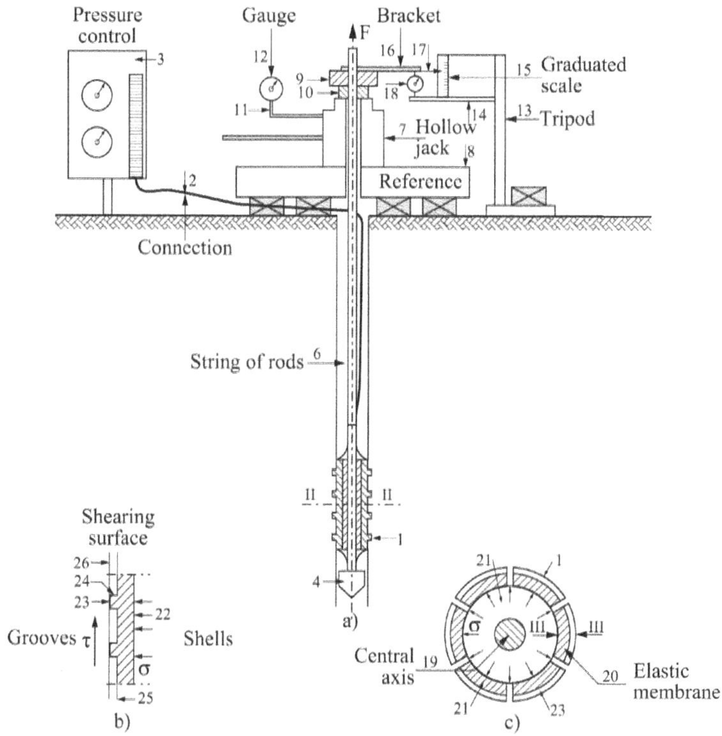

Figure 8.17. *Diagram of the principle of the Philiponnat phicomètre [AFN 97]*

The pressure of the fluid of the jack can be read on a gauge (12 – Figure 8.17) and provides the values of the axial force. The movement of the string of rods, and therefore of the probe, can be read on a fixed scale (15 – Figure 8.17). We assume that the probe shears a constant surface that corresponds to the external envelope of the grooves (26 – Figure 8.17).

8.3.2. *Practical realization of the test*

The procedures used for this test are the following:

– the probe is leave at the ground level, outside the borehole to determine the pressure of inertia needed to obtain a volume of water injected into the probe (after 1 min of applying the pressure) and the unique external diameter of the probe;

– a borehole is drilled in the conditions needed for the pressuremeter test;

– the probe is installed at the desired level;

– a pressure p_1 is applied to the fluid (water) in the probe and the volume of the water injected is measured after 1 min of applying the pressure;

– a shearing force is applied at a constant speed of 2 mm per minute;

– the axial efforts are measured every 30 s. The axial efforts increase linearly, and then stabilize at a maximal value (Figure 8.18). This plateau corresponds to the rupture of the crown of the soil, which surrounds the phicometric probe;

– the measurements of the axial effort are continued at a constant radial pressure, until a maximal movement of 90 mm is achieved;

– a variant of this operating mode involves measuring the axial effort until a movement of only 1 cm is achieved, which results in a nearly continuous line of rupture;

– the threshold value is that of a shear of τ_1, which corresponds to a normal constraint of σ_{r1};

– next, a pressure of p_2 is applied, greater than p_1, and the procedure is repeated.

8.3.3. *Interpretation*

If D_1 is the external diameter of these grooves when a pressure p_1 is applied, the value of the sheared surface [8.13] can be calculated, with h being the length of the probe. The diameter D is defined from the calibration curve linking the volume of water injected to the diameter of the probe. If F is the maximum axial force of shearing applied, then the shearing constraint is [8.14]:

$$S = \pi . D_1 . h \qquad [8.13]$$

$$\tau_1 = {}^{F}/_{\pi . D_1 . h} \qquad [8.14]$$

The applied pressure p_1 is corrected from the calibration pressure p_i corresponding with the same volume injected, as well as the thickness of the membranes, the metallic lantern and the height of the teeth e_p. This provides the radial constraint applied to the soil, using [8.15]:

$$\sigma_1 = (p_1 - p_i).\frac{(D_1 - e_p)}{D_1} \quad [8.15]$$

There is, therefore, a torque of value (τ_1, σ_{r1}) in the Mohr–Coulomb plane, which corresponds to a rupture for a chosen constraint level (Figure 8.18).

All of the (τ_1 , σ_{r1}) torques help to draw a Coulomb rupture curve, and to determine the cohesion c and an internal friction angle of the soil Φ (Figures 8.19 and 8.20).

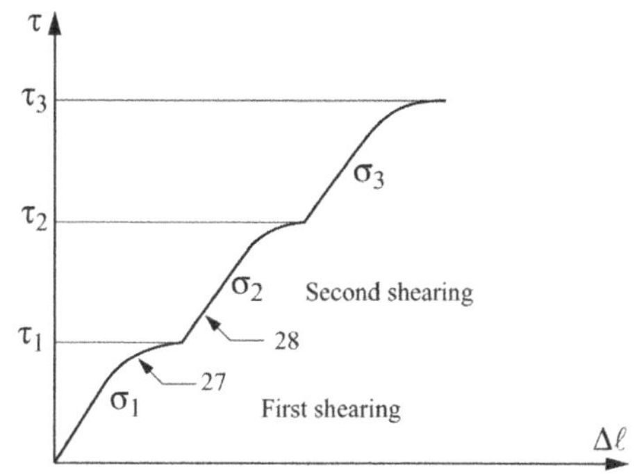

Figure 8.18. *Experimental shearing curves obtained with the phicomètre*

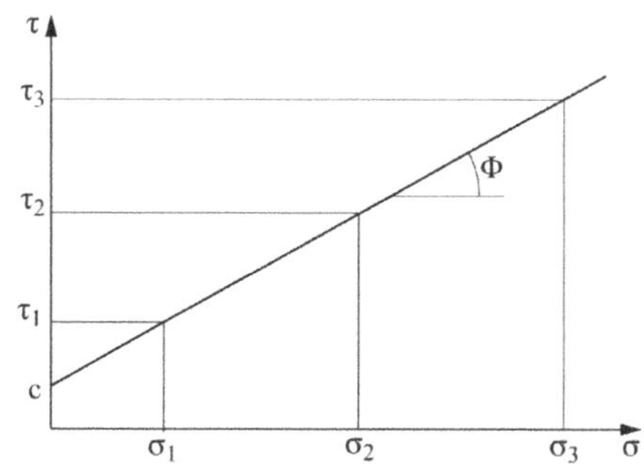

Figure 8.19. *Determination of the consistency c and the internal friction angle Φ with the phicomètre*

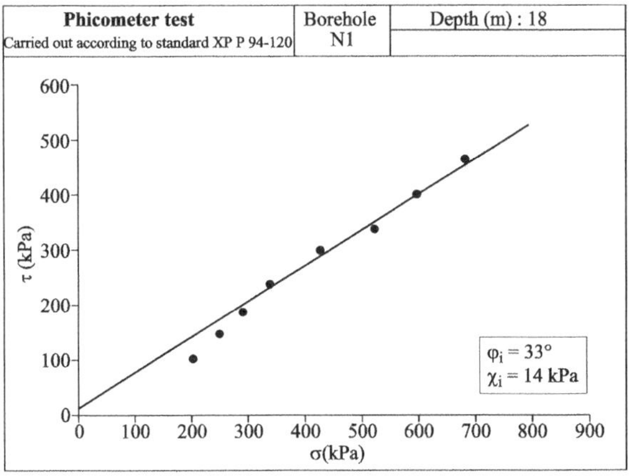

Figure 8.20. *Example of a phicomètre result, determination of c and Φ in situ*

8.3.4. *Advantages and disadvantages of the phicometric test*

8.3.4.1. *Disadvantages*

The disadvantages are as follows:

– it is not possible to test soils with a grain size greater than the gaps between grooves (D_{max} smaller than 2 or 3 cm);

– a crown of soil is sheared that has already been sheared in part during the application of the previous constraint. The cohesion c and the internal friction angle Φ that are obtained have been modified;

– we are limited to small soil covers (less than 10 m), as there cannot be any form of friction along the rods when the shearing force F is applied;

– the deformation characteristics of the soil are not measured.

8.3.4.2. *Advantages*

The advantages are as follows:

– this is a test that is quick and easy to carry out;

– it is reasonably priced (2–3 times that of a pressuremeter test);

– the cohesion and friction angle obtained are "*in situ*".

9

Pressuremeter Tests (PMT, SBP) and Variants

Many civil engineering and construction projects require an in-depth knowledge of the mechanical characteristics of foundation soil. Previous tests have shown that direct shearing (box, vane test and "phicomètre") allows for the measurement of cohesion as well as the soil's internal friction angle. However, they also give no indication about the moduli of soil deformation.

Ménard [MER 55] patented the pressuremeter on 19 January 1955. It is a device that possesses a dilatable probe that is placed in a borehole. This device can determine:

– the soil shear deformation modulus E_M;

– the pressure limit p_{LM} which is linked to the resistance of soil shearing;

– the settlement and bearing capacity under the foundations, by using the tables and graphs of empirical correlations of E_M and p_{LM};

– the cohesion or the internal friction angle, or the association of cohesion-angle of friction from a more thorough interpretation.

Other devices have been developed from the principle of the functioning of the pressuremeter such as the CamKometer, the self-boring pressuremeter, the dilatometer and the "Géomécamètre" which will be introduced in this chapter.

9.1. Ménard pressuremeter test (PMT)

The Pressuremeter Ménard Test (PMT) is covered by French standards [AFN 91a, AFN 91b] and american [AST 94b]. It is described in detail by Baguelin *et al.* [BAG 78] and is regularly featured in international conferences.

9.1.1. *Principle of the test*

The pressuremeter is composed of a dilatable probe placed in a borehole, tubes for transmitting fluids (air and water), a pressure and volume controller, and a reserve of compressed gas (Figures 9.1 and 9.2). The principle of the test consists of inflating a dilatable probe in a borehole and linking the pressure of the inflation to a measurement of the deformability of the soil, the pressuremeter modulus E_M and a measurement of the pressure limit p_{LM}.

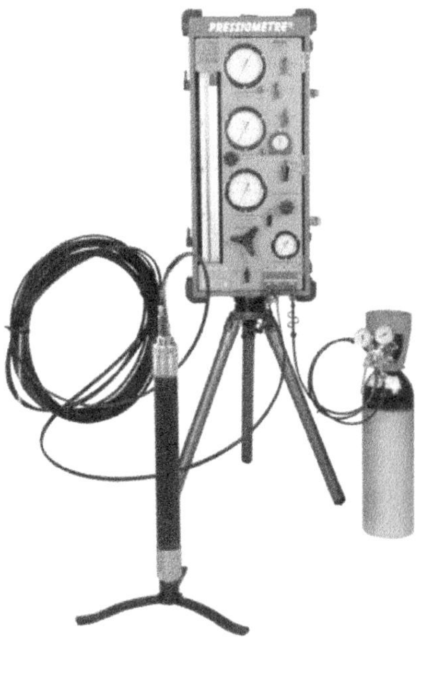

Figure 9.1. *Ménard pressuremeter*

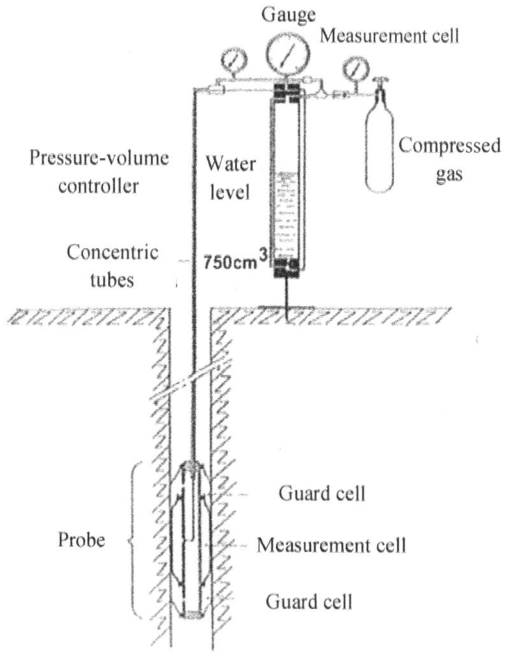

Figure 9.2. *Diagram of the principle of the pressuremeter*

9.1.2. *Execution of the test*

9.1.2.1. *Equipment*

The flexible pressuremeter probe (Figure 9.3) is placed in a borehole at chosen level. It is composed of metallic cylinder (Figure 9.3, no. 1; Figure 9.4, no. 1) with a diameter that is compatible with the diameters of the common boring tools (generally a diameter of 60 mm). It is covered with a rubber membrane (Figure 9.3, no. 3; Figure 9.4, no. 3). The central part of this probe is inflated with water and forms the measurement cell of length l_s (Figure 9.3, no. 2; Figure 9.4, no. 2). The extremities of the probe are inflated with compressed gas and constitute the guard cells of length l_g. The purpose of these guard cells is to distribute the pressures in the soil in a uniform manner in order to obtain a substantially constant pressure in the measurement cell. The probe is linked to the pressure and volume controller at the surface by a flexible, coaxial tube made from Rilsan, which transmits the water pressure through the central tube, and the gas pressure through the exterior tube. This device allows us to reduce in the largest possible way the parasite deformations of the flexible connection cable which could be taken for the deformation of the probe itself. A variant of the flexible connection consists of linking the probe to the pressure and volume controller by twin tubes placed besides one another, one transmitting the pressure of the water, the other transmitting the pressure of the gas.

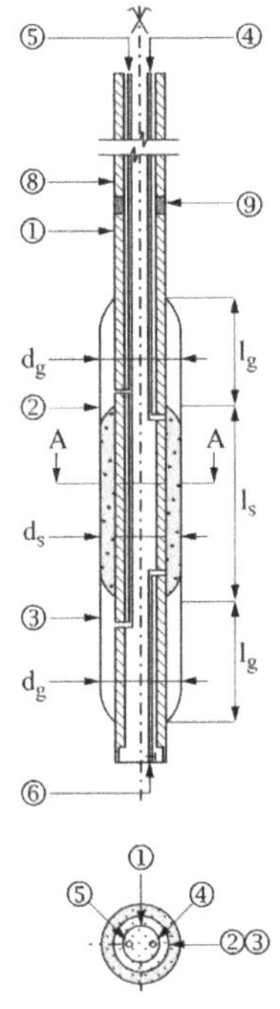

Figure 9.3. *Diagram of the flexible sheath pressuremeter [AFN 91]*

The split tube pressuremeter (Figure 9.4) is placed under the cover of a steel tube with an exterior diameter ds, possessing six equidistant slits over a longitudinal length lm, in such a way that relation [9.1] may be verified. The opening of each slit must be less than or equal to 0.4 mm at all times during the test. After dilation, the split tube must return to its initial state:

$$l_m = Max\{1{,}3.(l_s + 2.l_g); 800mm\} \quad [9.1]$$

The volume and pressure controller must be placed on the surface. The pressure is fed into it by a pressurized gas bottle (Figure 9.2) which is often carbonic gas. A pressure regulator regulates the pressure of the gas.

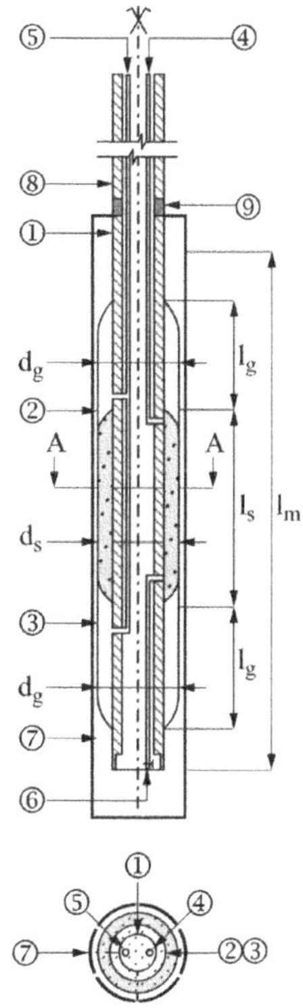

Figure 9.4. *Diagram of the split probe pressuremeter [AFN 91]*

This pressure is monitored on a gauge and it is transmitted in its entirety to the surface of the tank containing the water that is fed into the pressuremeter probe. Through the intermediary of a coupling block (Figure 9.2, no. 12), this initial pressure is equally transmitted into the two guard cells, after having been reduced by the differential pressure regulator (Figure 9.2, no. 19). The pressure of the gas in the guard cells is monitored by the differential gauge (Figure 9.2, no. 20). The venting valves allow for the reduction of the pressure of the gas first in the guard cells, followed by the measurement cells. The filler valve completes the level of the liquid so that the test may begin with a level close to zero for the first boring test. The boundary valve disconnects the probe from the pressure and volume gauge for transport.

The pressuremeter probes have normalized dimensions (Table 9.1). Generally, bare probes (Figure 9.3) are used in fine and slightly consistent soils, and once the boring has already been carried out. Lantern probes (Figure 9.4) are generally used in coarse soils and once the probing has been done by direct pile driving of the probe into the soil.

<table>
<tr><th colspan="4"></th><th>Reference</th><th>Unit</th><th>Value</th><th>Tolerance</th></tr>
<tr><td rowspan="4">Flexible sheath probe</td><td rowspan="2">Central cells</td><td colspan="2">Length</td><td>l_s</td><td>mm</td><td>210</td><td>+ 5
0</td></tr>
<tr><td colspan="2">External diameter</td><td>d_s</td><td>mm</td><td>58</td><td>± 2</td></tr>
<tr><td rowspan="2">Guard cells</td><td colspan="2">Length</td><td>l_g</td><td>mm</td><td>120</td><td>+ 15
– 15</td></tr>
<tr><td colspan="2">External diameter</td><td>d_g</td><td>mm</td><td>58</td><td>± 2</td></tr>
<tr><td rowspan="6">Split tube probe</td><td rowspan="2">“Short”</td><td>Central cells</td><td>Length</td><td>l_{sc}</td><td>mm</td><td>210</td><td>+ 2
0</td></tr>
<tr><td>Guard cells</td><td>Length</td><td>l_{gc}</td><td>mm</td><td>200</td><td>± 5</td></tr>
<tr><td rowspan="2">“Long”</td><td>Central cells</td><td>Length</td><td>l_{sl}</td><td>mm</td><td>370</td><td>+ 5
– 5</td></tr>
<tr><td>Guard cells</td><td>Length</td><td>l_{gl}</td><td>mm</td><td>110</td><td>± 5</td></tr>
<tr><td rowspan="2">Split tube</td><td colspan="2">External diameter</td><td>d_{sf}</td><td>mm</td><td>59</td><td>± 5</td></tr>
<tr><td colspan="2">Length of the slits</td><td>l_m</td><td>mm</td><td>≥ 800</td><td></td></tr>
</table>

Table 9.1. *Dimensional characteristics of the pressuremeter probe*

9.1.2.2. *Execution of the boring*

The French standard for the Ménard pressuremeter test [AFN 91, AFN 91b] determines the conditions for carrying out the boring which must be adapted to the soil to be tested.

The types of materials must not destroy the structure of the soil. The materials used are indicated (Table 9.2) and are dependent on the nature of the soil.

The tests are carried out immediately after the execution of a drilling pass. The maximal authorized lengths for boring are indicated (Table 9.3). The length of a darilling pass depends on the nature of the soil.

	Preliminary boring									Return
	Rotation					Beating and others				
Nature of the terrain	HA	AD	CDH	CD IN	CAR	ROTOP	CAR BEA	CAR VS	ST MR	ST BEA/VS
Mud and soft clays	–	R°	–	O°	–	–	O SWD	–	–	–
Medium compact clays	R	R°	R	R°	–	O°	–	–	O	–
Compact clays, stiff marl			R	R°	R°	O°	–	–	–	–
Silt - over the water table	R	O°	R	O°	–	O°	O	O	O	–
- under the water table	–	R°	–	O°	O°	O°	–	–	O^D	–
Loose sands - over the water table	R	R°	O	O°	–	O°	–	–	O	–
- under the water table	–	R°	–	O°	–	O°	–	–	–	O
Medium compact and compact sands	R	R°	R	R°	–	R°	O	O	O^Δ	O^+
Rough ground: gravels, pebble, silex clay, etc.			O	O*		R^-	O	O	O	O^+
Altered rocks Soft rocks			R	R	O	R°	O	O		O^+

Legend:
R : Recommended
O : Tolerated
– : Not tolerated
(shaded) Inapt
* Rotation speed of < 60 tr/min, and tool diameter of less or equal to 1.15 d_s
+ Possibly, previous boring and small diameter ($d_t < d_s$)
° Injection with mud (pressure < 500 kPa; flow < 15 l/min)
In the case where boring is done by rotation, the pressure (at the head of the string of rods) transmitted to the tool must be less than 200 kPa.
Δ With certain setups (for example, impose a downward vertical gradient, perform tests while descending, elongate the spilt tube with a guard tube)

HA	Hand auger
AD	Auger with drill mud injection
CDH	Continuous dry helicoidal auger
CD	Core drill
CAR	Carottier
ROTOP	Rotopercussion
IN	With mud injection
BEA	Beating
SWD	Sunk thin wall core drill
ST	Split tube
STMR	Split tube with simultaneous material removal
VS	Vibrosinkage

Table 9.2. *Methods of execution prescribed for pressiometric boreholes*

Nature of the terrain	Maximal boring length done before the test (m)
Mud and soft clays	1** +
Medium compact clays	3 +
Compact clays, stiff marl	5
Silt: - over the water table - under the water table	 5 3 +
Loose sands: - over the water table - under the water table	 3 1** +
Medium compact and compact sands	5
Coarse soils: gravels, pebble, silex clay, etc.	5
Altered rocks, soft rocks	5
+ Tubing is recommended beyond 10 m depth ** Or interval between two consecutive tests	

Table 9.3. *Maximal length of a boring pass before the tests*

9.1.2.3. *Calibration*

Before carrying out the pressuremeter test, calibration tests must be done in order to remove the influence of the equipment on the test results:

– the gauges and sensors must be calibrated every 6 months;

– the probe is first covered solely by the membrane of the measurement cell. It is placed vertically, and then a purge of the gases contained in the measurement circuits may be performed. It is then dilated three times in open air while injecting at least 600 cm^3 of liquid. The pressure volume controller is set at zero;

– the probe, still covered solely in the membrane of the measurement cell, is placed under pressure increasing by steps equal to 1/10° of its resistance p_m. Each pressure is maintained for 60 s. The conventional resistance limit of the membrane p_m is the pressure corresponding to a volume injected of 600 cm^3;

– every month or every 200 tests and during each period of frost, the gauges and sensors are checked in relation to the reference material;

– at the start of work session, each time the pressuremeter probe is configured, and each modification of the connection between the pressure and volume controller and the probe, the calibration tests described in the following must be carried out;

– the probe is dilated three times in open air by injecting 600 cm^3 of liquid, and the pressure and volume controller is reset to zero;

– the first correction is a test of the equipment dilatation. The probe is introduced into a steel calibration tube (Figure 9.5) and placed under pressure, increasing in increments of 250 kPa for the probes that can undergo a pressure of 2.5 Mpa, and by increments of 500 kPa for probes that can undergo a pressure of 5 Mpa. Each pressure is maintained for 60 s. The volume injected at the end of each step in traced depending on the pressure (Figure 9.6);

– the second correction is a test of resistance of the probe also called calibration of the probe or inertial test. The probe is placed in proximity to the volume pressure controller, therefore, not in the soil and in open air (Figure 9.9).

9.1.2.3.1. Expansion test: (dilatation of the equipment)

The pressuremeter probe exerts a pressure on the walls of the calibration tube. It measures the deformation by the volume of the water injected into the probe. The inherent expansion test allows us to subtract from the measurement of the volume found during the test the part of the inherent expansion (or dilation) corresponding the deformation of the equipment alone (crushing of the membranes and dilation of the pipes). For this, a probe is placed in a calibration tube (Figure 9.5). This tube is characterized by:

– an internal diameter d_i is less than or equal to 66 mm;

– thickness is more than or equal to 8 mm;

– a length l_c of the calibration tube is more than 1 m and l_c more than l_m, the total length of the slits of the lantern tube;

– an exterior diameter d_s of the probe which includes the housing (lantern tube, flexible sheath, jacket, tubes, etc.).

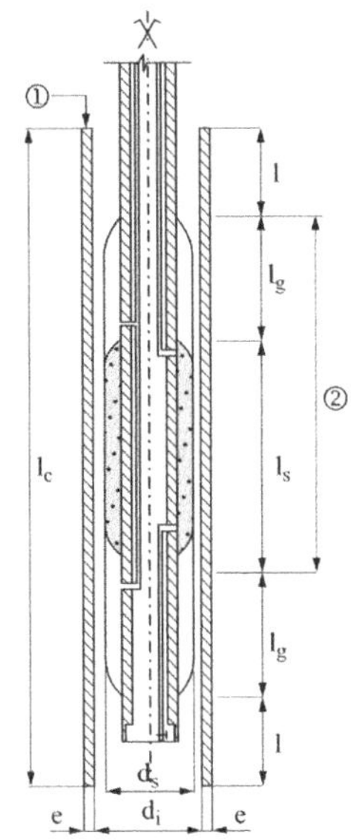

Figure 9.5. *The calibration tube*

The volume variation of the appartus is plotted versus the imposed pressure (Figure 9.6). From the inherent expansion test, (dilation of pipes), we can deduce:

– the initial volume of the central cell V_s, which is given in relation [9.2], with V_c being the volume injected into the probe to put it in contact with the calibration tube (Figure 9.6):

$$V_S = \pi.\frac{d_i^2}{4}.l_S - V_c; \qquad [9.2]$$

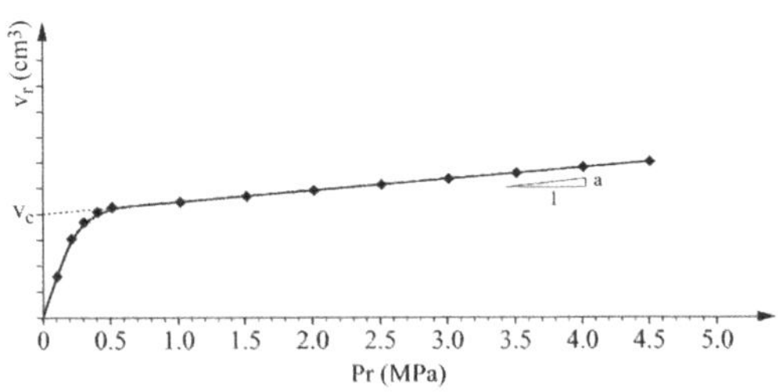

Figure 9.6. *The expansion curve of a pressurementer probe in a calibration tube*

– the coefficient of dilation of the tubes, a, which is the slope of the linear part of the dilation curve (Figure 9.6). This coefficient must have a value less than 6 cm^3/MPa for a device equipped with 50 m of tubing. It is recommended to use coaxial connecting tubes to determine the pressuremeter moduli that are greater than 100 MPa (Figure 9.8) rather than side-to-side piping. Coaxial tubing is generally characterized by an external diameter of 10 mm (Figure 9.7) and an internal diameter of 8 mm. It allows the transmission of air pressure to the guard cells. The coaxial central tube has an external diameter of 6 mm and an internal diameter of 4 mm. It allows the transmission of water pressure to the measurement cells.

For a given test pressure, the volume of inherent expansion must be subtracted from the measured volume.

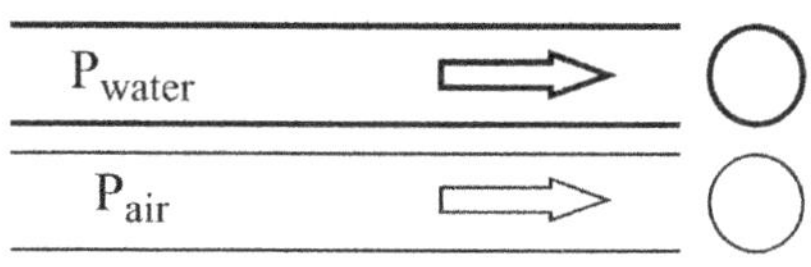

Figure 9.7. *Side-by-side connective tubing*

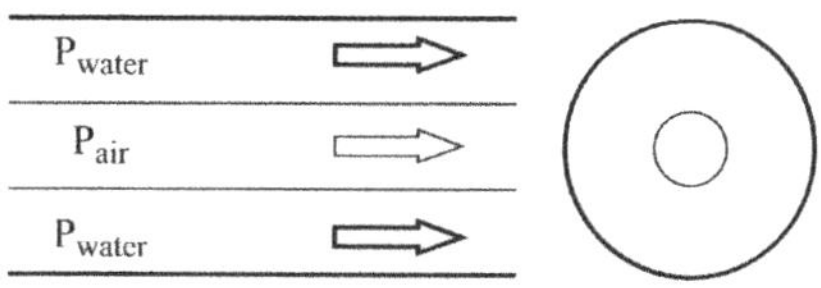

Figure 9.8. *Coaxial connective tubing*

9.1.2.3.2. Inherent probe resistance test (probe calibration or inertial test)

The probe is placed close to the pressure and volume controller, out of the soil and in open air (Figure 9.9).

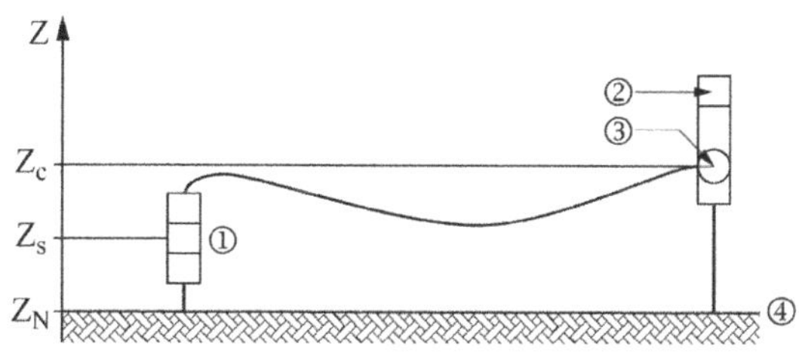

Figure 9.9. *Placement of the probe and the pressure-volume controller during the inherent resistance test (calibration or inertia)*

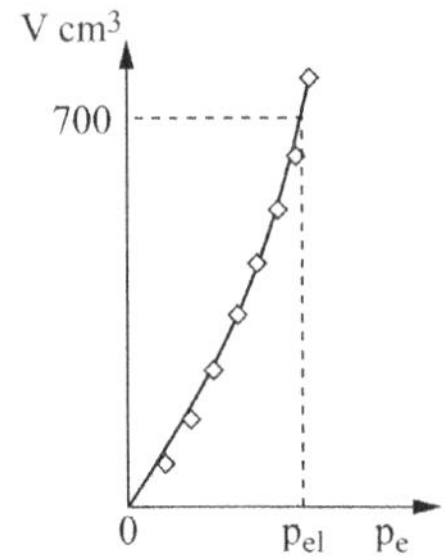

Figure 9.10. *Inherent probe resistance curve (calibration or inertia)*

The probe is inflated and the volume is noted at 60 s, which corresponds to the intervals of increasing pressure. Ten pressure intervals are carried out until a volume of water injected is equal to 1.2 V_s, with V_s the volume of the probe. The pressure curve p_e, which depends on the volume V is then traced (Figure 9.10). The measurement of pressure p_r carried out during the probe resistant test, are corrected using the relation [9.3], if necessary (when the probe is not placed at the same level of the pressure-volume controller).

$$p_e = \gamma_w . (z_c - z_s) + p_r \qquad [9.3]$$

The curve obtained is called calibration curve (Figure 9.10). Due to convention, the inherent resistance limit of the probe p_{el} is the pressure corresponding to the volume of water injected of 1.2 times that of V_s as determined by relation [9.2].

For a given volume, the calibration pressure is subtracted from the test pressure.

9.1.2.4. *Standard test*

It is described in standard [AFN 91a]. We must verify that the pressure in the guard cells p_k always remains less than that of the pressure in the measurement cell p_s. In order to do so, the difference of pressure, read by the pressure-volume controller, between the air and the water must verify in equation [9.4] with p_m the inherent resistance of the membrane of the central measurement cell:

$$p_s - 3. p_m \leq p_k \leq p_s - 2. p_m \qquad [9.4]$$

A practical rule is to set:

– from 0 to 10 m of depth, the difference p_s-p_k is of 100 kPa;

– from 10 to 20 m of depth, the difference p_s-p_k is of 0 kPa;

– from 20 to 30 m of depth, the difference of p_s-p_k is of –100 kPa;

– from 30 to 40 m of depth, the difference of p_s-p_k is of –200 kPa;

– from 40 to 50 m of depth, the difference of p_s-p_k is of –300 kPa, etc.

The way to place the probe is linked to the nature of the soil and the presence or absence of water table. In any case, it is necessary to:

– either lower the probe into a borehole of diameter d_t made previously and adapted to the diameter of the probe d_s so that the relation may be satisfied:

$$d_t \leq 1{,}15.\, d_s; \qquad [9.5]$$

– or directly sink the probe equipped with a lantern tube into the soil by driving.

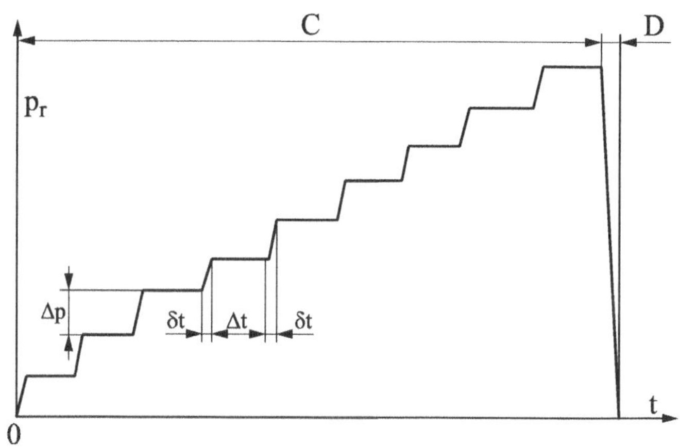

Legend:

p_r Pressure
t Time
Δt Length of a pressure interval
Δp Pressure interval step
δt Length from one step to the next
C Loading of the terrain
D Unloading of the terrain

Figure 9.11. *Loading program of the Ménard pressuremeter test [AFN 91a]*

Once the probe is placed in the borehole, the intervals of increasing pressure must be applied, each interval must be maintained for 60 s. The numbers of intervals of pressure are at least 10, the increasing intervals of pressure being at most equal to one-tenth of the limit pressure estimated from the data of the borage. The time allowed to the increase of pressure, between two different steps of constant pressure, should be 10s or less (Figure 9.11).

The test is over if one of the following conditions arises:

– 5 Mpa of pressure is reached;

– the volume of liquid V equal to V_s+2 .V_l has been injected;

– three intervals of pressure have been applied beyond the creep pressure p_f and four at least before the creep pressure.

The following conditions of execution must be upheld:

– the pressure steps between intervals are equal in the pseudo-elastic phase and can be reduced by half in the creep phase;

– in the same boring, two tests cannot be carried out at distances of less than 0.75 m;

– the precision of the depth of the probe is 10cm or the 1/100° of the length of the connecting rods;

– the precision of volume measurement is of 2 cm^3 or 1% of the value measured;

– volume deformations are noted at 15, 30 and 60 s;

– for each pressure interval, the measured pressures p_r are corrected, from calibration pressure p_e obtained for volume V, from the pressure of the water column (Figure 9.12) between the volume pressure controller and the probe using relation [9.6];

– for each pressure interval, the measured volumes V_r are corrected at 60 s, by dilatation of the tubes according to relation [9.7]:

$$p = p_r + \gamma_w \cdot (z_c - z_s) - p_e \quad [9.6]$$

$$V = V_r - a . p_r \quad [9.7]$$

We then trace the corrected pressuremeter curve (Figure 9.13) and the creep curve (Figure 9.14).

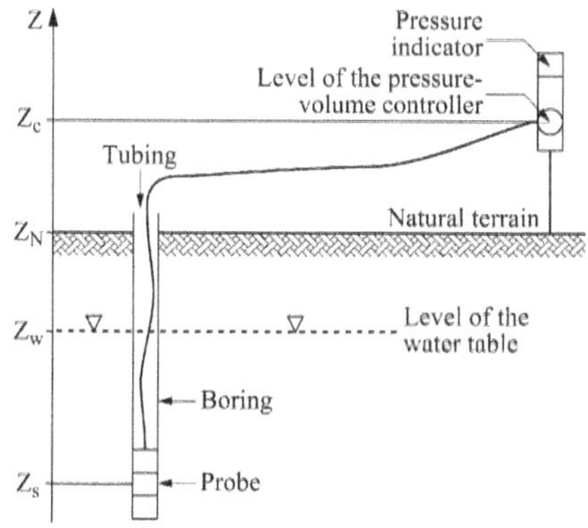

Figure 9.12. *Location of the probe and the pressure-volume controller during the test [AFN 91]*

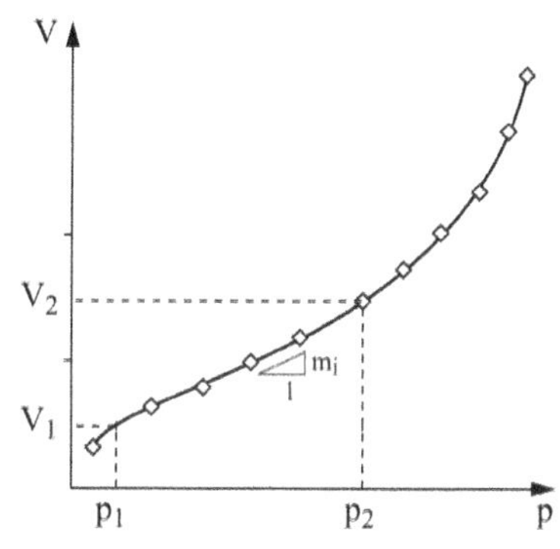

Figure 9.13. *Example of a corrected pressiometric curve [AFN 91]*

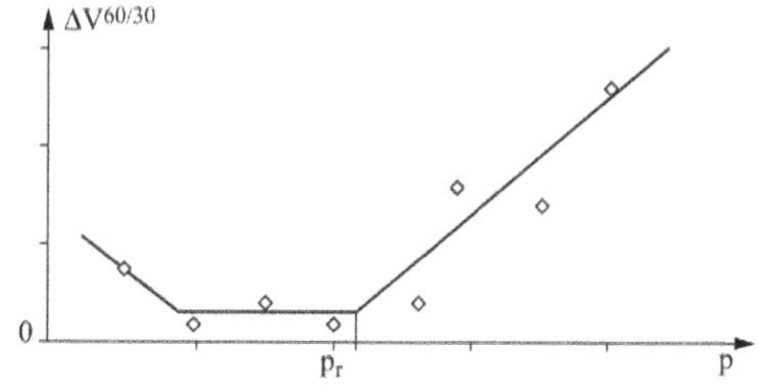

Figure 9.14. *Example of a pressiometric creep curve [AFN 91]*

9.1.2.5. *Cyclical test*

This is described by the standard [AFN 99b]. The principle of this test is to determine the elasticity modulus of the soil in the loading part of a loading-unloading cycle.

The loading program is shown in Figure 9.15. It consists of:

– unloading the probe from a pressure p_{rD} (Figure 9.15, D) which is immediately higher than the pressure p_2 at the end of the pseudo elastic part or phase;

– finishing the unloading down to a pressure p_{rE} (Figure 9.15, 2) which is immediately higher than pressure p_1 from the beginning of the pseudo (Figure 9.15, 3) elastic phase;

– reloading (Figure 9.15) in increments according to the modus operandi of the test without a cycle [AFN 91a] until the pressure prD (Figure 9.15, R);

– continuing the reloading until the limit specified in the test without a cycle [AFN 91a].

The resulting curve has a cyclic part above the virgin loading curve (Figure 9.16).

9.1.3. *Normalized interpretation of the standard and cyclical tests*

The test is carried out by calculating the characteristics of soil deformation, the Ménard pressuremeter modulus E_M and a resistance characteristic p_{LM}.

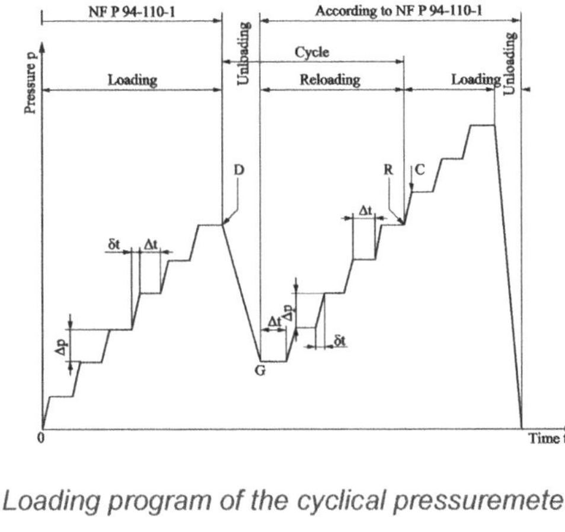

Figure 9.15. *Loading program of the cyclical pressuremeter test [AFN 99b]*

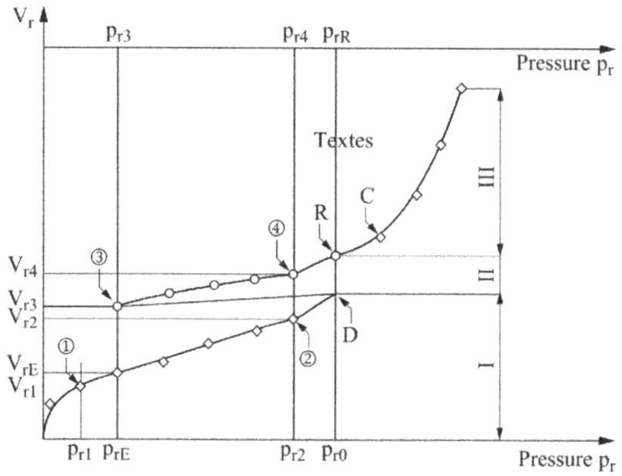

Figure 9.16. *Example of a pressuremeter curve in function of p [AFN 99b]*

9.1.3.1. *Pressiometric modulus E_M*

The pressiometric modulus corresponds to the pseudo-linear range of the pressuremeter curve between points (p_1, V_1) and (p_2, V_2) of the curve (Figure 9.13). Between these two points, the behavior of the ground is considered to be elastic, and the modulus of the corresponding elastic body becomes the pressiometric modulus determined by relation [9.8].

$$E_M = 2.(1+\vartheta).\left(V_S + \frac{V_1+V_2}{2}\right).\frac{p_2-p_1}{V_2-V_1} \quad [9.8]$$

9.1.3.2. *Ménard reloading pressiometric modulus E_R*

The pressuremeter modulus corresponds to the reloading line on the pressuremeter curve between points (p_{r3}, V_{r3}) and (p_{r4}, V_{r4}) of the reloading curve (Figure 9.16). Between these two points, the behavior of the soil is elastic which allows us to determine the reloading pressiometric modulus, determined by relation [9.9]. The behavior of the soil during the unloading-reloading cycle is practically reversible and we can consider that the modulus E_R is the horizontal, elastic modulus of the soil measured by pressuremeter.

$$E_R = 2.(1+\vartheta).\left(V_S + \frac{V_{r3}+V_{r4}}{2}\right).\frac{p_{r4}-p_{r3}}{V_{r4}-V_{r3}} \quad [9.9]$$

9.1.3.3. *Pressure limit* p_{LM}

The theoretical pressure limit corresponds to the pressure for which the probe reaches an infinite volume. This pressure is by definition impossible to reach, and as such we have chosen to measure the conventional pressure limit, which corresponds to the pressure at which the central cell of the probe has doubled its volume. The volume of the probe is denoted as V_s [9.2]. During the test, the probe comes into contact with the soil at point (p_1, V_1) of the experimental curve (Figure 9.13). The initial volume of the probe when the test begins is, therefore, V_s+V_1, and the injected volume V_l into the probe when it doubles its volume is thus given by [9.10]:

$$V_l = V_S + 2.V_1 \qquad [9.10]$$

The conventional pressure limit p_{LM}, which corresponds to the doubling of the volume of the probe, is the pressure for a volume V_l determined by relation [9.10].

If volume V_l cannot be achieved, the conventional limit pressure is estimated by the following two methods:

– the inverse linear method, which considers that the curve of the inverse of the volume as a function of the pressure becomes linear for the strongest pressures. The inverse of volume becomes null for the theoretical pressure limit. The regression line can be determined by the least squares of the relation (p, V^{-1}) of which the slope is A and the intercept at the origin B. The conventional pressure limit thus becomes:

$$p_{LM} = -\frac{B}{A} + \frac{1}{[A.(V_S+2.V_1)]}; \qquad [9.11]$$

– the hyperbolic method, which considers that the pressuremeter curve is a hyperbole branch for strong pressures. Under these conditions, the pressuremeter curve becomes linear [9.12] with the value of the abscissa being relation [9.13], and the value of the ordinate being relation [9.14]. Coefficients C and D are obtained by the linear regression of the least squares:

$$Y = C.X - D \qquad [9.12]$$

$$X = \frac{V^2 - V_1^2}{p - p_1} \qquad [9.13]$$

$$Y = \frac{p.V^2 - p_1.V_1^2}{p - p_1} \qquad [9.14]$$

$$p_{LM} = \frac{[p_1.(V_1^2+D)+C.(V_l^2+V_1^2)]}{(V_l^2+D)} \qquad [9.15]$$

The conventional pressure limit p_{LM} when volume V_l is not reached is the lowest of the pressure limits calculated by [9.11] and [9.15]. If the difference between these two relations goes beyond 20% of [9.15], the test does not allow the determination of p_{LM}.

9.2. Self-drilling pressuremeter test (SBP)

9.2.1. *Principle of the test*

The Self Boring Pressuremeter Test (SBP) is an improvement of the Ménard probe. It was patented by J. Jezequel and J. Touzé [JEZ 68] for the *Laboratoire Central des Ponts et Chaussées.*

By carrying out the tests with the Ménard pressuremeter probe, it became apparent that the quality of the drilling had a strong influence on the value of the pressuremeter modulus. In order to counter this inconvenience, Jezequel developed a self-drilling pressuremeter (Figures 9.17 and 9.18). This device is constructed to carry out its own drilling, and as a result to affect the soil the least possible. To do this, a cutter is fitted at the base of the device, in such a manner that the diameter is the exact size necessary for the pressuremeter to pass (Figures 9.19–9.21). A deagglomerating tool destroys the structure of the soil, and an injection fluid carries the mucking fluid (boring mud and deagglomerated soil) to the surface [REI 05].

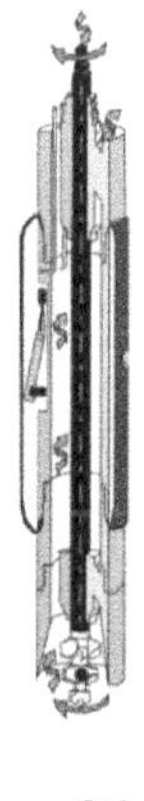

Figure 9.17. *Diagram of the principle of the self-drilling pressuremeter PAF2000 [REI 05]*

Figure 9.18. *Example of the self-drilling pressuremeter PAF2000 [REI 05]*

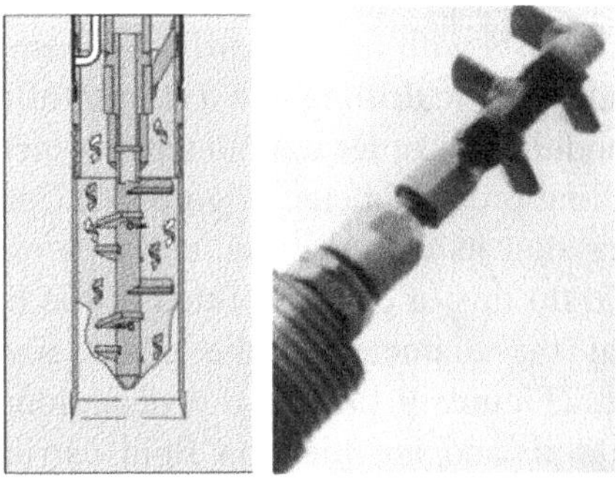

Figure 9.19. *Staged deagglomerating tool [REI 05]*

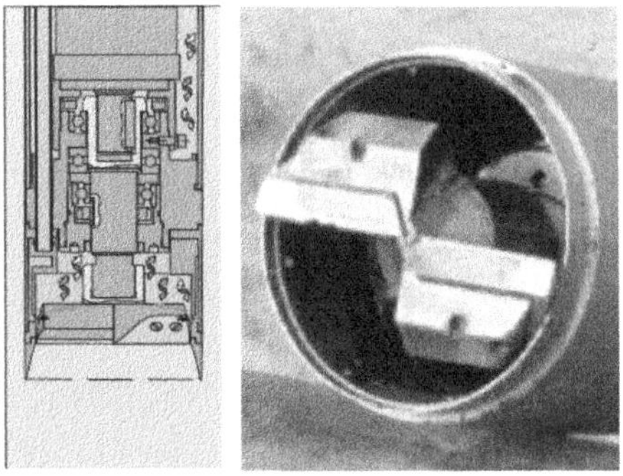

Figure 9.20. *Butterfly deagglomerating tool [REI 05]*

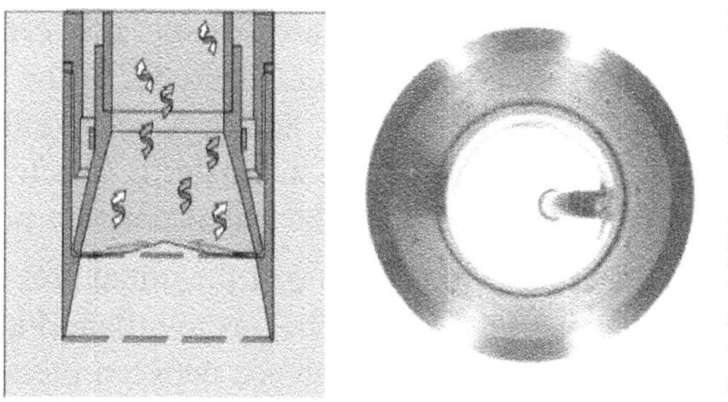

Figure 9.21. *Peripheral jet deagglomerating tool [REI 05]*

The Camkometer is a UK/USA variant of the self-drilling pressuremeter which was developed by C.P. Wroth and J.M.O Hughes in 1972 (Figure 9.22). This device has an external diameter of 2.5 inches (6.35 cm) or 3.25 inches (8.25 cm). It is inflated with gas and possesses a single cell. The radial deformation of the cell is recorded by three movement sensors placed at 120°. The device also possesses a cell that measures interstitial pressure.

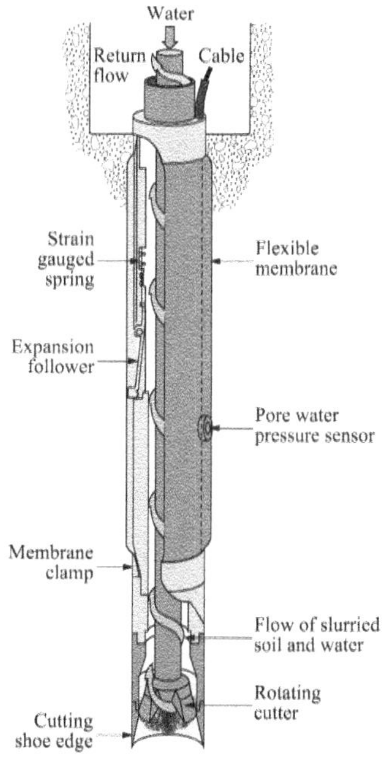

Figure 9.22. *The Cambridge Institute Camkometer*

It can be noted that if the initial phases are different between the self-drilling pressuremeter and the Ménard pressuremeter (Figure 9.23), the following phases of the creep pressure are fairly similar and produce parallel curves. The difference between these two types of pressuremeter is sensitive to the level of P_0, which is systematically smaller than the initial pressure p_1 for the Ménard pressuremeter (PMT), and close to the value of the initial pressure p_0 for which the self-boring pressuremeter probe (SBP) begins to inflate (Figure 9.24). It can also be noted that there is good correspondence between pressure p_0 measured by SBP and that measured by Glötz pressure sensors (Figure 9.25).

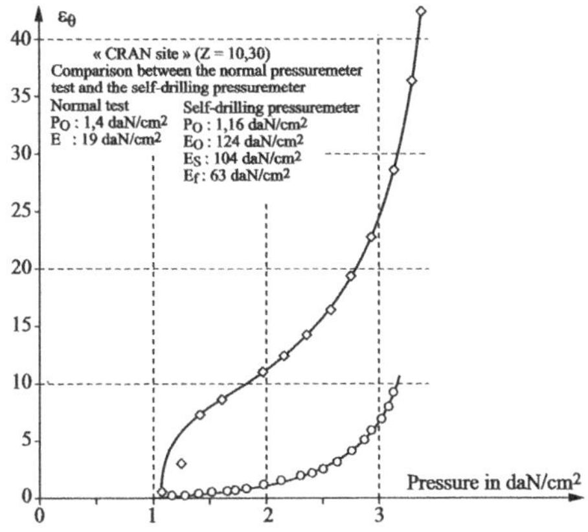

Figure 9.23. *Comparison of the expansion curves between Ménard pressuremeter and the self-drilling pressuremeter*

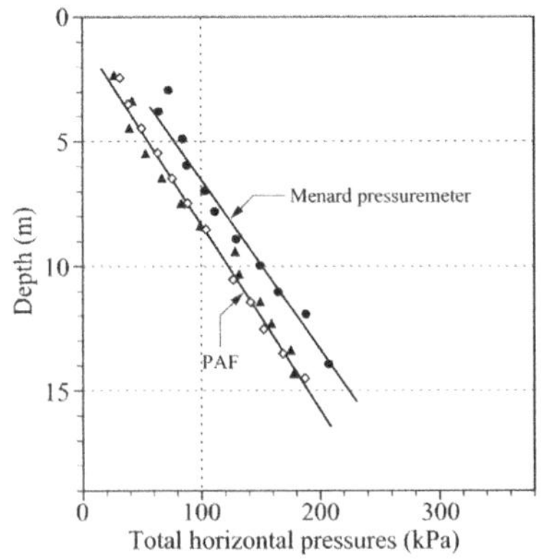

Figure 9.24. *Comparison of the measurement of the total horizontal pressure between the Ménard pressuremeter and the self-drilling pressuremeter*

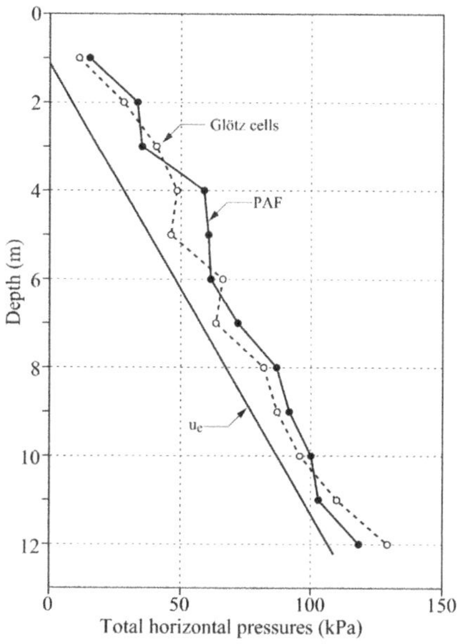

Figure 9.25. *Comparison of the measurement of total horizontal pressure between the self-drilling pressuremeter and the Glötz pressure sensors*

9.2.2. *Execution of the test*

We proceed in the following manner:

– the pressuremeter probe is put in place by self-drilling at a chosen level;

– once the test level has been reached, the drilling, rotation and injection cease and the evolution of the pressure in the measurement cell must be followed until stabilization. In practice, the volume is read for time values of 0, 1, 2, 5, 10, 20, 30 min, etc.;

– the stabilization having been reached, we proceed to the expansion test by dilating the probe to a controlled deformation and by applying a regular speed of deformation of 2% per minute until a maximal deformation of 25%.

The following three criteria must be upheld:

– the pressure p_i in the probe at the moment when the drilling is interrupted and the pressure p_0 in the probe at the end of relaxation are higher than the hydrostatic pressure in the soil and the depth of the depth;

– the difference p_i- p_0 is less than or equal to Δp_m which is obtained in the first increment of volume;

– the curves of relaxation and expansion show a monotone shape, without singular points;

– the tests cannot be further that 1.5 m apart along the same borehole.

9.2.3. *Evaluation of the tests*

From the tests, we can deduce:

– the initial pressure p_0, considered equal to the horizontal at rest pressure;

– the pressures p_5 and p_{20} corresponding to a variation of the volume by 5 and 20%;

– the shearing moduli G_{p2} and G_{p5} for the variations of the volume by 2 and 5% calculated by [9.16];

– the identification coefficient of the soils β such as [9.17];

– the undrained cohesion c_u by the Baguelin *et al.* method [BAG 72].

$$G = V_S \frac{\Delta p}{\Delta V} \quad [9.16]$$

$$\beta = \frac{p_{20} - p_5}{p_{20} - p_0} \quad [9.17]$$

9.3. The dilatometer

9.3.1. *Description*

The dilatometer is a device (Figure 9.26) that has a single cell inflated or oil, as opposed to the pressuremeter, which has a single central cell inflated with water and guard cells inflated with air. It is intended to test rocks during drilling. It possesses a thick, rubber membrane that can resist the strong pressures that are applied. Its diameter is 70 or 95 mm and its length is 1 m.

The thickness of the membranes is 1.5 cm. The device possesses three movement sensors halfway along the membrane, each sensor separated by 120° angles. The sensors are placed at 75 mm intervals. This layout allows us to determine the anisotropy of the rocks. A pressure sensor is placed in the body of the device and can measure the pressure inside the dilatometric probe at all times. The information from the sensors is transmitted to the surface by an electric cable. The measurement of movement has a precision of 10 μm with which has an interval of measurement of 25 mm maximum.

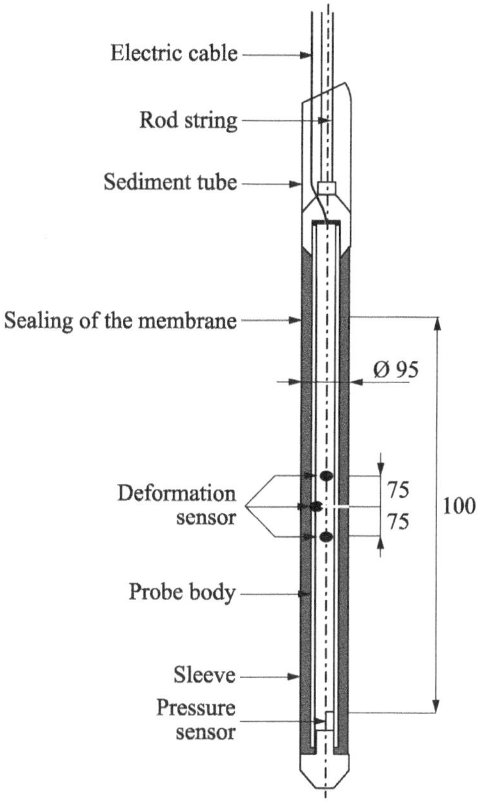

Figure 9.26. *Dilatometer probe*

Putting the sleeve under pressure can be done using compressed air or oil. Air inflation can be used up to 10 MPa. This method is the most practical, and is to be used unless security measures prohibit it. The dilatometer can be inflated with oil up to 20 MPa.

A computer allows us to measure deformation and pressure information gathered from the probe. An electronic cable carried by a reel links the measuring case to the probe.

The probe is lowered into the borehole (Figure 9.28) using tubular rod strings, which are also used as an airtight conduct for the compressed air or oil. The rods have an exterior diameter of 25 mm and a normal length of 2 m. They are equipped with special sockets that can connect to the next rod without rotation.

Figure 9.27. *HyperPAC device (Apageo, 2015)*

An orientation device is placed at the tip of the string of rods, and at the beginning of the test. This allows us to note the orientation of the sensors in the dilatometric probe in relation to the magnetic North. This device can measure the horizontal anisotropy of the rocks in relation to the geographic North.

An evolution of the pressuremeter was carried out in order to make the dilatometer's pressure range between 25 and 50 MPa. HyperPac (Apageo, 2015) uses a probe inflated with water (Figure 9.27) in a protective sheath. The cycle is controlled automatically. The precision of the measurement is

0.8 mm^3 for a probe volume of 600 cm^3, which corresponds to a precision of 7.10^{-7} which is at least the equivalent of the dilatometer's.

9.3.2. *Interpretation of the results*

Calculation of the elastic moduli is done according to the Lamé formulas, [9.18] of axial symmetry, with υ Poisson's coefficient, D_0 the initial diameter of the dilatometer, ΔD the variation of the diameter of the dilatometer and Δp the variation of the pressure applied to the drilling.

$$E = (1 + \upsilon)\frac{\Delta p.D_0}{\Delta D} \quad [9.18]$$

Movement of the arms can be significantly different (Figures 9.29–9.31) which allows us to find the elastic anisotropy of the rocks (Figure 9.32). Often, three cycles of unloading and reloading are carried out upon which the elasticity modulus of the rock is measured. It is frequent to have differences between the soil moduli measured with the pressuremeter and the moduli measured with a dilatometer.

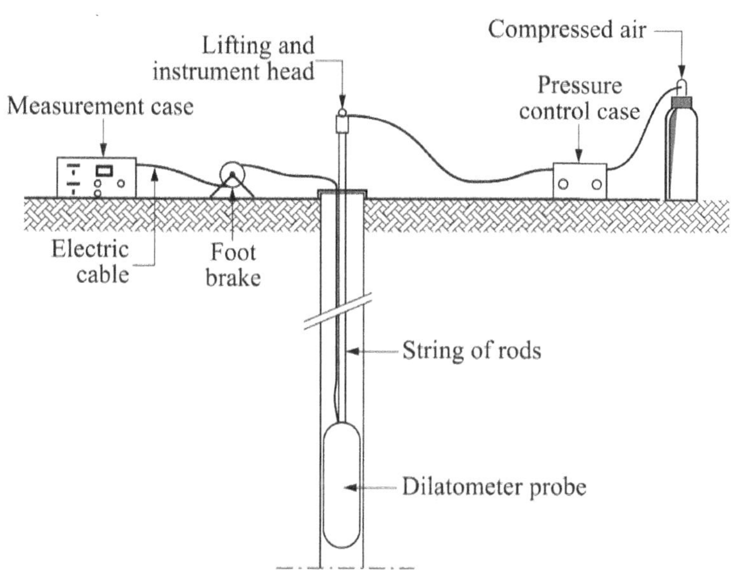

Figure 9.28. *Complete dilatometer device during boring*

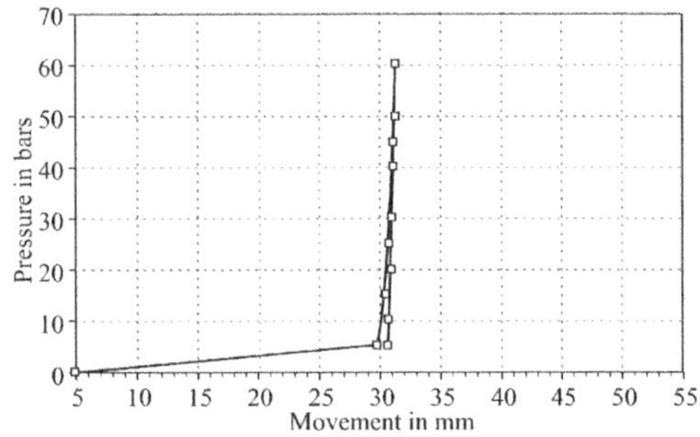

Figure 9.29. *Example of a dilatometer result: response of sensor 1 – rue du Havre – depth 23.7 m*

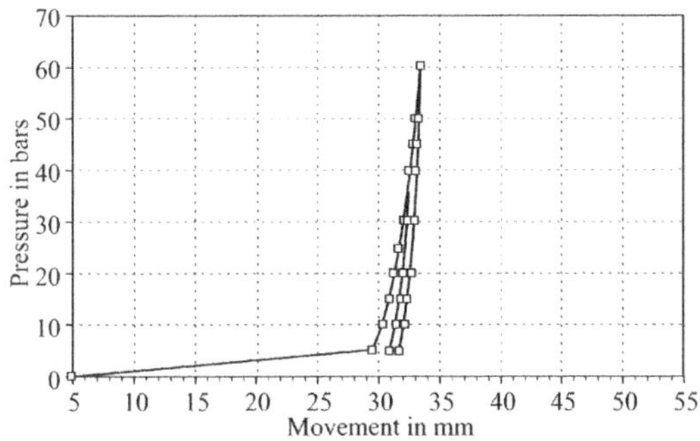

Figure 9.30. *Example of a dilatometer result: response of sensor 2 – rue du Havre – depth 23.7 m*

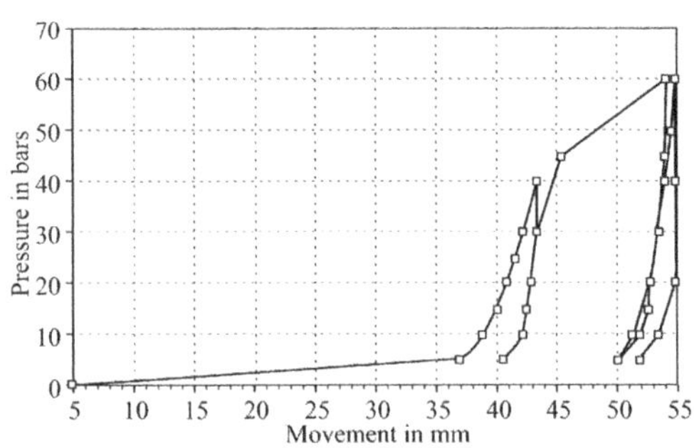

Figure 9.31. *Example of a dilatometer result: response of sensor 3 – rue du Havre – depth 23.7 m*

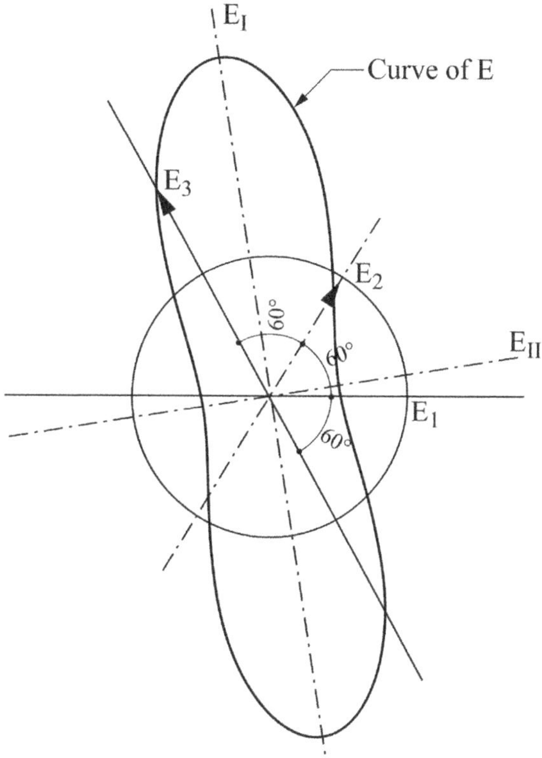

Figure 9.32. *Interpretation of the dilatometer tests in an elastic medium with a transverse anisotropy*

9.4. The "Géomécamètre"

9.4.1. *Principle of the test*

This device is an improvement of the pressuremeter. The "Géomécamètre" patent (Figure 9.33) has been filed [MON 99]. The pressuremeter test [MEN 57] can be considered as a shearing test between the radial stress and the circumferential stress in the horizontal plane, with the normal vertical stress at the plane of shearing [BAG 78]. The theoretical limit of the pressuremeter test is linked to the fact that the vertical stress is linked to the weight of the Earth above the test. The principle of the "Géomécamètre" is to vary this vertical stress by hydraulic flow from top to bottom (Figure 9.34).

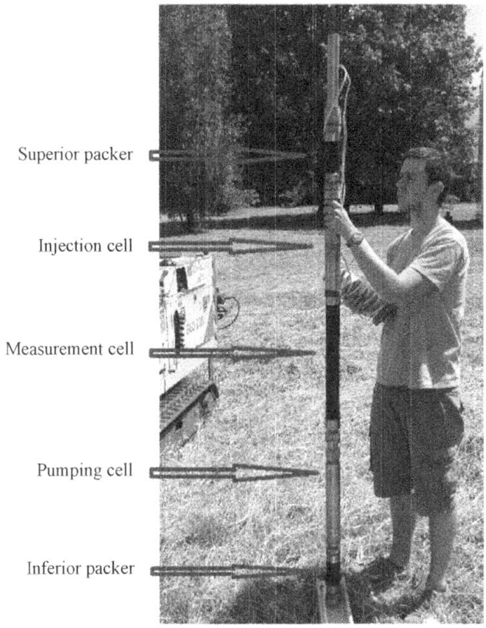

Figure 9.33. *The "Géomécamètre"*

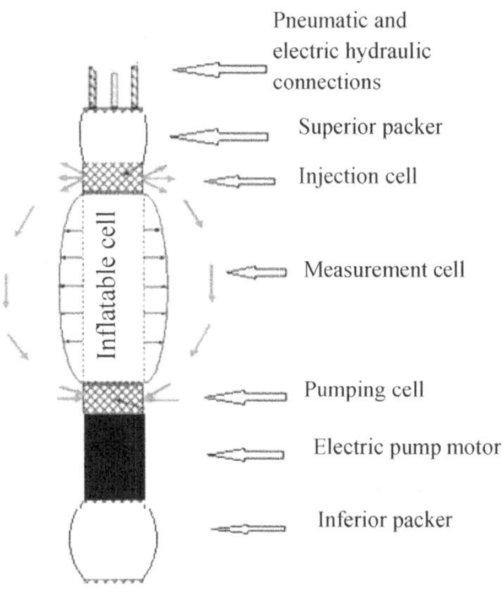

Figure 9.34. *Operating principle of the "Géomécamètre"*

9.4.2. *Modus operandi*

The execution of the drilling and the calibration (expansion test and resistance test) are identical to the modus operandi of the Ménard pressuremeter test.

The "Géomécamètre" test is a device that uses hydraulic flow around the inflatable probe. The hydraulic charge decreases along the length of the water's trajectory when the water moves downward from the head to the toe of the probe. Force is applied to the soil's particles in the direction of the drainage. In the "Géomécamètre" test, these forces are similar to the action of gravity's pull. This action can be assimilated to an action in the direction of the water's flow obtained by relation [9.20].

When the coefficient of permeability is more than 10^{-8} m/s, dilation of the probe does not influence the interstitial pressure [CAM 93, FRA 86] and the effective pressure can be used, but if the permeability coefficient is less than 10^{-10} m /s, the test cannot take place because the soil does not allow for sufficient drainage. A numerical model of the test [SEN 99] shows that, in a unit of volume of soil, the force applied by the hydraulic flow can be estimated through relation [9.21]. In the "Géomécamètre" test, water is injected into the soil by an injection cell at injection pressure (p_i). After circulation in the soil, water is pumped across a pumping cell with a pump with a controlled value of water pressure (p_P).

The probe has a diameter of 60 mm and a length of 1.80 m (Figure 9.33).

Before the test, the following points are carried out:

– an expansion test according to pressuremeter standards;

– a resistance test according to pressuremeter standards;

– a pump calibration to know the characteristic pumping curve.

The measurement must be done in the following manner:

– we carry out a preliminary boring according to the modus operandi of pressuremeter standards;

– we position the probe at the level chosen for the test;

– We inflate the packers and the measurement cell until pressure p_l is reached at contact with the probe in the borehole.

– We apply the injection load H_i chosen for the test and regulate the power of the pump in order to reach the pump charge H_p chosen for the test. The hydraulic gradient can then be imposed on the test.

– We take note every minute of the hydraulic conditions Q_i, H_i, p_i, Q_p, H_p, p_p, the turbidity of the water extracted.

– We must wait for the stabilization of the flow Q_p, which must remain stable for 5 min.

– We must carry out an expansion of the measurement cell of the "Géomécamètre" according to the standards of the cyclical pressuremeter [AFN 99b].

– The probe is unloaded and the device positioned 1 m below the previous level.

– We restart this measurement cycle two more times with increasing hydraulic gradients at each testing to obtain three expansion tests with the "Géomécamètre" under three increasing vertical stress.

9.4.3. *Interpretation*

The hydraulic gradient is obtained using the difference between the hydraulic injection charge H_i and the pump charge H_p. For the Terzaghi hydraulic gradient [9.19] which defines soil piping, the water flow is vertical upward the surface in such a way that the unit weight of the soil decrease to zero. In the Geomechameter test the water flow is vertical downward in such a way that the unit weight of the soil increase. The increase of the effective vertical stress [9.22] is associated with the hydraulic gradient i in the soil at the level of the median plane of the length of the probe. The effective vertical stress placed at the level of the probe [9.23] and simulated depth Z_{Sim} [9.24] are indicated.

The three tests can be interpreted in drained conditions with dilatance, under three vertical stress σ_v' – different by the method developed in the following Chapter [MON 12] – in order to measure cohesion and the friction angle of the soil.

$$i_{crit} = (\gamma_{sat} - \gamma_w)/\gamma_w \quad [9.19]$$

$$F = i.\gamma_w.\Delta V \quad [9.20]$$

$$i = (H_i - H_p)/l_e \quad [9.21]$$

$$\delta\sigma'_v = i.\gamma_w.l_e/2 \quad [9.22]$$

$$\sigma'_v = \gamma'.z_{sim} = [(\gamma' + i.\gamma_w).l_e]/2 + \sigma'_{v0} \quad [9.23]$$

$$z_{sim} = [(1 + i.\gamma_w/\gamma').l_e]/2 + z_0 \quad [9.24]$$

$$k = \frac{Q.\gamma_w}{2.\pi.l.(p-p_0)}.\left[Ln\frac{l}{r_0} + \frac{1}{2}\right] \quad [9.25]$$

$$C_v = \frac{k.E_{oed}}{\gamma_w} = \frac{k.E.(1-\upsilon)}{(1+\upsilon).(1-2.\upsilon)} \quad [9.26]$$

The measurement of the elasticity modulus E with the hypothesis of a constant Poisson coefficient close to 0.3 allows us to determine the oedometric modulus. As the "Géomécamètre" is able to measure the permeability k of the soil [9.25], it is possible to determine the coefficient of consolidation C_v [9.26]. The measurement of the elasticity modulus is carried out on an unloading reloading cycle in around 10 min, the measurement of permeability is also done in a few minutes. The consolidation coefficient C_v can thus be measured without waiting for final consolidation.

An example of an experimental curve with the "Géomécamètre" is shown in Figure 9.35.

9.5. Theoretical interpretation of the pressuremeter test

9.5.1. *Cohesive soil: the Baguelin et al. (1972) interpretation*

Baguelin *et al.* [BAG 72] propose a method of interpretation of the pressuremeter test in undrained soil. The soil is not supposed to undergo any variation in volume. This method is more complete than Ladanyi method [LAD 72] and is based on the principle of derivation of the pressuremeter curve. Deformations are used with notations [9.27] and [9.28] in deformation (with u the radial displacement at the point of radius r) and σ_r in radial stress and σ_θ in circumferential stress:

$$\varepsilon_r = \frac{du}{dr} \quad [9.27]$$

$$\varepsilon_\theta = \frac{u}{r} \quad [9.28]$$

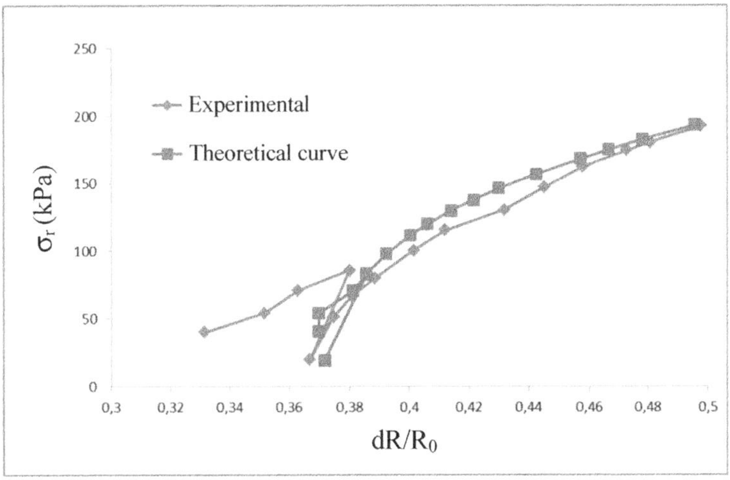

Figure 9.35. *Example of an expansion curve obtained with the "Géomécamètre"*

9.5.1.1. *Deformation conditions*

If we consider a zone in the soil in the angle dθ delimited by the radii r_0 and r prior to deformation, and by the radii ρ_0 and ρ after the deformation (Figure 9.36), we can establish the relation between variable r and ρ through [9.29], [9.30]:

$$\rho_0 = r_0 + u_0 \quad [9.29]$$

$$\rho = r + u \quad [9.30]$$

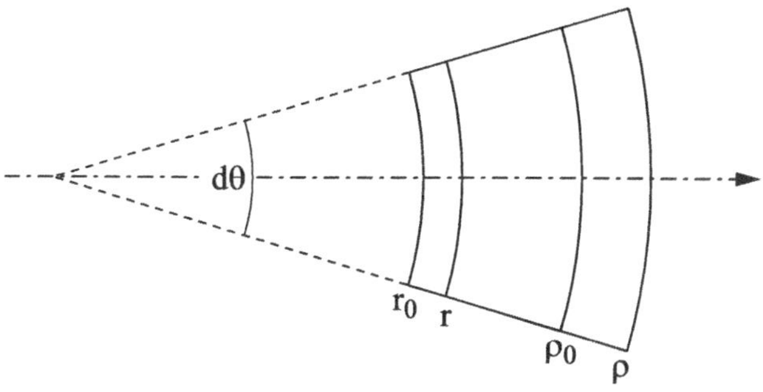

Figure 9.36. *Deformation of a circular section of the ground in the horizontal plane*

The volume of the circular sector is constant as the shearing is done without volume variation:

$$d\theta.(\rho^2 - \rho_0^2) = d\theta.(r^2 - r_0^2) \quad [9.31]$$

$$(\rho^2 - r^2) = (\rho_0^2 - r_0^2) = a^2 \quad [9.32]$$

Using relations [9.29] and [9.30], we obtain:

$$(r + u)^2 - r^2 = a^2 \quad [9.33])$$

$$2.r.u + u^2 = a^2 \quad [9.34]$$

Using equations [9.27] and [9.28] linking deformations to movements, movement u can be replaced by ε_θ.r, which results in:

$$2.r^2.\varepsilon_\theta + r^2.\varepsilon_\theta^2 = a^2 \quad [9.35]$$

$$r^2.\left[(1 + \varepsilon_\theta)^2 - 1\right] = a^2 \quad [9.36]$$

By deriving [9.36] in relation to r, we get [9.37] which can be transformed to [9.38]:

$$2.u + 2.r.\frac{du}{dr} + 2.u.\frac{du}{dr} = 0 \quad [9.37]$$

$$(1 + \varepsilon_r).(1 + \varepsilon_\theta) = 0 \quad [9.38]$$

which is translated into the following equations [9.39]–[9.40]:

$$\varepsilon_r = -\frac{\varepsilon_\theta}{(1+\varepsilon_\theta)} \quad [9.39]$$

$$\varepsilon_\theta = -\frac{\varepsilon_r}{(1+\varepsilon_r)} \quad [9.40]$$

9.5.1.2. *Conditions under constraint*

The condition for equilibrium in the horizontal plane is expressed by small deformation by relation [9.41] and by a large deformation for [9.42]:

$$\sigma_r - \sigma_\theta - r.\frac{d\sigma_r}{dr} = 0 \quad [9.41]$$

$$\sigma_r - \sigma_\theta - \rho.\frac{d\sigma_r}{d\rho} = 0 \quad [9.42]$$

9.5.1.3. *Relation under deformation constraint*

We assume that the relation between shearing τ and deformation ε_θ is an experimental curve that is a function of the circumferential deformation [9.43]. Taking into account the relation of equilibrium [9.42] for large deformation, we can deduce [9.44] and [9.45]:

$$\tau = \frac{\sigma_r - \sigma_\theta}{2} = f(\varepsilon_\theta) \quad [9.43]$$

$$\tau = \frac{\sigma_r - \sigma_\theta}{2} = -\frac{\rho}{2}.\frac{d\sigma_r}{d\rho} = f(\varepsilon_\theta) \quad [9.44]$$

$$d\sigma_r = -2.f(\varepsilon_\theta).\frac{d\rho}{\rho} \quad [9.45]$$

Taking into account relation [9.30] between movement and radius, we can write [9.46] and [9.47]:

$$\rho = r.(1 + \varepsilon_\theta) \quad [9.46]$$

$$\frac{d\rho}{\rho} = \frac{dr}{r} + \frac{d\varepsilon_\theta}{1+\varepsilon_\theta} \quad [9.47]$$

Now, going back to equation [9.35], and deriving it in relation to r, we obtain [9.48] and then [9.49]:

$$2.r.dr.[(1+\varepsilon_\theta)^2-1]+r^2.2.\left(1+\varepsilon_\theta\right).d\varepsilon_\theta=0 \quad [9.48]$$

$$\frac{dr}{r}+\frac{(1+\varepsilon_\theta).d\varepsilon_\theta}{(2+\varepsilon_\theta).\varepsilon_\theta}=0 \quad [9.49]$$

This equation can be introduced into equation [9.47], which results in [9.50] and [9.51]:

$$\frac{d\rho}{\rho}=-\frac{(1+\varepsilon_\theta).d\varepsilon_\theta}{(2+\varepsilon_\theta).\varepsilon_\theta}+\frac{d\varepsilon_\theta}{(1+\varepsilon_\theta)} \quad [9.50]$$

$$\frac{d\rho}{\rho}=-\frac{d\varepsilon_\theta}{\varepsilon_\theta.(1+\varepsilon_\theta).(2+\varepsilon_\theta)} \quad [9.51]$$

This equation can be introduced into relation [9.45] and this results in:

$$d\sigma_r=2.f(\varepsilon_\theta).\frac{d\varepsilon_\theta}{\varepsilon_\theta.(1+\varepsilon_\theta).(2+\varepsilon_\theta)} \quad [9.52]$$

$$\sigma_r=\int_0^{\varepsilon_\theta} f(\varepsilon_\theta).\frac{d\varepsilon_\theta}{\varepsilon_\theta.(1+\varepsilon_\theta/2).(1+\varepsilon_\theta)}=F(\varepsilon_\theta) \quad [9.53]$$

But, this integral is known experimentally. It is the evolution of the radial pressure in relation to movement during the pressuremeter test. Inversely, if $F(\varepsilon_\theta)$ according to [9.53], we get expressions [9.54] and then [9.55]:

$$\frac{dF(\varepsilon_\theta)}{d\varepsilon_\theta}=f(\varepsilon_\theta).\frac{d\varepsilon_\theta}{\varepsilon_\theta.(1+\varepsilon_\theta/2).(1+\varepsilon_\theta)} \quad [9.54]$$

$$f(\varepsilon_\theta)=\varepsilon_\theta.(1+\varepsilon_\theta/2).(1+\varepsilon_\theta).\frac{dF(\varepsilon_\theta)}{d\varepsilon_\theta}=\tau \quad [9.55]$$

Relation [9.55] shows that if we derive the pressiometric curve, and multiple the result by $\varepsilon_\theta.(1+\varepsilon_\theta/2).(1+\varepsilon_\theta)$, this results in the undrained soil shearing curve, on which the following can be measured (Figure 9.37):

– the undrained peak cohesion at the maximum of the shearing curve;

– the residual undrained cohesion at the interval following the shearing peak,

– the shearing modulus G is obtained by considering the elasticity equations [9.56] in the unloading loading cycle (Figure 9.39).

$$\tau = \frac{E}{2.(1+\upsilon)}.(2.\varepsilon_\theta) = 2.G.\varepsilon_\theta \qquad [9.56]$$

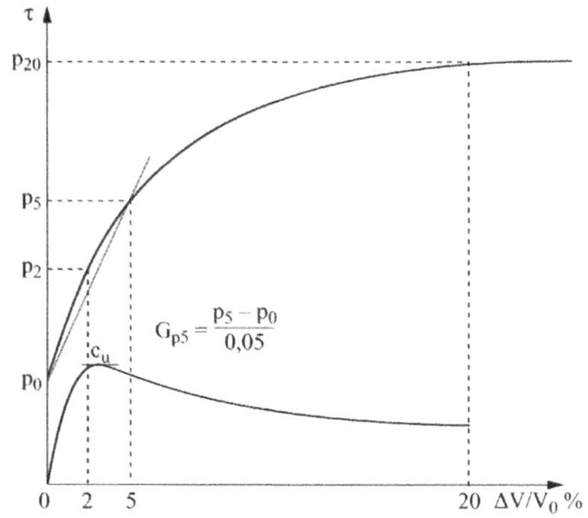

Figure 9.37. *Interpretation of the self-drilling pressuremeter test and the shearing curve*

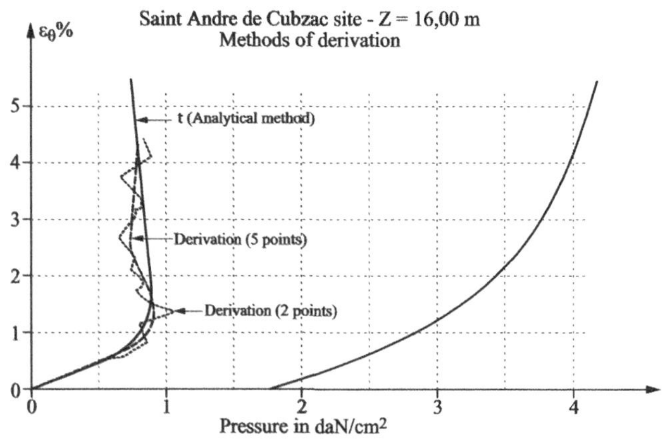

Figure 9.38. *Influence of the derivation mode on the results*

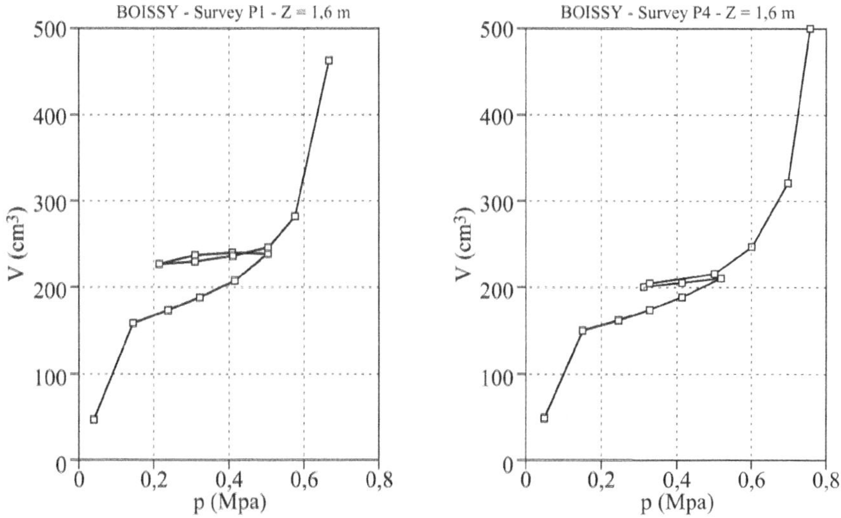

a) L-type test – unloading in steps

b) R-type test – quick unloading

Figure 9.39. *Determination of the shearing modulus with the pressuremeter [COM 01]*

The Baguelin *et al.* [BAG 72] method presents several advantages:

– it allows us to deduce the curve of soil shearing from the pressuremeter curve;

– it allows us to measure the peak undrained consistency and the residual undrained consistency;

– it allows us to determine the shearing modulus G in the unloading loading cycle.

However, it also presents several disadvantages:

– it is limited to impermeable soils such as clay and cannot allow for variation of soil volume;

– it is sensitive to the method of deriving the pressuremeter curve that is not continuous. We recommend using a second-degree polynomial regression of (parabola) by using five regression points rather than the experimental curve directly from two experimental points (Figure 9.38);

– it leads to an overestimate that can be quite considerable of the peak undrained consistency. It is better to use the residual consistency;

– it was developed for the use of the self-drilling pressuremeter test. It is sensitive and cannot be used for the Ménard pressuremeter.

9.5.2. *Cohesive soil: Monnet and Chemaa (1995) interpretation*

Baguelin *et al.*'s theory [BAG 71] is too sensitive to the device used and to the derivation method to be easily used. Furthermore, it does not take into account the decrease in soil volume that occurs at the beginning of the pressuremeter test.

Other theories have been presented, such as that of Gibson *et al.* [GIB 61] on pressuremeter testing in clays. This article is a weighty contribution to the knowledge of equilibrium around the pressuremeter. However, this study suffers from simplification. One hypothesis of plane deformation is made in the vertical direction, with no movement allowed along this axis. Equilibrium is studied only on the horizontal plane and vertical forces are not taken into account. The possibility of a rupture occurring between the circumferential and vertical constraints in such a way that the radial constraint becomes intermediary was envisaged by Wood *et al.* [WOO 77] but no formulation of this was proposed.

Monnet and Chemaa's theory [MON 95] offers answers to these unresolved questions. It also allows us to take into account the contraction that occurs at the beginning of the pressuremeter test. It encompasses the case in which the vertical constraint constitutes a direction of plastic deformation, and the case in which the vertical constraint has no role in the plastic flow.

9.5.2.1. *Hypotheses*

The soil is considered to have standard elasto-plastic behavior (Figure 9.40). The elastic part is linked to elasticity constants E (Young's modulus) and ν (Poisson's coefficient) which allows us to define the Lamé coefficients λ and μ:

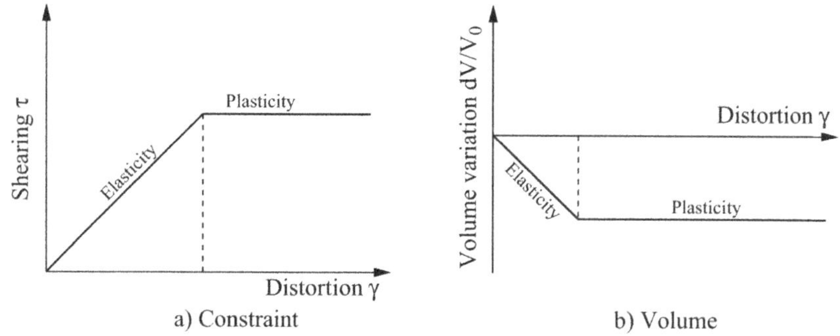

Figure 9.40. *Theoretical behavior of consistent soil a/ to shearing, b/ to volume variation*

The behavior's plastic part is linked to the undrained consistency c_u. The standard plastic flow [9.57] uses the scalar ξ, which defines the length of the plastic deformation. The function of charge F(σ) is a Mohr-Coulomb relation [9.58] which cancels itself in terms of plasticity.

$$d\varepsilon^p = \xi.\frac{dF(\sigma)}{d\sigma} \quad [9.57]$$

$$F(\sigma) = \sigma_1 - \sigma_3 - 2.c_u \quad [9.58]$$

Three different soil zones are taken into consideration. If we start from the drilling axis (Figures 9.41 and 9.42), we find:

1) A plastic zone in which plasticity is produced between the circumferential constraint σ_θ and radial stress σ_r, or in other terms between radii a (drilling radius) and b (external radius from the plastic zone).

2) According to the stress conditions that will be elaborate, a second plastic zone can evolve between circumferential stress σ_θ and vertical stress σ_z, or in other words between the radii b (external radius to the first plastic zone) and c (external radius to the second plastic zone) (see Figure 9.41) or otherwise this zone does not exist (Figure 9.42). Depending on the horizontal pressure conditions of soil at rest, this plastic zone can disappear.

3) An elastic zone is created beyond.

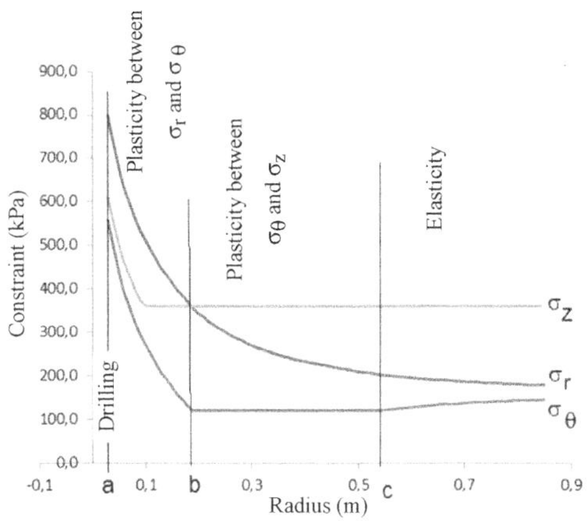

Figure 9.41. *State of the constraints around the pressuremeter, along the radius with vertical plasticity [MON 95]*

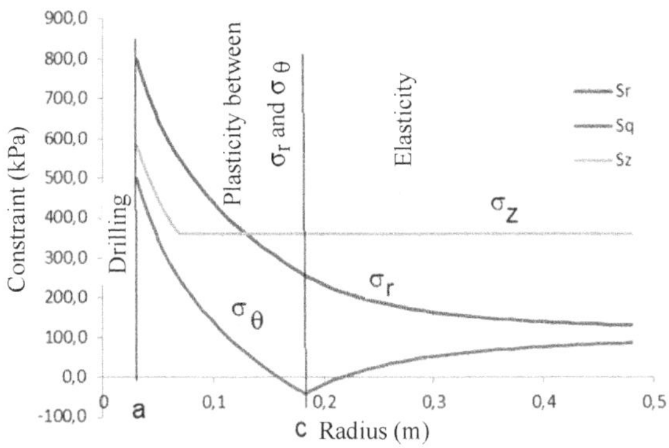

Figure 9.42. *State of the constraints around the pressuremeter, along the radius without vertical plasticity [MON 95]*

9.5.2.2. *Equilibrium conditions*

In the horizontal plane, the condition for equilibrium for a small deformation is [9.59], and in the vertical plane [9.60]:

$$\sigma_r - \sigma_\theta - r.\frac{d\sigma_r}{dr} = 0 \qquad [9.59]$$

$$\frac{d\sigma_z}{dz} = \gamma \qquad [9.60]$$

9.5.2.3. *Equilibrium of the first plastic zone between radii a and b*

This plastic zone is delimited by the drilling radius a and by the radius b which is the external limit for the zone considered. The limit equilibrium is produced between the circumferential constraint σ_θ and the radial constraint σ_r according to the Mohr–Coulomb criterion [9.58], which allows us to express [9.61].

The condition for horizontal equilibrium [9.61] can be written as function of the cohesion and integrated between the values a of the drilling radius where pressure p is applied, and any radius r which is internal to the first plastic zone where the radial constraint σ_r is applied. We can, therefore, deduce relation [9.62]:

$$\sigma_r - \sigma_\theta = 2.c_u = -r.\frac{d\sigma_r}{dr} \qquad [9.61]$$

$$Ln\left(\frac{r}{a}\right) = \frac{1}{2.c_u}.(p - \sigma_r) \qquad [9.62]$$

The standard plasticity condition [9.57]. shows that the ratio between the radial ε_p and circumferential ε_θ increments of plastic deformation is constant [9.63]. This equation can be integrated and the constant C_1 appears. C_1 does not depend on the movement of u and radius r [9.64]:

$$\frac{d\varepsilon_r^p}{d\varepsilon_\theta^p} = -1 \qquad [9.63]$$

$$\varepsilon_r^p = -\varepsilon_\theta^p + C_1 \qquad [9.64]$$

We assume that the elastic part of the deformation is negligible in this zone, and we use the relation between the deformations and movements [9.27], [9.28] to find equation[9.65]. This can be integrated between values a and r of the radius to find equation [9.65]. This relation can be integrated between the values a and r of the radius, which allows to find the equation [9.66]

$$\frac{du}{dr} = -\frac{u}{r} + C_1 \qquad [9.65]$$

$$Ln\left(\frac{r}{a}\right) = \frac{1}{2} Ln\left[{(2.u_a/a - C_1)}\Big/{(2.u/r - C_1)} \right] \quad [9.66]$$

From relations [9.62] and [9.66], we find the relation linking the radial stress, the drilling movement and movement u for the radius r [9.67] for the first plastic zone:

$$Ln\left[{(2.u_a/a - C_1)}\Big/{(2.u/r - C_1)} \right] = \frac{1}{c_u}.(p - \sigma_r) \quad [9.67]$$

9.5.2.4. *Equilibrium of the second plastic zone between radii b and c*

Under certain conditions, the second plastic zone appears between the vertical stress and the circumferential stress. The condition for vertical equilibrium [9.60] allows us to calculate the vertical stress at depth z without force applied to the surface: [9.68]. In the case where plasticity appears in this zone, the Mohr–Coulomb criterion [9.58] gives a constant difference between the circumferential and vertical stress. This leads to a constant value of the circumferential stress [9.69].

Equilibrium condition [9.59] can be written as a function of c_u and can be integrated between the values c and r of the radius. We use the notation $\sigma_{\rho\chi}$ for the radial stress at the limit with the elastic domain, for the value of radius c. Then we have the radial stress for any radius r so that [9.70]:

$$\sigma_z = \gamma.z \quad [9.68]$$

$$\sigma_\theta = \gamma.z - 2.c_u \quad [9.69]$$

$$\frac{(\gamma.z - 2.c_u + \sigma_{rc})}{(\gamma.z - 2.c_u + \sigma_r)} = \frac{r}{c} \quad [9.70]$$

The standard plasticity condition [9.57] provides the ratio between the increments of circumferential and vertical plastic deformations, as well as the increment of radial plastic deformation [9.71], with the integration constants C_2 and C_3. Thus, we find the value of the movement as a function of the radius [9.72]. This equation, as well as formula [9.70] provides the relations between the stress and the movements in the second plastic zone

[9.73], with u_c being the movement for radius c, which is the external limit of the second plastic zone and the beginning of the elastic zone.

$$\frac{d\varepsilon_z^p}{d\varepsilon_\theta^p} = -1 \text{ and } d\varepsilon_r^p = 0 \qquad [9.71]$$

$$u = C_3.r + C_4 \qquad [9.72]$$

$$\frac{(\gamma.z-2.c_u+\sigma_{rc})}{(\gamma.z-2.c_u+\sigma_r)} = \frac{(u-C_4)}{(u_c-C_4)} \qquad [9.73]$$

9.5.2.5. *Elastic equilibrium beyond c*

The analysis of the elastic zone is well known, and can be found in [CAS 78] and [BAG 78]. With the Lamé notations, the radial and circumferential constraints are written [9.74], [9.75]. The condition for horizontal equilibrium [9.59] and relations[9.27], [9.28] give the second-order differential equation which can be integrated with constant C_5 [9.76] to find the movement with a movement of zero for an infinite radius.

$$\sigma_r = (1 + 2.\mu).\varepsilon_r + \lambda.(\varepsilon_\theta + \varepsilon_z) \qquad [9.74]$$

$$\sigma_\theta = (1 + 2.\mu).\varepsilon_\theta + \lambda.(\varepsilon_r + \varepsilon_z) \qquad [9.75]$$

$$u = C_5/r \qquad [9.76]$$

Relation [9.76] can be introduced in the elastic equations [9.74], [9.75]. For an infinite radius, the radial constraint is linked to the vertical constraint by the coefficient of at rest pressure K_0, so that the relations become relations in terms of stress and deformation strain:

$$\sigma_r = 2.\mu.\frac{C_5}{r^2} + K_0.\gamma.z \qquad [9.77]$$

$$\sigma_\theta = -2.\mu.\frac{C_5}{r^2} + K_0.\gamma.z \qquad [9.78]$$

$$\varepsilon_r = \frac{C_5}{r^2} \qquad [9.79]$$

$$\varepsilon_\theta = -\frac{C_5}{r^2} \qquad [9.80]$$

9.5.2.6. *Creep pressure for two plastic zones*

The case of two plastic zones is shown in Figure 9.41. In this case, a plastic flow occurs between the vertical and circumferential axes. The particular case where only the first plastic zone would exist will be examined in the following sections.

The limit between the elastic and plastic zones is located at the value c of the radius. We know the difference $2.c_u$ between the vertical and circumferential constraints. The circumferential stress can be written as a function of the consistency c_u by the Mohr–Coulomb criterion, and as a function of C_5 by the elastic equations. Taking into consideration the relations of elasticity [9.77] and [9.78], we find the value of the constant C_5 and $\sigma_{\rho\chi}$, which is both constant and independent of the loading p applied by the pressuremeter to the bore wall:

$$C_5 = \frac{c^2}{2.\mu}.[2.c_u - \gamma.z.(1 - K_0)] \qquad [9.81]$$

$$p_f = 2.c_u + \gamma.z.(2.K_0 - 1) = \sigma_{rc} \qquad [9.82]$$

9.5.2.7. *Pressiometric curve for two plastic zones*

Relation [9.82] gives the creep pressure in the case where two plastic zones appear. Relations [9.27] and [9.79] can be used to find the radial movement at limit C [9.83] and the strain at radius C which is equal to the integration constant C_3 [9.84]:

$$u_{rc} = \frac{c}{2.\mu}.[2.c_u - \gamma.z.(1 - K_0)] \qquad [9.83]$$

$$\varepsilon_{rc} = \frac{[2.c_u - \gamma.z.(1-K_0)]}{2.\mu} = C_3 \qquad [9.84]$$

At the limit of the first plastic zone of value b of the radius, the radial stress takes the value of the vertical stress (Figure 9.41). This condition allows us to calculate the value of the first plastic radius b [9.85]. By using equation [9.70] we find the value c [9.86] and finally the value of $\varepsilon_{\theta b}$ [9.87]:

$$b = a.e^{[(p-\sigma_z)/(2.c_u)]} \qquad [9.85]$$

$$c = \frac{b.c_u}{[2.c_u - \gamma.z.(1-K_0)]} \qquad [9.86]$$

$$\varepsilon_{\theta b} = \frac{\gamma.z.(K_0 - 1)}{2.\mu} \qquad [9.87]$$

At the limit between the first and second plastic zones for radius b, equation [9.64] is applied. The value of the radial, plastic deformation can be equaled at constant C_3, the circumferential deformation given by equation [9.87], which allows us to find the value of constant C_1 [9.88]. The distribution of stress in the first plastic zone [9.67] can be used between the values a and b of the radius and thus we find the general relation [9.89] which is the theoretical expression of the pressuremeter curve in the case where two distinct zones of plasticity appear:

$$C_1 = -\frac{c_u}{\mu} \qquad [9.88]$$

$$Ln\left[\frac{u_a}{a} + \frac{c_u}{2.\mu}\right] = \frac{p}{c_u} - \frac{\gamma.z}{c_u} + Ln\left[\frac{\gamma.z.(1-K_0)}{2.\mu} + \frac{c_u}{2.\mu}\right] \qquad [9.89]$$

The term $c_u/(2.\mu)$ is very small. By considering it negligible, we find the linear relation between the logarithms of movement u_a at drilling, and pressure p is applied of which the slope is the inverse of cohesion. Measurement of the slope allows us to determine the undrained cohesion c_u (Figure 9.43). This result has previously been found by Gibson and Anderson [GIB 61].

9.5.2.8. *Limit pressure for two plastic zones*

The conventional pressure limit is reached when the probe doubles in volume, which corresponds to a doubling of section [9.90], to a constant relation between the initial and deformed radii [9.91], and to a particular deformation [9.92] of drilling u_a/a equal to $(\sqrt{2} - 1)$. This particular value of radial deformation can then be introduced into the theoretical expressions of the pressuremeter curve. We thus find the analytical expression of the conventional pressure limit in the two cases of plastification:

$$S_{LM} = 2.S_1 \qquad [9.90]$$

$$R_{LM} = \sqrt{2}.R_1 \qquad [9.91]$$

$$\varepsilon_{\theta LM} = \frac{u_a}{a} = \frac{R_{LM}-R_1}{R_1} = \sqrt{2} - 1 \qquad [9.92]$$

In the case where two plastic zones appear, equation [9.89]. allows us to determine the value of the pressure limit through equation [9.93]:

$$p_{LM} = \gamma.z + c_u.Ln\left[\frac{c_u+2.\mu.(\sqrt{2}-1)}{c_u+\gamma.z.(1-K_0)}\right] \qquad [9.93]$$

9.5.2.9. *Creep pressure for a plastic zone*

The case of a single plastic zone is shown in Figure 9.42. In this configuration, there is only one plastic zone of radius c. Constraints at radius c can be expressed by relations [9.77] and [9.78]. By applying the condition of plasticity [9.61], we can find the constant C_5 and the deformation $\varepsilon_{\theta c}$ [9.94] as well as the creep pressure[9.96].

Relation [9.96] gives the creep pressure in the case where there is one plastic zone. We find the Gibson *et al.* formula [GIB 61]. Relation [9.61] applied for a radius c and the radial constraint σ_{rc} defined by relation [9.96] gives the value of the plastic radius c [9.95].

$$\varepsilon_{\theta c} = -\varepsilon_{rc} = \frac{c_u}{2.\mu} = \frac{C_5}{c^2} \qquad [9.94]$$

$$c = a.e^{[(p-c_u-K_0.\gamma.z)/(2.c_u)]} \qquad [9.95]$$

$$p_f = c_u + K_0.\gamma.z = \sigma_{rc} \qquad [9.96]$$

9.5.2.10. *Pressuremeter curve for a single plastic zone*

The elastic deformations are negligible compared to the plastic deformations. Formula [9.64] in which we introduced the values of the deformations [9.94] allows us to find the new constant C_1 in this state of equilibrium [9.97]. Formula [9.67] applied between radii a and c gives the analytical expression of the pressuremeter curve [9.98]:

$$C_1 = 0 \qquad [9.97]$$

$$Ln\left[\frac{u_a}{a}\right] = \frac{p}{c_u} - \frac{K_0.\gamma.z}{c_u} - 1 + Ln\left[\frac{c_u}{2.\mu}\right] \qquad [9.98]$$

We find a linear relation between movement logarithms u_a at drilling, and the applied pressure p of which the slope is the inverse of the cohesion. The measurement of the slope allows us to determine the undrained cohesion c_u.

9.5.2.11. *Condition for the development of vertical plasticity (case of two plastic zones)*

The distinction between the two possible behaviors is made on the value of the radial constraint of radius c. For the second case of behavior when a single plastic zone exists, the radial stress of plastic radius c exceeds the value of the vertical stress under compression (or that the cohesion is strong enough to reach the indicated limit value [9.99]):

$$c_u \geq (1 - K_0).\gamma.z \qquad [9.99]$$

The case of shearing with two plastic zones can only occur when the soil coefficient at rest is K_0 smaller than 1, which corresponds to a Poisson coefficient of less than 0.5 and to a clay above groundwater.

9.5.2.12. *Limit pressure for a single plastic zone*

In the case of a single plastic zone, we use equation [9.98], which allows us to determine the limit pressure through equation [9.100].The relation corresponds to formula [9.101] proposed by Hughes and Withers [HUG 74], the difference being that the value of the deformation of the conventional pressure limit for these authors is of 0.5, when it is really $(\sqrt{2} - 1)$:

$$p_{LM} = K_0.\gamma.z + c_u.\left\{1 + Ln\left[\frac{E(\sqrt{2}-1)}{c_u.(1+\upsilon)}\right]\right\} \qquad [9.100]$$

$$p_{LM} = K_0.\gamma.z + c_u.\left\{1 + Ln\left[\frac{E}{2.c_u.(1+\upsilon)}\right]\right\} \qquad [9.101]$$

In these two relations, the limit pressure is not proportional to the depth of the test. It is proportional to the undrained cohesion, which seems logical in a consistent soil. However, these relations are considerably different from that proposed by CTR4 [AMA 91] which are the following:

$$p_{LM} = 5{,}5.c_u + K_0.\gamma.z \; pour \; (p_{LM} - K_0.\gamma.z) < 300kPa \qquad [9.102]$$

$$p_{LM} = 10.c_u - 250 \quad \text{when} \quad (p_{LM} - K_0.\gamma.z) > 300kPa \qquad [9.103]$$

The representation of pressuremeter curves on a semi-logarithmic scale (Figure 9.43) allows us to find a line with a slope that is the inverse of the undrained cohesion c_u. This result was also found by other researchers [GIB 61, AMA 91].

9.5.2.13. *Measuring consistency c_u, controlling its value*

Determination of undrained cohesion is done by the semi-logarithmic transformation of the pressuremeter curve (Figure 9.43). The measurement of the slope of the line gives the inverse of the cohesion.

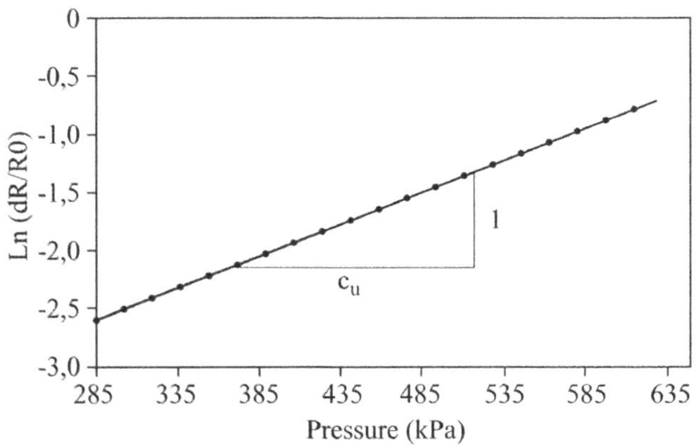

Figure 9.43. *Linear transformation of the pressuremeter curve for a test at of 13.20 m in the clay of Geneva [MON 95]*

Controlling the value of the cohesion is done by comparing (Figure 9.44) the experimental curve and the theoretical curve according to expressions [9.89] or [9.98]. In the calculation of the theoretical curve, we use the total stress of the ground, and the elastic modulus is considered to be equal to the elastic modulus of unloading and reloading. Comparison of the experimental and theoretical curves allows us to confirm the cohesion and the elastic modulus. Gaiapress software [GAL 12] makes the automation of this calculation possible.

Another manner of operating is to use the pressure limit [9.93] and to put (Figure 9.45) the shearing modulus G^e [9.104] into the graph as a function of the net pressure limit. The graph (Figure 9.45) gives the undrained cohesion. This determination is less precise than the previous method. An example of this is indicated for the CCIP project in Sparnacian clays (Figure 9.46) and gives similar results to those found in laboratory tests:

$$G^e = E^e/[2.(1+\nu)] \qquad [9.104]$$

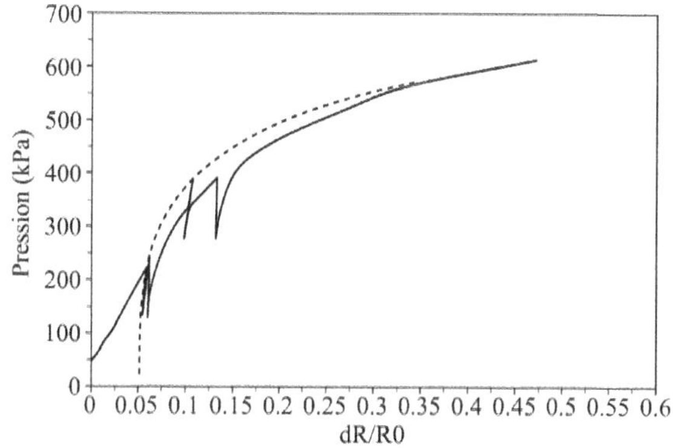

Figure 9.44. *Comparison of the experimental curve (solid line) and theoretical curve (broken line), test at 13.20 m, Geneva clay [MON 95]*

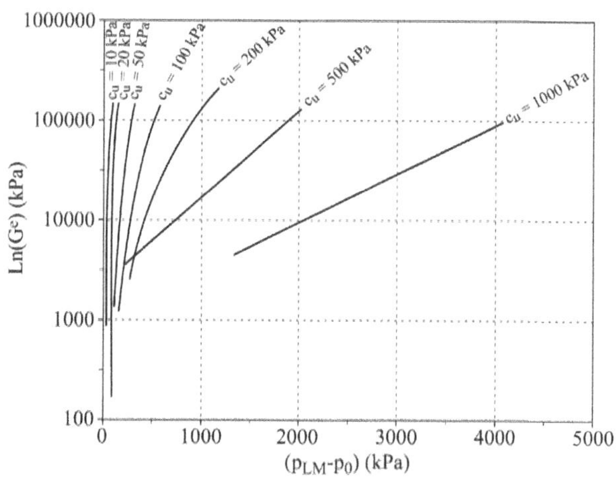

Figure 9.45. *Relation between the elastic modulus and the pressure limit as a function of the consistency [MON 13]*

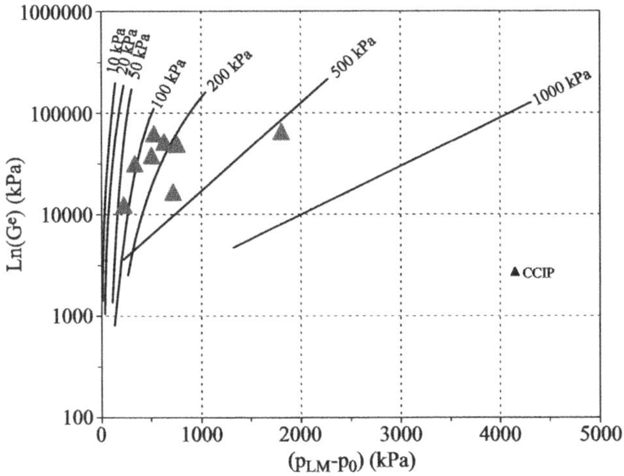

Figure 9.46. *Interpretation in consistency of the pressuremeter test of the CCIP project, Sparnacian clays, Paris [MON 13]*

9.5.3. *Granular soil: the Monnet and Khlif (1994) and Monnet (2012) interpretations*

One of the most interesting contributions using the pressuremeter was, among others, the work of Hughes *et al.* [HUG 77]. However, this study suffers of two simplifications. First, a hypothesis of plane deformation is taken in the vertical direction, and no movement is authorized along that axis. Furthermore, equilibrium is only studied on the horizontal plane, without taking into account vertical constraints. However, the possibility of a rupture occurring between circumferential and vertical stress was envisaged by Wood and Wroth [WOO 77], but no formula was proposed.

Other interpretations are possible, such as the representation of the pressuremeter curve as a hyperbole [FAH 93]. A summary of the possible methods can be found in [VIA 15].

The studies undertaken by Monnet and Khlif [MON 94] and Monnet [MON 12] do not formulate a hypothesis on the shape of the pressuremeter curve. We will lay out the principle steps at first as well as the principle consequences of the equations presented.

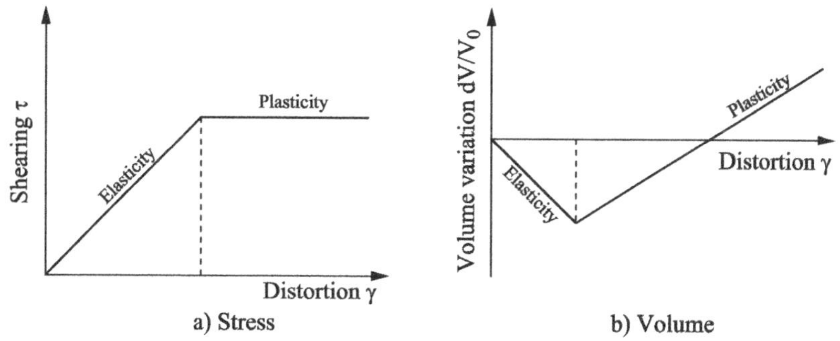

Figure 9.47. *Theoretical behavior of granular soil, a/ in deformation stress; b/ in volume variation*

9.5.3.1. *Hypotheses*

The stress are positive negative under compression. The soil is considered to have a non-standard elasto-plastic behavior (Figure 9.47). The elastic part is linked to the constants of elasticity E (Young's modulus) and ν (Poisson's coefficient), which allow us to define the Lamé coefficients λ and μ.

The plastic part of the behavior is linked to the two friction angles Φ (internal friction angle) and Φ_μ (interparticle angle of friction). In the case of the triaxial test of revolution, the internal friction angle Φ_μ is considered equal to the angle Φ_{cv} from the Rowe theory [ROW 62]. The non-standard plastic strain uses the scalar ξ which corresponds to the length of the non-standard plastic deformation given by relation [9.105] where H(s) is the plastic potential of the granular soil. The expression of plastic potential H(s) is shown [9.106] with the angle of dilatance of the soil ψ. The equilibrium limit is reached when Mohr–Coulomb's criterion is canceled out [9.107] when the soil is in plasticity, shown on the right in figures a/ and b/ (Figure 9.47). When the soil is in elasticity, shown on the left in figures a/ and b/ (Figure 9.47), Mohr–Coulomb's criterion [9.107] becomes negative:

$$d\varepsilon^p = \xi.\frac{dH(\sigma)}{d\sigma} \qquad [9.105]$$

$$H(\sigma) = \sigma'_1 - \sigma'_3 - sin\psi.(\sigma'_1 + \sigma'_3) \qquad [9.106]$$

$$F(\sigma) = \sigma'_1 - \sigma'_3 - sin\phi.(\sigma'_1 + \sigma'_3) \qquad [9.107]$$

9.5.3.2. *Dilatance*

The expression found by Frydman *et al.* [FRY 73] allows us to link the stress and plastic deformations in a tridimensional state. Notations used are shown for the average constraint σ_{oct} in relation [9.108], the average shearing τ_{oct} in relation [9.109], the average volume deformation ε_{oct} by relation [9.110], and the average distortion deformation ε_d by relation [9.111]. All these values correspond to the state of constraint and deformation in the octahedral plane, which is the plane in which the majority of the plasticity criteria are expressed.

Frydman *et al.* [FRY 72] assume that the energy dissipated per unit of volume during loading [9.112] is equal to the energy dissipated through the rubbing of grains in the octahedral plane [9.113], with an average intergranular friction angle Φ_μ. Exponent p shows the plastic part of the deformations considered. Because they are equal, we can determine the relation linking the stress and the increments of plastic deformation [9.114]:

$$\sigma'_{oct} = \frac{1}{3} \cdot (\sigma'_1 + \sigma'_2 + \sigma'_3) \qquad [9.108]$$

$$\tau_{oct} = \frac{1}{3} \cdot \sqrt{(\sigma'_1 - \sigma'_2)^2 + (\sigma'_2 - \sigma'_3)^2 + (\sigma'_3 - \sigma'_1)^2} \qquad [9.109]$$

$$\varepsilon_m = \frac{1}{3} \cdot (\varepsilon_1 + \varepsilon_2 + \varepsilon_3) \qquad [9.110]$$

$$\varepsilon_d = \frac{1}{3} \cdot \sqrt{(\varepsilon_1 - \varepsilon_2)^2 + (\varepsilon_2 - \varepsilon_3)^2 + (\varepsilon_3 - \varepsilon_1)^2} \qquad [9.111]$$

$$dW^p = 3.\sigma'_{oct}.d\varepsilon^p_m + 3.\tau_{oct}.d\varepsilon^p_d \qquad [9.112]$$

$$dW^p_d = 3.\sigma'_{oct}.tg\phi_\mu.d\varepsilon^p_d \qquad [9.113]$$

$$\frac{\tau_{oct}}{\sigma'_{oct}} + \frac{d\varepsilon^p_m}{d\varepsilon^p_d} = tg\phi_\mu \qquad [9.114]$$

When perfect plasticity appears during the triaxial test, the relation between the main stress depends solely on the internal friction angle Φ' through equation [9.115]. The non-standard plasticity rule [9.105] allows us to determine the plastic deformation of volume during the triaxial test [9.116], as well as major [9.117]and minor [9.118] plastic deformations. The

Frydman *et al.* relation [FRY 72] can be used for plastic behavior at the failure of the triaxial test by using equations [9.114], [9.115], [9.117] and [9.118]. We thus find a relation between the three angles defining dilatance [9.119], which can be shown theoretically. This relation allows us to theoretically calculate dilatance as a function of the average interparticle angle of friction and the internal friction angle [9.120]:

$$\frac{\sigma_1'}{\sigma_3'} = \frac{1+sin\phi'}{1-sin\phi'} \qquad [9.115]$$

$$\varepsilon_m = \xi.\frac{1}{3}.\left(d\varepsilon_1^p + d\varepsilon_2^p + d\varepsilon_3^p\right) = \xi.\frac{2}{3}.sin\psi \qquad [9.116]$$

$$d\varepsilon_1^p = \xi.(1 - sin\psi) \qquad [9.117]$$

$$d\varepsilon_3^p = d\varepsilon_2^p = \xi.\frac{1}{2}\left(3.d\varepsilon_m^p - d\varepsilon_1^p\right) = -\xi.\frac{1}{2}.(1 + sin\psi) \qquad [9.118]$$

$$\frac{2.sin\phi'.\sqrt{2}}{(3-sin\phi')} - \frac{4.sin\psi}{(3-sin\psi).\sqrt{2}} = tg\phi_\mu \qquad [9.119]$$

$$sin\psi = \frac{3.\sqrt{2}.\left[\frac{2.sin\phi'.\sqrt{2}}{(3-sin\phi')} - tg\phi_\mu\right]}{4+\sqrt{2}.\left[\frac{2.sin\phi'.\sqrt{2}}{(3-sin\phi')} - tg\phi_\mu\right]} \qquad [9.120]$$

A relation between dilatance angle ψ and internal friction angle Φ' [9.121] was discovered by Rowe [ROW 62] which can be transformed into a dilatance equation [9.122]. Bolton [BOL 86] later found another dilatance equation through experimentation [9.123]:

$$\frac{1+sin\phi'}{1-sin\phi'} = \frac{1+sin\phi_{cv}}{1-sin\phi_{cv}}.\frac{1+sin\psi}{1-sin\psi} \qquad [9.121]$$

$$sin\psi = \frac{\frac{1+sin\phi'}{1-sin\phi'}.\frac{1-sin\phi_{cv}}{1+sin\phi_{cv}} - 1}{\frac{1+sin\phi'}{1-sin\phi'}.\frac{1-sin\phi_{cv}}{1+sin\phi_{cv}} + 1} \qquad [9.122]$$

$$0{,}8.\psi = \phi' - \phi_{cv} \qquad [9.123]$$

Comparison (Figure 9.48) of the theoretical theory of dilatance [9.120] with the Rowe [9.122] Bolton [9.123] expressions is shown. In this figure, interparticle friction angle Φ is considered equal to the friction angle in the critical state Φ_{cv} as has been observed by other authors [NEG 87, BIA 94].

The difference between theoretical dilatance [9.120] and Rowe's dilatance [9.122] is less than 4°. The theoretical expression is also close to Bolton's experimental expression [9.123] for the usual range of the friction angle (Φ= 20-45°) and the critical angle (Φ_{cv}= 20-30°) but gives a smaller value for the dilatance of friction angles Φ greater than 45° and for small critical angles (Φ_{cv}= 10°). The theoretical expression of dilatance [9.120] can be shown and will be used in the interpretation of the pressuremeter test.

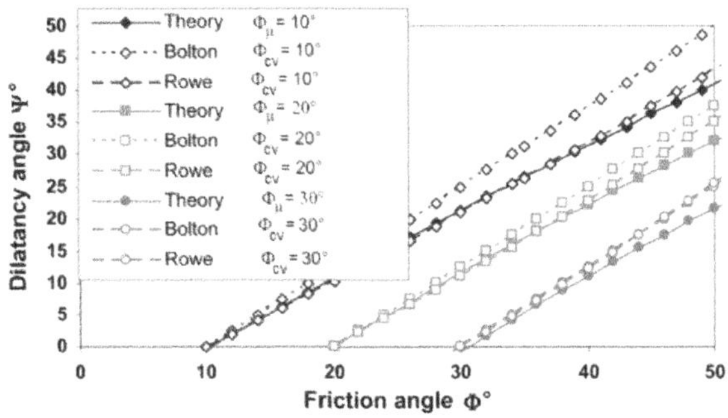

Figure 9.48. *Dilatance angle as a function of the friction in the presented theory, the Rowe theory [ROW 62], the empirical Bolton expression [BOL 86, MON 13]*

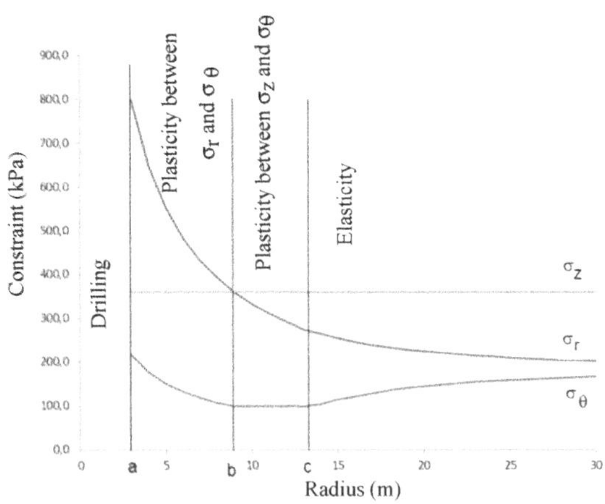

Figure 9.49. *State of stress around the pressuremeter, along the radius with two plastic zones [MON 94]*

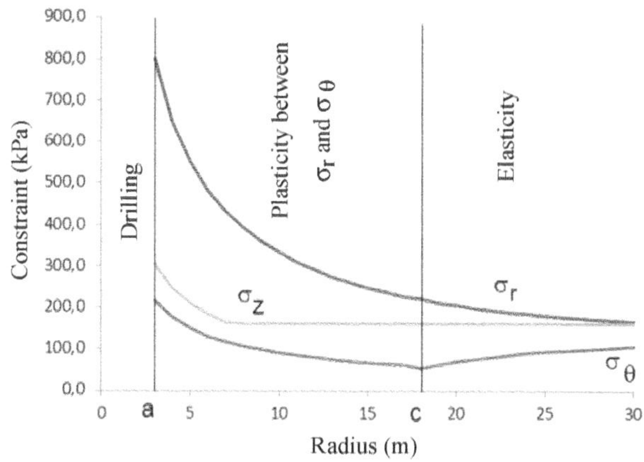

Figure 9.50. *State of stress around the pressuremeter, along the radius with a single plastic zone [MON 94]*

9.5.3.3. *Equilibrium conditions*

The conditions for equilibrium have been previously shown in the horizontal plane by relation [9.59] and in the vertical plane by relation [9.60].

Three different zones of the soil are considered. If we start at the drilling axis (Figure 9.49), we find:

– a plastic zone in which plasticity occurs between circumferential constraint σ_θ and radial constraint σ_r, that is between radii a (drilling radius) and b (external radius to the first plastic zone);

– according to stress conditions that will be elaborated, a second plastic zone can appear between circumferential stress σ_θ and vertical stress σ_Z, which is to say between radii b (external radius of the first plastic zone) and c (external radius of the second plastic zone) (Figure 9.49) or otherwise this zone does not exist (Figure 9.50);

– an elastic zone develops beyond this.

9.5.3.4. *First plastic zone for two plastic zones*

This plastic zone is limited by radius a of the drill and the pressuremeter, and by radius b the external limit of the considered zone. Maximum equilibrium occurs between circumferential stress σ_θ and radial stress r according to Mohr–Coulomb's criterion [9.107], which allows us to find the proportionality between the radial and circumferential constraints [9.124] with the notation N [9.125].

The equilibrium condition [9.59] can be written as a function of relation N of the stress and integrated between the value a of the drilling radius where pressure p is applied, and any radius r, internal to the first plastic zone where the radial stress σr is applicable. We can deduce relation [9.126]:

$$\frac{\sigma'_\theta}{\sigma'_r} = N \qquad [9.124]$$

$$N = \frac{(1-sin\phi)}{(1+sin\phi)} \qquad [9.125]$$

$$Ln\left(\frac{r}{a}\right) = \frac{1}{(N-1)}.Ln\left(\frac{\sigma'_r}{p}\right) \qquad [9.126]$$

The condition of non-standard plasticity [9.105], [9.106], [9.107] shows the relation between the increments of radial ε_p and circumferential ε_θ plastic deformation is constant [9.127]. Equation [9.127] can be integrated, and it shows the constant C_1 that does depend on movement u and radius r [9.129].

By assuming that the elastic part of the deformation is negligible in this zone, and by using the relations between deformations and movements [9.27], [9.28], equation [9.129] can be written in terms of movement and can be integrated between values a and r of the radius. The notation u_a corresponds to the radial movement at the radius a of the drilling, and notation u corresponds to radial movement at radius r. The expression of movement as a function of the radius is thus found [9.130].

From relations [9.126] and [9.130], the relation linking the radial stress, the movement during drilling and the movement u for radius r can be determined [9.131].

At the limit of the first plastic zone, the radial stress takes the value of the vertical stress (Figure 9.49). Under these conditions, expression [9.126] allows us to find the radius b of the first plastic zone [9.132]:

$$\frac{d\varepsilon_r^p}{d\varepsilon_\theta^p} = -n \qquad [9.127]$$

$$n = \frac{(1-\sin\psi)}{(1+\sin\psi)} \qquad [9.128]$$

$$\varepsilon_r^p = -n.\varepsilon_\theta^p + C_1 \qquad [9.129]$$

$$Ln\left(\frac{r}{a}\right) = \frac{1}{1+n}.Ln\left[\frac{\left(u_a/_a\right).(1+n)-C_1}{\left(u/_r\right).(1+n)-C_1}\right] \qquad [9.130]$$

$$Ln\left[\frac{\left(u_a/_a\right).(1+n)-C_1}{\left(u/_r\right).(1+n)-C_1}\right] = \frac{(1+n)}{(N-1)}.Ln\left(\frac{\sigma_r'}{p}\right) \qquad [9.131]$$

$$b = a.\left(\frac{\gamma\prime.z}{p}\right)^{\left[1/(N-1)\right]} \qquad [9.132]$$

9.5.3.5. *Second plastic zone*

Under certain conditions, a second plastic zone develops between the vertical constraint and the circumferential constraint. The vertical equilibrium condition [9.60] gives the vertical constraint at a depth of z, without any surface forces [9.133].

In the case where plasticity develops in this zone, the Mohr–Coulomb criterion [9.107] provides a constant N ratio between the circumferential and vertical stress [9.125]. This leads to a constant value of the circumferential stress [9.134].

The condition of equilibrium [9.59] can be written as a function of N and can be integrated between values c and r of the radius. The notation σ_{rc} is used for the radial constraint at the limit with the elastic domain, for the value c of the radius. The radial stress with any value of the radius r is such that relation [9.135] is verified:

$$\sigma'_z = \gamma'.z \qquad [9.133]$$

$$\sigma'_\theta = N.\gamma'.z \qquad [9.134]$$

$$\frac{(N.\gamma\prime.z-\sigma\prime_{rc})}{(N.\gamma\prime.z-\sigma_r)}=\frac{r}{c} \qquad [9.135]$$

The condition of non-standard plasticity [9.105] provides the ratio between the increments of the vertical and circumferential plastic deformations. The integration constants C_2 and C_3 are used to determine the radial and vertical deformations [9.136].

The elastic part of the deformation is neglected. By using relations [9.27] and [9.28] between deformation and movement and radial deformation in [9.136], the movement in the second plastic zone can be determined with integration constant C_4 [9.137].

This equation, as well as formula [9.135], provides the relation between the stress and the movements in the second plastic zone, with u_χ the movement for a radius of c, external limit of the second plastic zone and start of the elastic zone [9.138]:

$$\varepsilon_r^p = C_3 \quad et \quad \varepsilon_z^p = -n.\varepsilon_\theta^p + C_2 \qquad [9.136]$$

$$u = C_3.r + C_4 \qquad [9.137]$$

$$\frac{(N.\gamma\prime.z-\sigma\prime_{rc})}{(N.\gamma\prime.z-\sigma\prime_r)}=\frac{u-C_4}{u_c-C_4} \qquad [9.138]$$

9.5.3.6. *Elastic zone*

The analysis of the second plastic zone is well known, and can be found in [CAS 78] or [BAG 78]. This analysis is carried out with the previous notations and the constraints and the elastic deformations are identical to the ones for clay [9.74]–[9.80].

9.5.3.7. *Equilibrium of the second plastic zone*

The limit between the elastic and plastic zones is located at value c of the radius. The relation between the circumferential and vertical stress is equal to N. The circumferential stress can be written as a function of the relation N through the intermediary of the Mohr–Coulomb criterion, and as a function of C_5 by the elastic equations [9.139]. By taking into account formulas [9.77] and [9.78], the value of constant C_5, as well as stress σ_{rc}, can be found. It can be noted that $\sigma_{p\chi}$ is constant and independent of the loading p applied by the pressuremeter to the walls of the borehole.

The elasticity equation gives movement u_{cc} [9.141] at radius c and the corresponding deformations [9.142], [9.143] as well as C_3 [9.142].

Equation [9.138] can be applied between plastic radii b and c. With condition σ_{rb} equal to the vertical constraint, σ_{rc} equal to expression [9.140] and b given by equation [9.132], we finally obtain the equation of the second plastic radius c [9.144]:

$$C_5 = c^2.(K_0 - N).\frac{\gamma'.z}{2.\mu} \quad [9.139]$$

$$\sigma'_{rc} = (N - 2.K_0).\gamma'.z \quad [9.140]$$

$$u_c = c.(N - K_0).\frac{\gamma'.z}{2.\mu} \quad [9.141]$$

$$\varepsilon_{rc} = (K_0 - N).\frac{\gamma'.z}{2.\mu} = \varepsilon_r^p = C_3 \quad [9.142]$$

$$\varepsilon_{\theta c} = (N - K_0).\frac{\gamma'.z}{2.\mu} \quad [9.143]$$

$$c = a.\frac{N-1}{2.(N-K_0)}.\left(\frac{\gamma'.z}{p}\right)^{\left[1/(N-1)\right]} \quad [9.144]$$

9.5.3.8. *Pressuremeter curve for two plastic zones*

The limit between the elastic and plastic zones is located at the value c of the radius.

In the second plastic zone, the radial stress is intermediate between the vertical and circumferential constraints (Figure 9.49). The radial deformation is, therefore, constant. At the limit between the first and second zones, the value of the radial deformation can be calculated with equation [9.130] and is equaled for constant C_3 in [9.142], which gives equation [9.145]. The value of constant C_1 can be determined by [9.146].

Relation [9.137] on movements in the second plastic zone taken between the values b and c of the radius gives the ratio between corresponding movements u_b and u_c. The condition of equilibrium [9.138] of the second plastic zone gives the circumferential deformation at radius b [9.148] as a function of u_c.

The conditions of elastic equilibrium [9.141] provide the movement u_c at radius c, which can be introduced into the previous equation, giving the circumferential deformation at radius b [9.149].

The distribution of the constraints [9.130] in the first plastic zone can be used between values a and b of the radius, the theoretical expression of the pressuremeter curve in the case where there are two plastic zones is determined by [9.150]:

$$\frac{1}{1+n}\cdot\left[\left({}^{u_a}/_a\right).(1+n)-C_1\right].\left(\frac{a}{b}\right)^{(1+n)} = (K_0-N).\frac{\gamma\prime.z}{2.\mu} \quad [9.145]$$

$$C_1 = \frac{n.\left({}^{u_a}/_a\right).(1+n).\left(\frac{\gamma\prime.z}{p}\right)^{\delta}+(1+n).(N-K_0).\frac{\gamma\prime.z}{2.\mu}}{1+n.\left(\frac{\gamma\prime.z}{p}\right)^{\delta}}. \quad [9.146]$$

$$\delta = \frac{(1+n)}{(1-N)} \quad [9.147]$$

$$\varepsilon_{\theta b} = \frac{u_b}{b} = \frac{u_c}{b} + C_3.\left[1-\frac{N-1}{2.(K_0-N)}\right] \quad [9.148]$$

$$\varepsilon_{\theta b} = \frac{u_b}{b} = (1-K_0).C_3.\frac{\gamma\prime.z}{2.\mu} \quad [9.149]$$

$$Ln\left[\left(\frac{u_a}{a}\right).(1+n)-C_1\right] = \delta.Ln(p)-\delta.Ln(\gamma'.z)$$
$$+Ln\left[(1-K_0).\gamma'.z.\frac{1+n}{2.\mu}-C_1\right] \quad [9.150]$$

9.5.3.9. *Plastic zone for a single plasticity between σ_r and σ_θ*

In this configuration, the second plastic zone disappears, and the radius b that determined the limit between the two plastic zones has no more reason to be. Only drilling radius a, and radius c which limits the plastic and elastic zones remaino.

The stress at radius c can be expressed by the elastic relations [9.74]–[9.80] and by the plasticity relation [9.124] which allows us to determine the radial deformation [9.151]. The value of the constraints at plastic radius c is, therefore, given by equations [9.152] and [9.153].

The equilibrium relation in the plastic zone [9.126] can be applied for radius c and the radial constraint σ_{rc} defined by relation [9.153], which provides the value of the plastic radius c:

$$\frac{C_5}{c^2} = \frac{K_0.\gamma\prime.z.(1-N)}{2.\mu.(1+N)} = \varepsilon_{rc} = -\varepsilon_{\theta c} \qquad [9.151]$$

$$\sigma'_{rc} = \frac{2.K_0.\gamma\prime.z.}{(1+N)} \qquad [9.152]$$

$$\sigma'_{\theta c} = \frac{2.K_0.\gamma\prime.z.N}{(1+N)} \qquad [9.153]$$

$$c = \left[\frac{2.K_0.\gamma\prime.z}{p.(1+N)}\right]^{\left[1/(N-1)\right]} \qquad [9.154]$$

9.5.3.10. *Pressuremeter curve for a single plastic zone*

As in the previous case, we neglect the elastic deformations in front of the plastic deformations, and formula [9.129] in which we introduced the values of the deformations [9.151] allows us to find the new constant C_1 in this state of equilibrium [9.155].

Formula [9.131] applied between radii a and c gives the formula of the pressuremeter curve with a single plastic zone [9.156] with a value δ given by [9.147].

$$C_1 = \frac{K_0.\gamma\prime.z.(1-N).(1-n)}{2.\mu.(1+N)} \qquad [9.155]$$

$$Ln\left[\left({}^{u_a}/_{a}\right).(1+n) - C_1\right] = \delta.Ln(p) - \delta.Ln\left(\frac{2.K_0.\gamma\prime.z.}{(1+N)}\right) + Ln\left[\frac{K_0.\gamma\prime.z.(1-N).(1+n)}{2.\mu.(1+N)}.-C_1\right] \qquad [9.156]$$

9.5.3.11. *Condition for the development of a vertical plasticity*

The distinction between the two possible behaviors is made on the value of the radial constraint for radius c. For the second case of behavior, for it to be that a single plastic zone exists the radial constraint at plastic radius c must surpass the value of the vertical constraint under compression

($\sigma_{rc} > \sigma_z$) or even condition [9.158]. This relation was previously found by Wood *et al.* [WOO 77] to have a single plastic zone:

$$\frac{2.K_0.\gamma'.z.}{(1+N)} \geq \gamma'.z \qquad [9.157]$$

$$K_0 \geq \frac{1}{1+sin\phi} \qquad [9.158]$$

9.5.3.12. *Creep pressure for the case of two plastic zones*

In this case, the circumferential stress enters into plasticity with the vertical stress and the corresponding radial stress is given by relation [9.152], which gives the creep pressure [9.159]:

$$p'_f = \frac{2.K_0.\gamma'.z.}{(1+N)} \qquad [9.159]$$

9.5.3.13. *Limit pressure for the case of two plastic zones*

When the volume injected corresponds to the initial volume, then the radial deformation u_a/a is equal to $\sqrt{2}-1$ and the pressure is equal to the conventional pressure limit. This value of the radial deformation can be introduced into relation [9.150] to find the expression of the pressure limit [9.160].

In this relation, the limit pressure is proportional to the depth of the test, which is logical in granular soil. However, this relation is considerably different from [9.161] by Amar *et al.* [AMA 91], which seems to set at a depth of dry terrain of 14–15 m.

$$p'_{LM} = \gamma'.z.\sqrt[\delta]{\frac{\left[\left((\sqrt{2}-1).(1+n)-C_1\right).2.\mu\right]}{[(1-K_0).(1+n).\gamma'.z-2.\mu.C_1]}} \qquad [9.160]$$

$$p'_{LM} = 250.2^{\left[(\phi-24)/4\right]} + K_0.\gamma'.z \qquad [9.161]$$

9.5.3.14. *Creep pressure for the case of a single plastic zone*

Plasticity occurs when the radial constraint reaches the limit value of σ_{rc}. Under these conditions, creep pressure becomes [9.162]:

$$p'_f = \frac{2.K_0.\gamma'.z.}{(1+N)} \qquad [9.162]$$

9.5.3.15. *Limit pressure for the case of a single plastic zone*

When the injected volume corresponds to the initial volume, the radial deformation u_a/a is equal to $\sqrt{2}-1$, and the pressure is equal to the conventional pressure limit. If we now consider equation [9.156], it can be transformed into [9.163]. In this relation, the pressure limit is proportional to the depth of the test, which is logical for a coarse soil:

$$p'_{LM} = \frac{2.K_0.\gamma'.z.}{(1+N)} \sqrt[\delta]{\frac{\left[\left((\sqrt{2}-1).(1+n)-C_1\right).2.\mu.(1+N)\right]}{[K_0.\gamma'.z.(1-N).(1+n)-2.\mu.C_1.(1+N)]}} \qquad [9.163]$$

9.5.3.16. *Measurement of friction angle Φ and controlling its value*

Determining friction angle Φ is done by logarithmic transformation of the pressuremeter curve. In relations [9.150] and [9.156], term C_1 is very small and can be neglected in front of $u_a/a.(1+n)$. Under these conditions, relations [9.150] and [9.156] show a linear relation between pressure logarithms and radial deformation (Figure 9.51). This property was previously found by Hughes *et al.* [HUG 77]. Knowing the geological nature of the soil and its plasticity allows us to determine the interparticle angle of friction Φμ by statistical analysis, with this value not depending on density but solely on the state of the contact with the grains (Figure 9.53 and 9.54). Table 9.4 indicates the usual values of interparticle angle of friction Φ_μ in relation to the plasticity index of the soil and its *Guide Technique de Référence*, GTR class [GUI 00] class. The measurement of slope δ is, therefore, a function of internal friction angle Φ with N by [9.125], of intrparticle angle of friction Φ_μ with n by [9.128] and dilatance ψ itself a function of Φ and Φ_μ by [9.120]. The measurement of δ associated with the knowledge of intergranular friction Φ_μ allows us to determine the friction angle Φ.

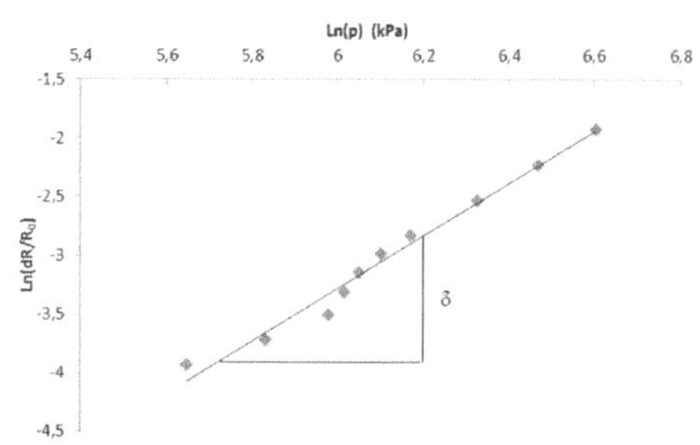

Figure 9.51. *Linear transformation of the pressuremeter curve; test at 9 m depth in the Lyon grave [MON 94]*

Monitoring this value is done by comparing (Figure 9.52) the experimental curve and the theoretical curve according to expressions [9.150] or [9.163]. In this calculation of the theoretical curve, we use the weight of the land soil under effective stress, and the elastic modulus is equal to the elastic modulus of unloading and reloading. The comparison between the experimental and theoretical curves allows us to validate the friction angle Φ and the elastic modulus. Calculating the theoretical curve (Figure 9.52) is done with only five parameters (E = 40MPa; ν = 0,3; σ_Z = 160 kPa, Φ = 34 ;Ψ = 4°). Gaiapress software [GAI 12] automates this calculation.

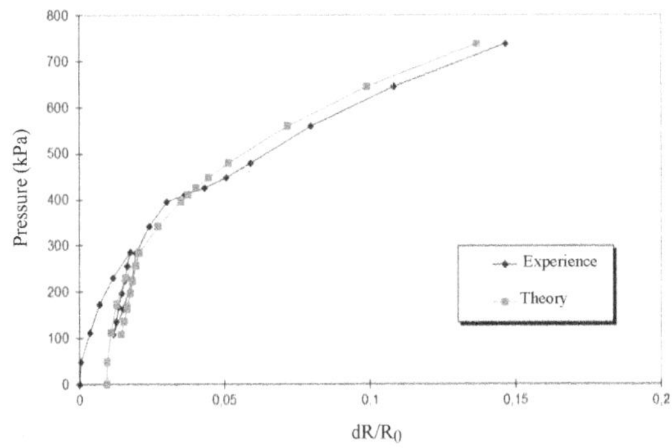

Figure 9.52. *Comparison of the experimental and theoretical curves: test at 9 m depth in the Lyon grave*

Another way of proceeding is the use of the pressure limit [9.163] and to place in the graph (Figure 9.55) the ratio of the shearing modulus G^e [9.104] to the effective horizontal pressure p'_0 as a function of the ratio of the effective limit pressure to the effective horizontal pressure p'_0. The graph (Figure 9.55), therefore, provides the friction angle for non-plastic soils. This determination is less precise than the previous method. An example of its use is shown for Lyon's metro station, in the Bellecour gravels (Figure 9.56) and gives similar results to direct shearing tests *in situ*.

9.5.3.17. *Conclusions*

We can draw several lessons from the previous equations:

– the relations show that plasticity appears early around the pressuremeter and before the creep pressure observed experimentally;

– the pressiometric modulus, taken on a virgin loading, is highly affected by an error caused by plasticity. It is recommended to use the elastic modulus deduced from the loading unloading cycle carried out during the transformed test;

– the linear relation as a logarithm can only be traced if there are enough points beyond the creep. The pressuremeter test must be carried out so that to obtain 4 or 5 creep points;

– if the creep is well marked out, and for granular soils, the theoretical interpretation allows us to find the internal friction angle of the soil in place (Figure 9.52).

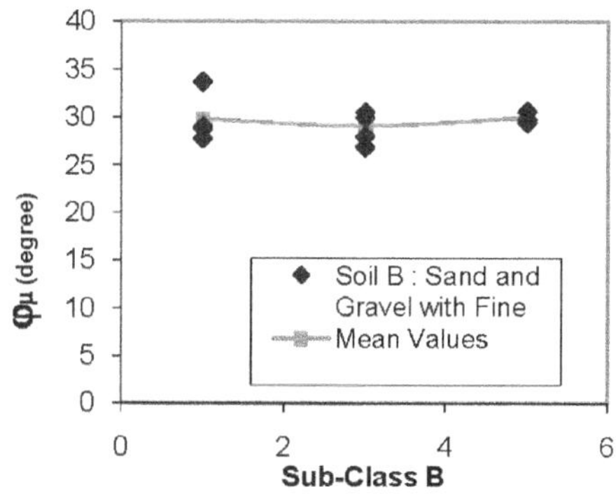

Figure 9.53. *Relation between the interparticle angle of friction and the GTR B subclasses [MON 02]*

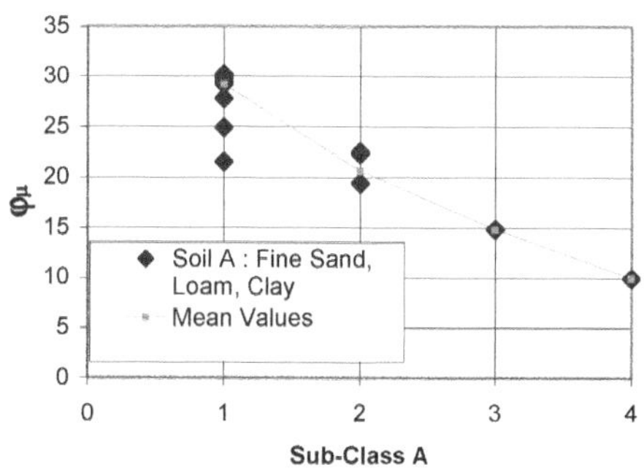

Figure 9.54. *Relation between the interparticle angle of friction Φ_μ and the GTR A subclasses [MON 02]*

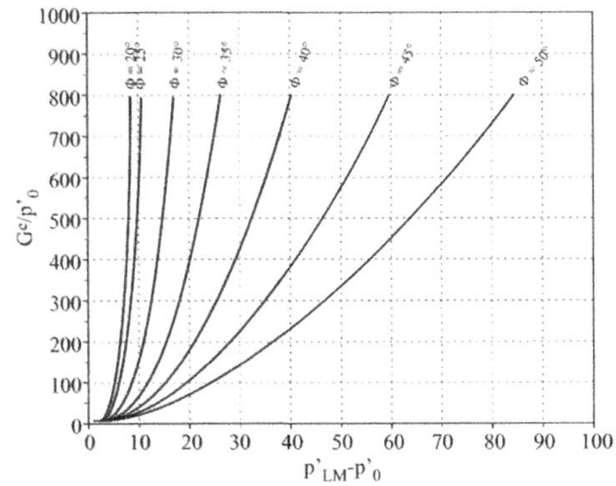

Figure 9.55. *Relation between the elastic modulus and the limit pressure as a function of the internal friction angle [MON 13]*

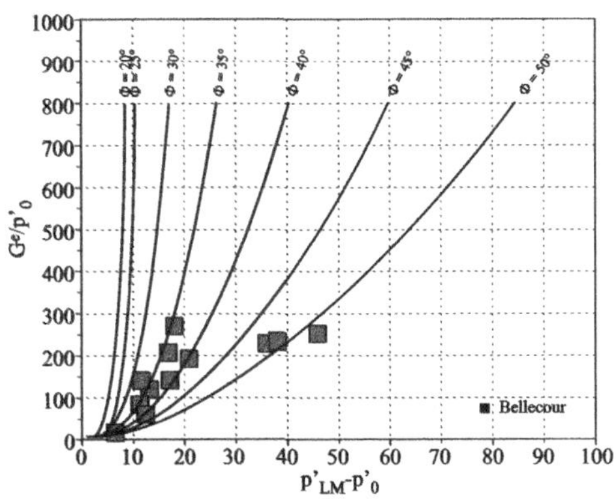

Figure 9.56. *Friction interpretation of the pressuremeter tests of Place Bellecour [MON 13]*

Class of soil	Ip (%)	Φ_u°	Φ_{cv}°
Sand, gravel	–	29°	29,4°
Silt, class A1	0-12	28°	28,3°
Low plasticity clay, A2	12-25	20°	20°
Plastic clay, A3	25-40	15°	15°
Very plastic clay, A4	> 40	10°	10,1°

Table 9.4. *Average value of intergranular friction [MON 13]*

10

Water Tests in Soils

Knowledge of the hydrodynamic characteristics of a soil, i.e. its water storage capacity, void ratio and permeability, is necessary to solve the following problems:

– stability of installations exposed to water (dykes, dams, natural slopes, etc.);

– design of catching constructions (wells, galleries, etc.);

– preliminary estimate of drainage flows of an excavation (trench drying, etc.);

– impact of catchments works on the area (lowering of the water table, etc.);

– long-term evolution of a catchment.

To deal with these problems, *in situ* water tests are carried out; these are more representative of the real operating conditions of the installation than those carried out in the laboratory. As a result, either one of the following is carried out:

– a punctual test (Lefranc and Lugeon);

– a test involving an important sector (pumping and drawdown).

A detailed description of these tests can be found in [CAS 80] and [CAS 05].

10.1. Punctual water tests

10.1.1. *Infiltrometer test*

The single or double ring infiltrometer test measures permeability at the surface of the soil for medium permeability soils (k $\geq 10^{-9}$ m/s) for open-type tests [AFN 98], to low permeability soils (k $\geq 10^{-10}$ m/s) for closed-type tests [AFN 99a].The depth of investigation is approximately 10 cm.

10.1.1.1. *Principle of the test*

The test involves injecting water at the surface from a single or double ring, which is either in the open air for the open-type (Figure 10.1), or under pressure for the closed-type (Figure 10.2).

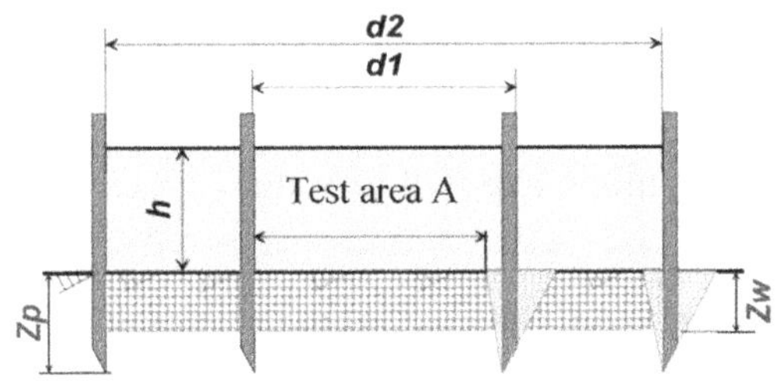

Figure 10.1. *Open-type double ring infiltrometer test [BER 02]*

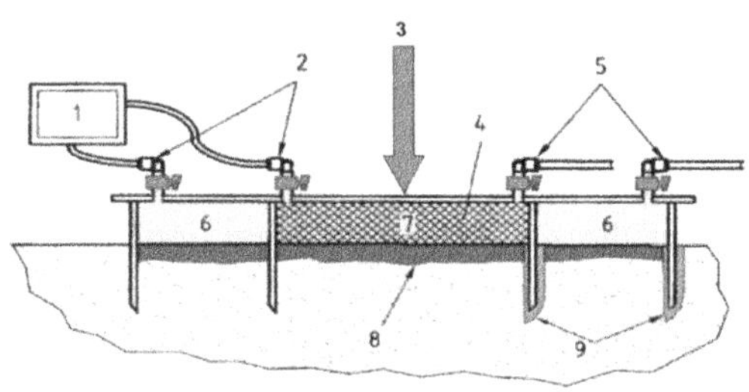

Figure 10.2. *Closed-type double ring infiltrometer test [BER 02]*

The double ring device is equipped with a guard ring with the same hydraulic load as the internal ring. This external ring maintains the vertical flow of the internal ring and limits the risk of leaks under the internal ring. An example of a double ring infiltrometer test is shown in Figure 10.3.

Figure 10.3. *Example of an open-type double infiltrometer test – Panda II [BER 02]*

10.1.1.2. *Execution of the two tests*

10.1.1.2.1. Execution of the open-type test

This test is carried out in the following manner:

– the test area is leveled;

– the ring is sunk using jacks (Figure 10.4, top);

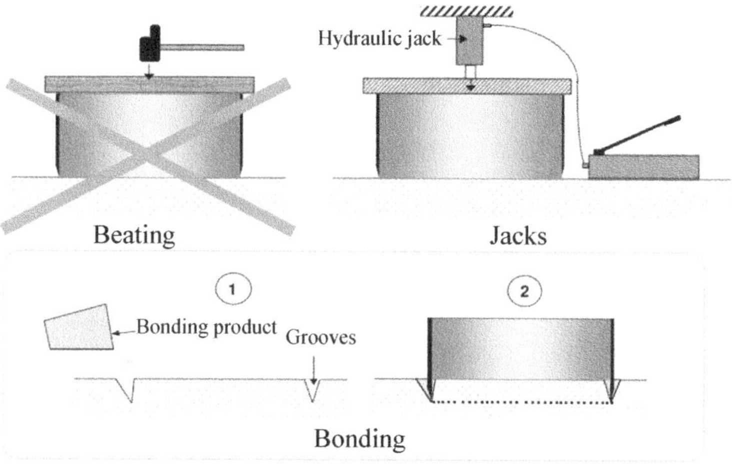

Figure 10.4. *Installing the rings [BER 02]*

– otherwise, grooves are made for the insertion of the ring, and this is sealed using a bonding product (Figure 10.4, bottom);

– the soil is saturated using a constant load of water over the timeframe indicated (Figure 10.5) and from a height limited to between 10 and 30 cm;

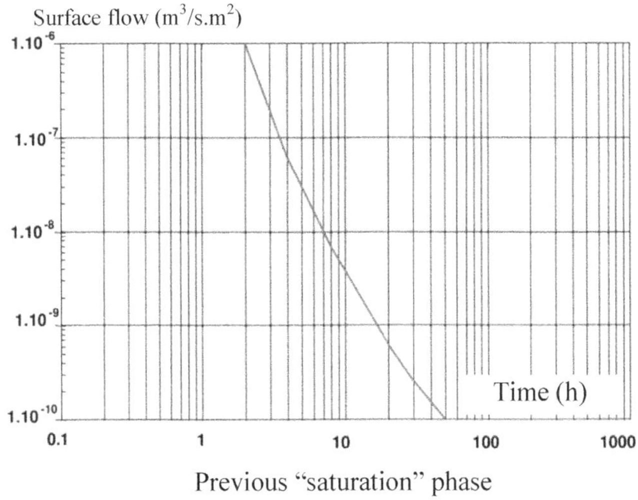

Figure 10.5. *Duration of the saturation phase [AFN 98]*

– when the permanent flow state is established, the water supply is maintained and the constant flow obtained during the indicated timeframe is noted (Figure 10.6);

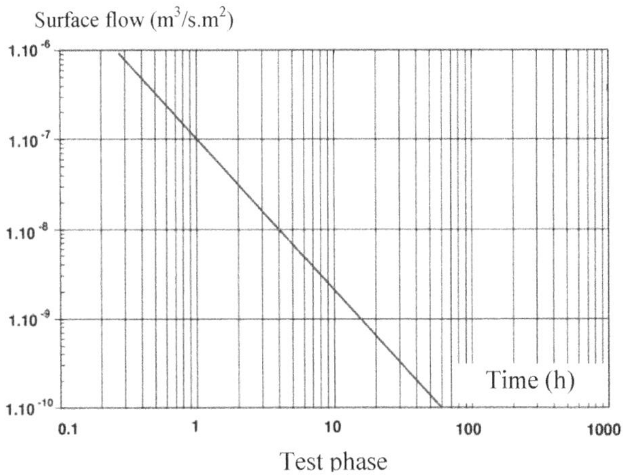

Figure 10.6. *Duration of the test phase [AFN 98]*

– at the end of the test, the device is taken apart;

– the soil under the device is sampled to determine the thickness z_w of the saturated zone.

10.1.1.2.2. Execution of the closed-type test

This test is executed in the following way:

– preparation and setup are identical to the open test;

– the soil is saturated with a constant load of water h_1 over the duration indicated (Figure 10.7) and from a height limited to 1 m;

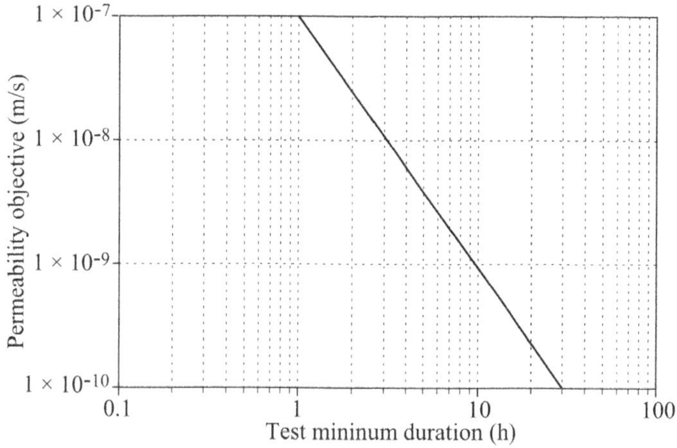

Figure 10.7. *Duration of the watering phase [AFN 99a]*

– a permanent rate is obtained and we note ΔV_1 and Δt_1 to have a flow of Q_1;

– a second constant load of water h_2 is applied, greater than h_1 and the same process is followed to measure ΔV_2 and Δt_2 so as to obtain flow Q_2;

– a third constant load of water h_3 is applied, greater than h_2 and the same process is followed to measure ΔV_3 and Δt_3 so as to obtain flow Q_3;

– the device is disassembled;

– some soil is sampled to measure depth z_w of the saturation front.

10.1.1.3. *Interpretation*

10.1.1.3.1. Interpretation of the open-type test

Darcy's law [10.1] provides the relation between the flow rate v, permeability k and the hydraulic gradient i. By replacing v with ratio Q/A between flow Q and the straight section of the infiltrometer A, we obtain relation [10.2]. On the terrain, we have observed that the depth of the saturated zone is z_w (Figure 10.1). The load being equal to zero on the saturation front, we can deduced the hydraulic gradient i using relation [10.3], and the surface permeability using relation [10.4]:

$$v = k.i \qquad [10.1]$$

$$Q = k.A.i \qquad [10.2]$$

$$i = \frac{h+z_w}{z_w} \qquad [10.3]$$

$$k = \frac{Q.z_w}{A.(h+z_w)} \qquad [10.4]$$

10.1.1.3.2. Interpretation of the closed-type test

The hydraulic gradient of each load step is calculated following relation [10.3]. The graph is traced linking flow Q_i of each step to the corresponding gradient i_i (Figure 10.8). The value of the intersect to the ordinate on a bi-logarithmic scale provides the permeability coefficient k of the ground.

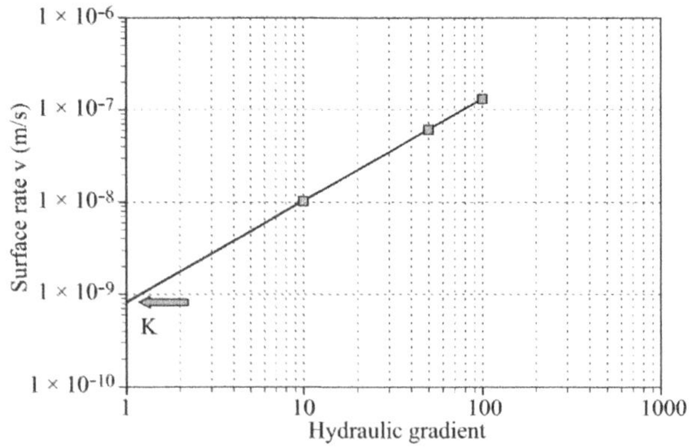

Figure 10.8. *Determination of the permeability coefficient for a gradient of 1 [AFN 99a]*

10.1.2. *Lefranc permeability test*

The Lefranc test is covered by standard NFP94-132 [AFN 00b]. It is an injection test (or pumping test) of water into the ground using gravity (Figure 10.9), or with an injection pump by boring (Figure 10.10) into a soil underneath the water table.

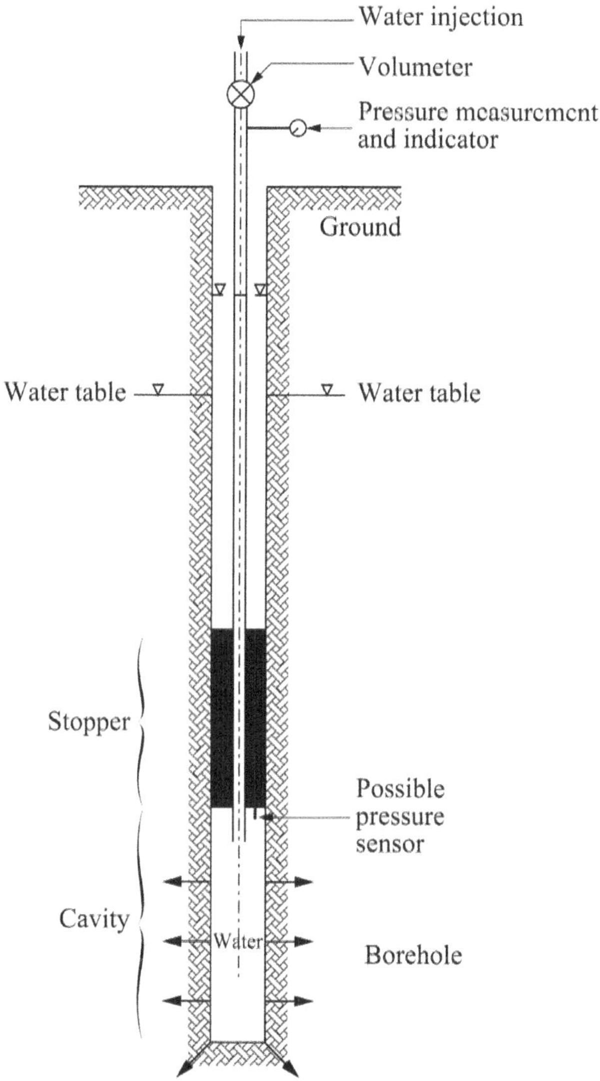

Figure 10.9. *Principle of the Lefranc test*

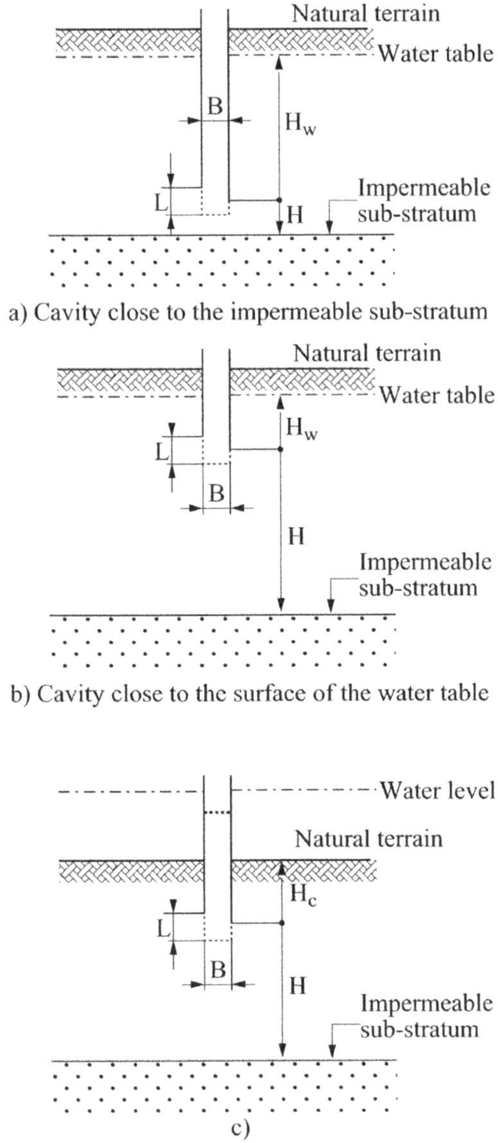

Figure 10.10. *Notation used for the Lefranc test [AFN 00b]*

10.1.2.1. *Principle of the test*

Inside the permeable soil and under the water table, we build a cavity (of any shape) that is invariable. The dimensions of the cavity must be known.

This is done at the bottom of a borehole and the cavity is arranged so that filtration only takes place in the walls, and not through the borehole. Water is removed (for depths of less than 9 m) or injected, preferably, from the cavity. The hydraulic head is set at a fixed quantity. Under these conditions, the percolation flow is written [10.5], with Q the removed or injected punctual flow, k the permeability coefficient of the soil, h the load variation caused by removal or injection, B the diameter of the pumping or injection cavity and m the shape coefficient of the cavity.

The shape coefficients can take on the values shown in (Table 10.1). If the location is close to the aquifer domain (surface of the table and impermeable wall), m is replaced in formula [10.5] by m_0 shown in formula [10.6], and in Table 10.2, with the given notation (Figure 10.10) with H_w the distance from the cavity under the level of the water table, H the distance from the cavity to the impermeable wall and L the length of the cavity.

Shape of the cavity C = L / B	Value of m
Elongated ellipsoid C > 10	$\frac{2.\pi.C}{Ln(2.C)}$
Ellipsoid 1.2 <= C <=10	$\frac{2.\pi.C}{Ln\left(C+\sqrt{C^2+1}\right)}$
Spherical 0.7 <= C < 1.2	$2.\pi.\sqrt{C+{}^1/_4}$
0.3 <= C <0.7	$\pi.\sqrt{2.C+{}^1/_2}$
C < 0.3	$\frac{\pi.\sqrt{1-4.C^2}}{2.ArcTan\left(-2.C+\sqrt{4.C^2+1}\right)}$
Plane, the cavity corresponds to the end of the tubed borehole	2

Table 10.1. *The values of m for different cavity shapes*

$$Q = m.k.h.B \quad [10.5]$$

$$\frac{1}{m_0} = \frac{1}{m} - \frac{B}{8.\pi.H_W} + \frac{B}{8.\pi.H} \quad [10.6]$$

Type of cavity	Value of m_0
Cavity far from the limits of the aquifer	$m = m_0$
Cavity close to the impermeable substratum (Figure 10.10(a))	$1/m = 1/m_0 = B/(8.\pi.H)$
Cavity close to the surface of the water table (Figure 10.10(b))	$1/m = 1/m_0 = B/(8.\pi.H_w)$
Cavity close to the surface of the soil located under the water table (Figure 10.10(c))	$1/m = 1/m_0 = B/(8.\pi.H_c)$

Table 10.2. *The values of m_0 as a function of the limits of the aquifer*

10.1.2.2. *Execution of the tests*

The following conditions must be respected:

– the tests are done under the water table;

– the positions of the superior boring edge and the initial water level are measured with a precision of 1 cm;

– the position of the test is measured with a precision of 10 cm;

– the water level must be constant during the measurement of a flow;

– the water level is variable during the observation of the rising or falling of the level of the water table within the borehole;

– the installation of the cavity is controlled as it is essential to the proper execution of the test.

To create the cavity, we must operate in the following manner:

– with a strainer, to stop the penetration of small elements. The strainer is located roughly 1.5 m above the end of the borehole (Figure 10.11(b)). The tube is closed off with a stopper;

– with a calibrated gravel filter (between 5 and 10 mm), more permeable than the soil (Figure 10.11(a)). The tube is elevated by a given height h, the diameter of this cavity then corresponds to that of the boring tube (Figure 10.11(c));

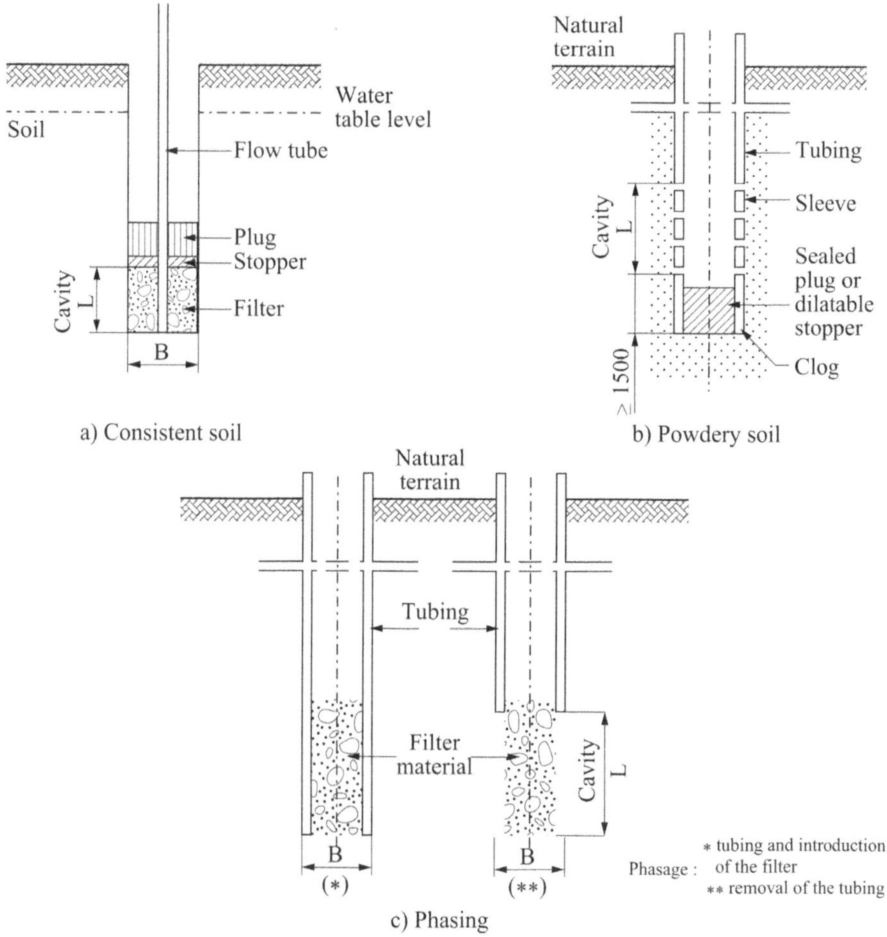

Figure 10.11. *Lefranc test, creation of the cavity: a) with a calibrated gravel filter; b) with a perforated sleeve; c) realization phasing [AFN 00b]*

– the level of the water table in the tube is left to rise. For this, the test is prepared in the evening, ready for the next morning;

– we check that the height of the pocket of test h is less than 10 diameters.

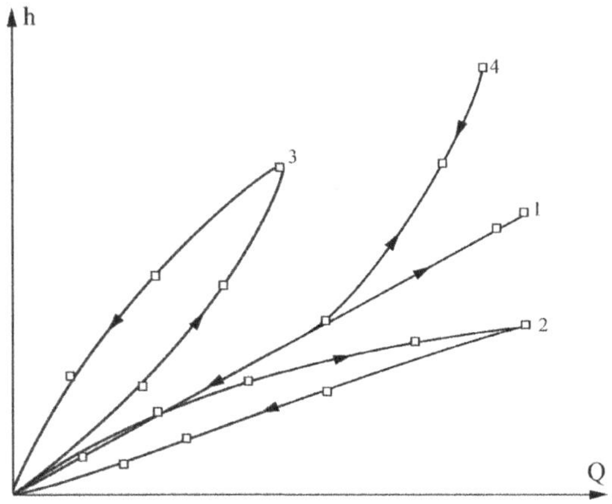

Figure 10.12. *Lefranc test, determination of permeability; clogging and unblocking effects [RAT 68]*

10.1.2.2.1. Execution of the test at a constant level

This test is represented in Figure 10.9. The test is carried out as follows:

– a constant hydraulic overload of the water table is established, by pouring a known volume of fresh water into the tubing (for very permeable terrain, we reduced the height of the hydraulic overload);

– a small stabilized flow is sent into the borehole so that the water level in the borehole increases by at least 10 cm during the first minute;

– we wait for the stabilization of the level and we mark down the flow and the level;

– we start over again for stabilized flows $2.Q_1$, $3.Q_1$, ...;

– we mark the static level after the feed is stopped;

– the total rising of the water cannot be more than 2 m;

– the flows are measured to the nearest 3%.

10.1.2.2.2. Execution of the variable load test

This type of test is less precise, and the interpretation is more complex as a result than for the constant level test. Measurements can be made using the Brillant device (Figure 10.13). This test is often carried out by pumping. The following operations are carried out:

– the pumping, or the injection, is stopped (a variation involves keeping a constant flow Q_a);

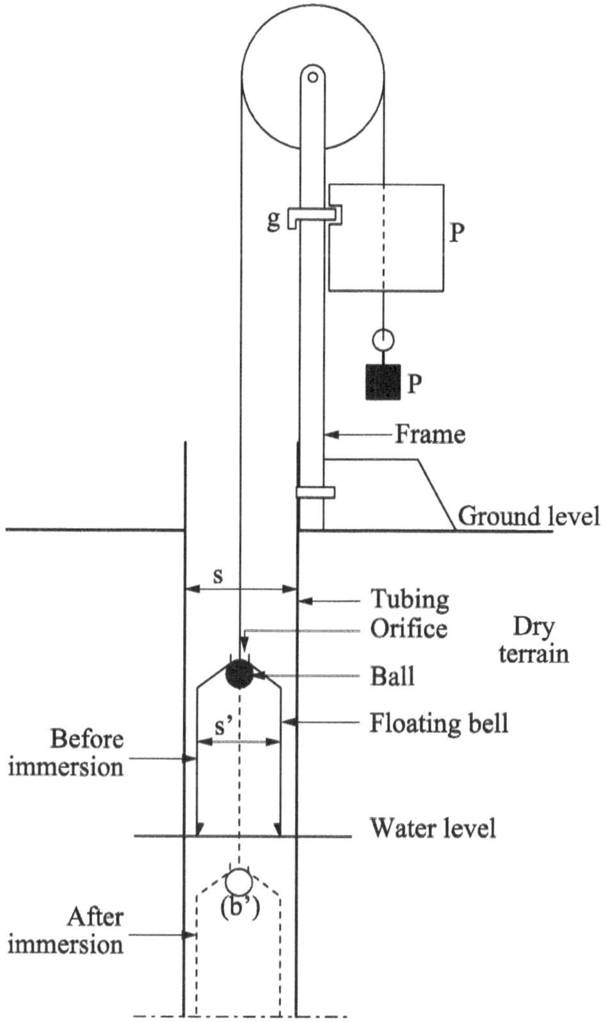

Figure 10.13. *Lefranc test, diagram of the Brillant device [CAS 80]*

– the water level is noted when the flow is stopped, then 30 s later, 1 min later and then every minute over a period of time equal to half of the duration of the constant (not 0) flow. The measurements are continued every 5 min after 20 min of measurement, and until three successive values are obtained that are not different by more than 1 cm between them. If this does not happen, the test is stopped after 1 h.

10.1.2.3. *Interpretation of the tests*

10.1.2.3.1. Interpretation of the constant level test

The hydraulic overload Δh is traced as a function of flow Q (Figure 10.12). The slope of the line obtained is equal to m.k.B, following relation [10.5] and allows us to determine permeability k. In general, we can obtain the following curves:

– an experimental line (Figure 10.12, (1)) with a proper test, allowing us to calculate the permeability k;

– after a proper start, the cavity is unblocked (Figure 10.12, (2));

– the strainer is gradually blocked (Figure 10.12, (3));

– the hydraulic gradient becomes too great and Darcy's law no longer applied (Figure 10.12, (4)).

10.1.2.3.2. Interpretation of the variable load test

In the case of the variable load test, the initial unloading (or overload for an injection) is written as H, and h is the value of the unloading at time t. It can be written that in the time interval dt, the level varies by dh in a borehole of section S. The flow that passes through the cavity under a residual drop of h is then written [10.7], which, considering [10.5], is transformed into equation [10.10]:

$$Q = Q_a - S.\frac{dh}{dt} \qquad [10.7]$$

$$Q_a - S.\frac{dh}{dt} = m.k.h.B \qquad [10.8]$$

$$\frac{Q_a}{S} - \frac{dh}{dt} = \frac{m.k.B}{S}.h \qquad [10.9]$$

$$h = \left(\frac{Q_a}{S} - \frac{dh}{dt}\right) \cdot \frac{S}{m.k.B} \qquad [10.10]$$

$$\frac{dh}{h} = -\frac{4.m.k}{\pi.B} . dt \qquad [10.11]$$

In the case where flow Q_a is equal to zero, the traced relation of the rate of the rising of the water in the borehole – with a circular section S whose value is $\pi.B^2/4$, the value of the unloading h at a time t – is linear following relation [10.11]. Relation [10.11] can be integrated as a natural logarithm [10.14] and as a common logarithm [10.15]. On a semi-logarithmic diagram, if we trace the relation between the natural logarithm of the ratio h/H over time t, we obtain a line (Figure 10.14) for which the slope is 4.m.k/π.B, which allows us to determine the value of the permeability k. Unfortunately, it can be noted that the points that should align are located on a curve, which causes the value of k to vary considerably. If this problem does occur, the relation between the ratio h/H and the inverse of time 1/t must be traced. This curve presents a point of inflexion, with a tangent at this point whose slope is V/(4.m.k). An alternative solution involves using the complete equation of the transitional system [10.16] and to calculate k by overlay the theoretical curve on the experimental points (Figure 10.15).

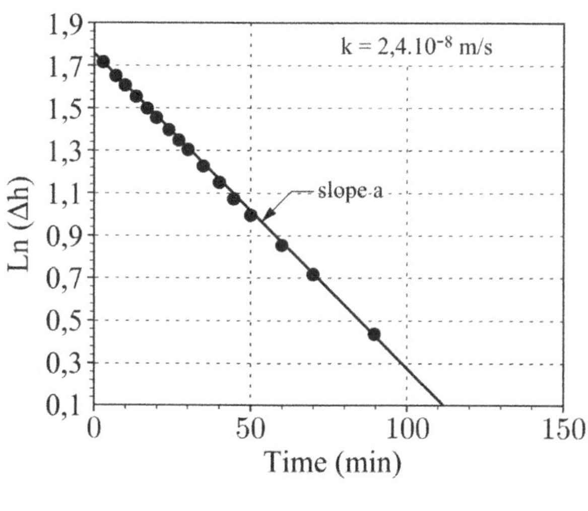

Figure 10.14. *Lefranc test, measurement of the variable load permeability [BER 02]*

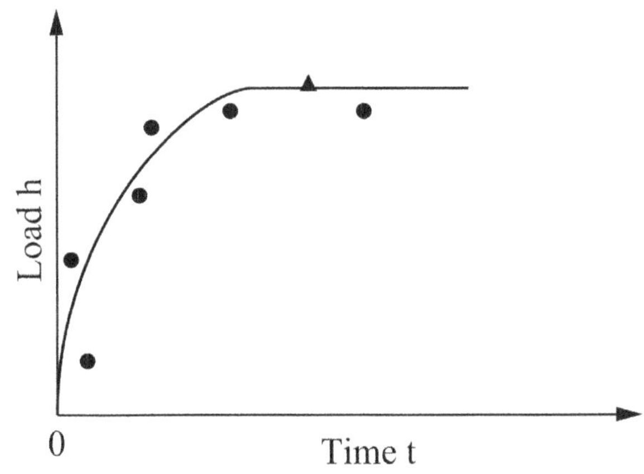

Figure 10.15. *Lefranc test, overlay of the theoretical curve on the experimental points [AFN 00b]*

In the case where the flow Q_a is not equal to zero (Figures 10.16 and 10.17), we can solve the equation graphically [10.10]. The experimental line intersects the axes at the coordinates shown in [10.12] and [10.13] and the slope of this line is equal to 4.m.k/π.B, allowing us to calculate k.

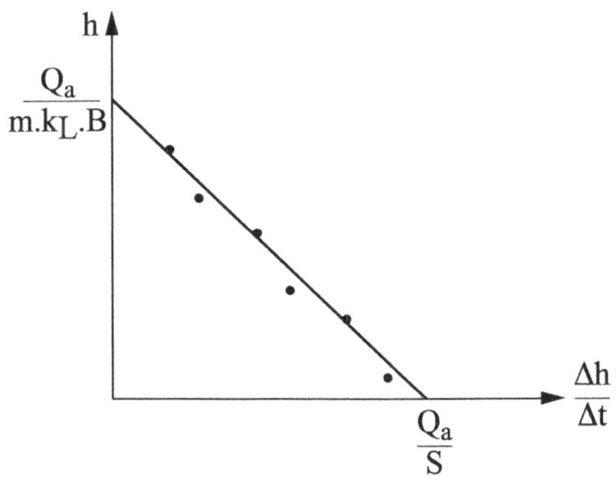

Figure 10.16. *Lefranc test, measurement of the permeability with a constant removal Qa [AFN 00b]*

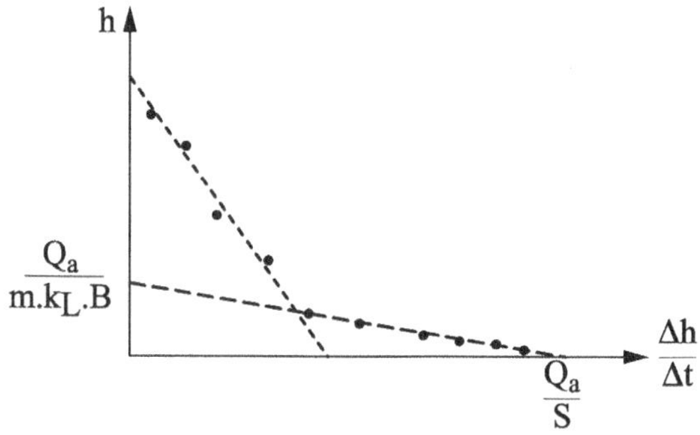

Figure 10.17. *Lefranc test, measurement of the permeability with a constant removal Qa and clogging [AFN 00b]*

$$h = \frac{Q_a}{m.k.B} \quad for \quad dh/dt = 0 \tag{10.12}$$

$$h = 0 \quad for \quad dh/dt = Q_a/S \tag{10.13}$$

$$Ln\left(\frac{h}{H_{ini}}\right) = -\frac{4.m.k}{\pi.B}.t \tag{10.14}$$

$$Log\left(\frac{h}{H_{ini}}\right) = -\frac{4.m.k}{2{,}303.\pi.B}.t \tag{10.15}$$

$$H(t) = \frac{Q_a}{m.k_L.B} - \left[\frac{Q_a}{m.k_L.B} - h_0\right].e^{\left(-\frac{t}{t_e}\right)} \tag{10.16}$$

$$t_e = \frac{S}{m.k_L.B} \tag{10.17}$$

10.1.3. *Permeability test in borehole current section*

The drilling permeameter was invented by Louis Ménard 1958 and the permeafor was invented by Paul Ursat in 1980.

10.1.3.1. *Principle of the test*

The drilling permeameter test involves injecting water into the cavity from a section of the borehole (Figure 10.18). This test assumes that the soil is consistent enough for the boring to take place along the length of the probe.

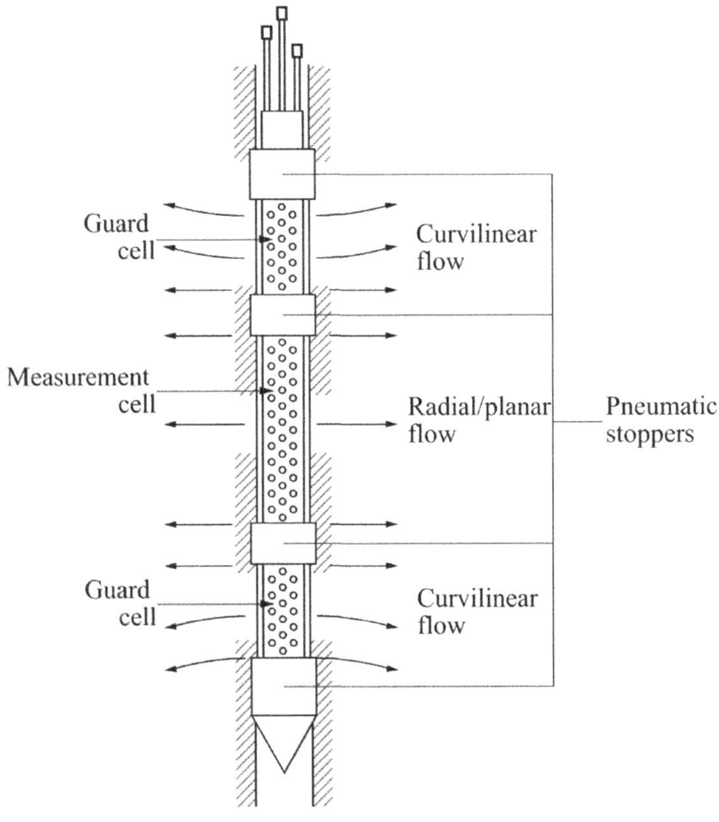

Figure 10.18. *Diagram of the principle of the permeametric probe [CAS 80]*

The permeafor test takes place by sinking the probe into the ground using jacks (Figure 10.19). It is not subject to the condition of the low consistency soil, but it is limited to low granulometry soils to avoid a failure of penetration.

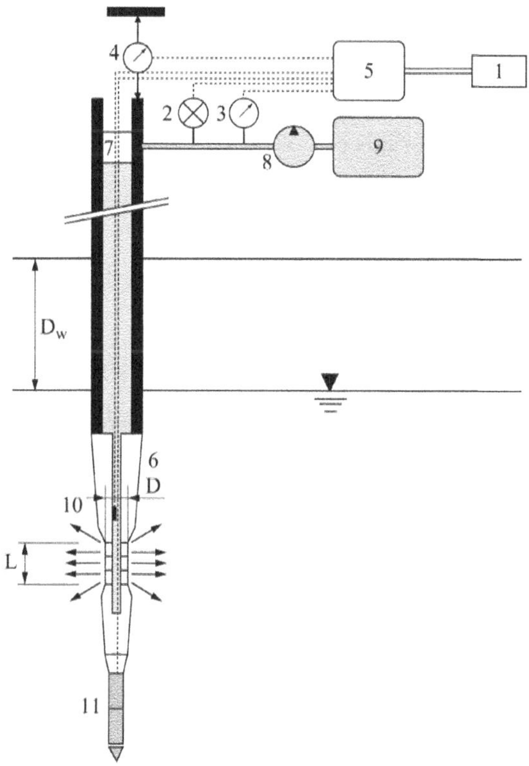

Figure 10.19. *Diagram of the principle of the permeafor [DOR 10]*

10.1.3.2. *Execution of the test*

10.1.3.2.1. Execution of the drilling permeameter test

The test can only be carried out in a terrain with enough consistency to allow the walls of the borehole to hold over a length at least equal to the length of the probe. Deaerated water is used for the test. The operations are carried out in the following order:

– filling of the reserves and various circuits;

– positioning of the probe at the chosen level;

– reading of the pressures p_0 written on the gauges of the two pneumatic stoppers;

– inflation of the stoppers at a pressure corresponding to half of the pressure limit of the soil p_{LM} with a maximal value of 0.5 MPa;

– water is injected at a constant pressure step and the volumes injected are read at equal time intervals (15 or 30 s);

– the measurements are continued until six points are obtained, corresponding to six equal flows;

– two other series of measurements are carried out, with two different increasing pressure steps;

– the water is depressurized and the pneumatic stoppers deflated;

– the pressure in the stoppers is noted, and must be equal or close to the initial pressure p_0.

10.1.3.2.2. Execution of the permeafor test

This test is carried out in terrains with low levels of granulometry for the sinking to be possible. The following operations are carried out:

– the depth of the water table D_w is measured;

– the probe is beaten into the ground down to the desired depth;

– during the beating, the hydraulic load H is applied to the injection strainer;

– every 20 cm, the beating is paused and a water test is carried out by exerting a constant hydraulic load H_i on the injection strainer over a time period of 5–10 s;

– the water flow Q_i is measured over the duration of the injection;

– the beating is resumed to reach the next level.

10.1.3.3. *Interpretation*

10.1.3.3.1. Interpretation of the drilling permeameter test

This test allows us to determine the permeability of the soil at a given boring level. We assume that around the injection cell, the current lines are horizontal (Figure 10.20). The flow is then revolving around the drilling axis. Near the borehole, the equipotential surfaces are revolving cylinders. However, far from the probe, the flow seems to come from a punctual source. Ménard suggested:

– a cylindrical network close to the probe;

– a spherical network far away;

– the connection zone is made of a ring with a medium size radius α.L with α being an empirical coefficient and L the length of the injection cell.

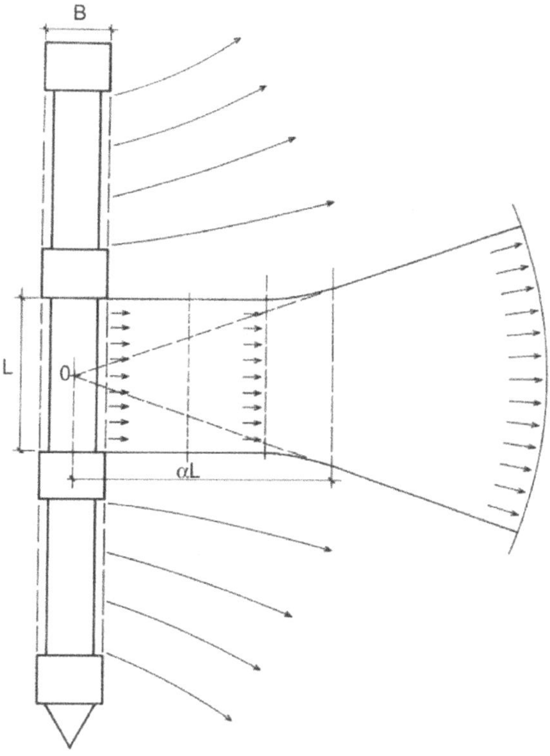

Figure 10.20. *Simplified flow network [CAS 80]*

In the cylindrical zone, loss of load between two equipotentials of radius r and r+dr is dh and corresponds to the hydraulic gradient i [10.18]. With Darcy's law, we can find the flow going through the equipotential of radius r [10.19], with the following conditions at the limits: on the borehole of radius r_0 a fixed load of $h_0 = p/\gamma_w$, on the equipotential of radius α.L, a load of $h_1 = p_1/\gamma_w$. This last relation can be integrated in [10.20].

In the spherical zone, the flow going through the equipotential surface of radius ρ is given by [10.21], with the following condition at the limits: on the sphere of radius e and the borehole of radius α.L a fixed load of $h_1 = p_1/\gamma_w$, at infinity a load of $h_\infty = p_0/\gamma_w$. By integrating this relation between α.L an infinity, we obtain relation [10.23].

Pressure p_1 can be eliminated between relations [10.20] and [10.22] and this results in relation [10.23]. Ménard proposed giving σ the value of 1, resulting in an error rate of less than 10% for the determining of the value of permeability k using relation [10.24].

Considering a launch of $\lambda = L/B$, the previous relation can be written following equation [10.25], with $B = 2.r_0$. We can note that ratio $(p - p_0/\gamma_w)$ has the dimension of a length. Under these conditions, equation [10.25] has exactly the same shape as the general relation of the Lefranc test [10.5] with form coefficient of the cavity m equal to [10.26]:

$$i = -\frac{dh}{dr} \qquad [10.18]$$

$$Q = -2.\pi.k.L.r.\frac{dh}{dr} \qquad [10.19]$$

$$\frac{Q}{2.\pi.k.L} Ln\left(\frac{\alpha.L}{r_0}\right) = \frac{p-p_1}{\gamma_w} \qquad [10.20]$$

$$Q = -4.\pi.k.\rho^2.\frac{dh}{dr} \qquad [10.21]$$

$$\frac{Q}{4.\pi.k.\alpha.L} = \frac{p_1-p_0}{\gamma_w} \qquad [10.22]$$

$$\frac{Q}{2.\pi.k.L}\left[Ln\left(\frac{\alpha.L}{r_0}\right) + \frac{1}{2.\alpha}\right] = \frac{p-p_0}{\gamma_w} \qquad [10.23]$$

$$k = \frac{\gamma_w.Q}{4.\pi.L.(p-p_0)}\left[1 + 2.Ln\left(\frac{L}{r_0}\right)\right] \qquad [10.24]$$

$$Q = \left[\frac{4.\pi.\lambda}{2.Ln(2.\lambda)+1}\right].\left(\frac{p-p_0}{\gamma_w}\right).k.B \qquad [10.25]$$

$$m = \frac{4.\pi.\lambda}{2.Ln(2.\lambda)+1} \qquad [10.26]$$

10.1.3.3.2. Interpretation of the permeafor test

The permeafor can be compared with a Lefranc test, but it is not able to reach a stabilized hydraulic system. It is, therefore, limited to the detection of variations of permeability (Figure 10.21) in an investigation volume corresponding to the size of the strainer (roughly 5 cm). Interpretation is limited to the calculation of the ratio between the flow Q and the static load applied H (a function of the height of the above ground reserve H_e, the depth of the water table D_w and the loss of load dH). H' is calculated using [10.27] and the result of the test is calculated with [10.28].

$$H' = H_e + D_w - dH \quad [10.27]$$

$$result = \frac{Q}{H'} \quad [10.28]$$

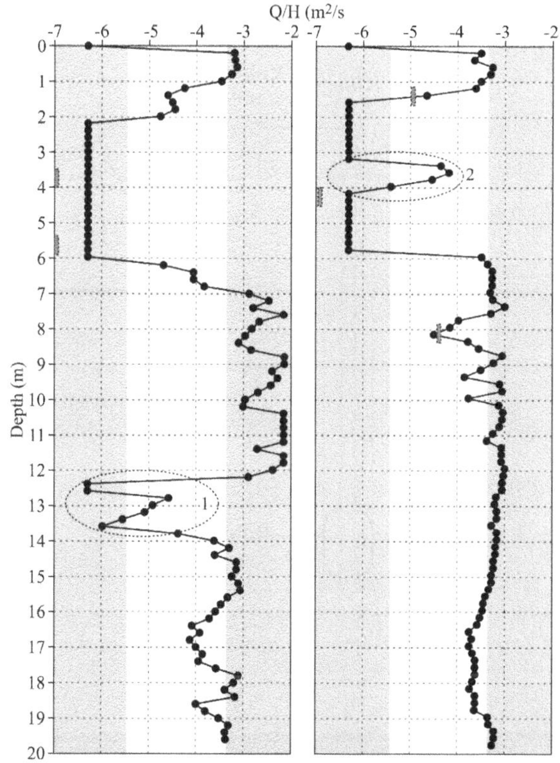

Figure 10.21. *Example of a permeafor result [FAR 92]*

10.1.4. *Lugeon permeability test*

The Lugeon test is covered by standard NFP94-131 [AFN 94]. It is an injection test in rocks or consistent soils able to support considerable water pressures during the test.

10.1.4.1. *Principle of the test*

The test involves injecting water in a cavity (Figure 10.22) by regulating the injection pressure p through a pressure pumping system. The test can be carried out below the level of the water table. Relation [10.29] allows us to determine the permeability coefficient k, as well as the injected or removed punctual flow Q, the shape coefficient m, the static injection pressure p, the boring diameter B and the specific weight of the water γ_w.

$$Q = \frac{m.k.p.B}{\gamma_w} \qquad [10.29]$$

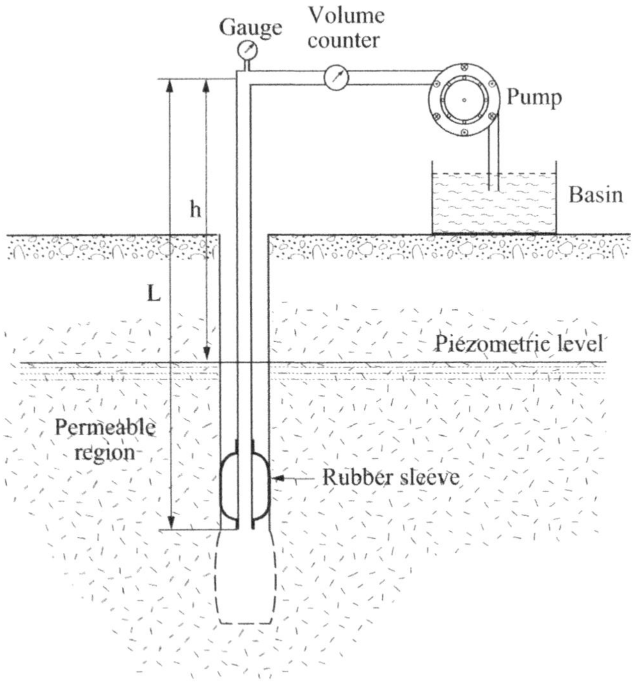

Figure 10.22. *Principle of the Lugeon test*

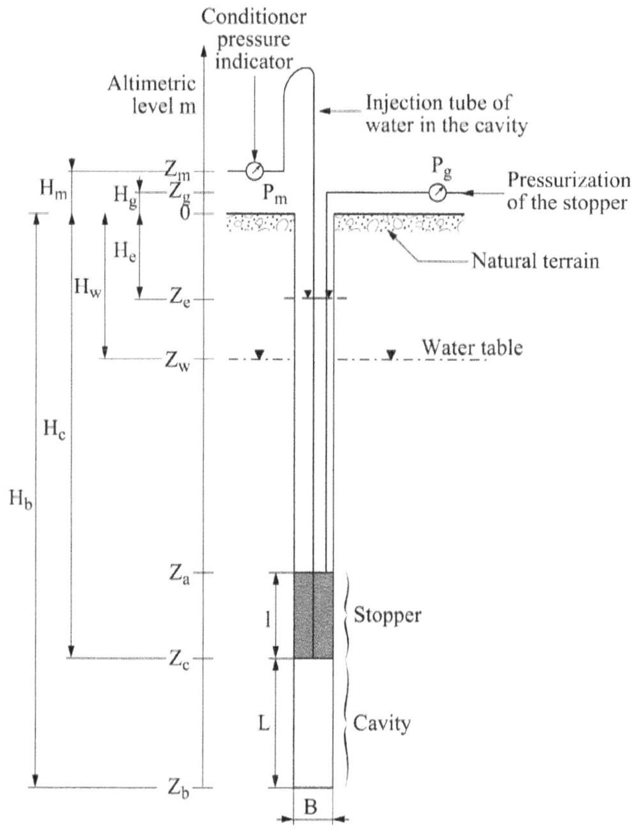

Figure 10.23. *Notation used for the Lugeon test [AFN 94]*

10.1.4.2. *Execution of the test*

10.1.4.2.1. Measurement system

The following conditions are respected:

– the position of the injection pocket is measured with a precision of 1 cm;

– stoppers of at least 1 m and with lengths greater or equal to 10 times the bore diameter delimit the injected zone;

– the obturator allows us to reach dilatation pressures of at least 2 MPa;

– its inflation pressure (deduced inherent resistance) must be greater by at least 500 kPa to the maximal pressure present in the injection cavity;

– the diameter of the test pocket is greater or equal to 0.7 times the bore diameter;

– the height of the test pocket is of 3–5 m;

– the injection pump allows for a flow of 100 l/min under a pressure of 1.1 MPa;

– the flowmeter must have a precision of at least 5%;

– the pressure sensor must have a precision of at least 5%;

– at the surface, a water reserve of fresh water of at least 1 m^3 is available;

– injection is done with a maximal pressure p_m (often of 1 MPa);

– the water level in the borehole is brought back up after stabilization (less than 2 cm of variation over 5 min) with a precision of 1 cm.

10.1.4.2.2. Building the cavity

In the absence of specific indications from the geotechnician, the height of the cavity is 3 m and its diameter is between 75 and 140 mm. The following conditions must be respected:

– the bore fluid is fresh water or air;

– the borehole is cleaned by washing before the test so as to eliminate particles caused by the perforation;

– isolation of the superior section of the boring is carried out using a dilatable stopper;

– to avoid the water bypassing the stopper, it must be dilate under a greater pressure p_g [10.30], after correction of its inherent resistance (p_e) by at least 0.5 MPa to the maximal water pressure that will be present in the test pocket:

$$p_g - p_e \geq p_0 + p_{max} + (z_m - z_g).\gamma_w \qquad [10.30]$$

where p_0 = 0.5 MPa;

– the water is left to flow out through the injection tube;

– during this phase, the dimensions of the cavity are determined (diameter B and length L).

10.1.4.2.3. Determining the pressure loss in the tubes

The tube is calibrated as follows before being lowered into the borehole:

– the injection water is passed through the tube by placing the free end at the same height as the conditioner-pressure indicator;

– 1 min after the flow becomes stabilized, the pressure Δp_c is noted, as is the corresponding flow Q;

– the operation is repeated for at least four flows;

– the calibration curve is traced $\Delta p_c = f(Q)$ with a minimum of five points for the range of usage of the pump;

– the loss of pressure in the tubing between the sensor and the test cavity is checked to make sure it is less than 25% of the value read on the pressure sensor.

10.1.4.2.4. Conducting the measurements

The following conditions must be respected:

– p_m is reached by five increasing steps, measured at the surface: 0.2, 0.4, 0.6, 0.8 and 1 MPa; alternatively, we can go down from p_m with four decreasing steps 0.7, 0.5, 0.3 and 0.1 MPa;

– the test pressure is maintained for 10 min, and the volume of water absorbed by the ground is measured every minute; the pressure is maintained with a variation of less than 5%; the precision of the pressure measurement is better than 3%;

– if, at the start of a step, the pressure is not constant to a degree of precision of 5% at least, the time from the moment where stabilization is achieved to the nearest 10% is not counted;

– for each step, the pressure of the circuit of the stopper and the water level in the borehole above the cavity must be measured;

– during each pressure step, pressure p_g in the stopper is checked to make sure it remains constant, and the water level in the borehole above the stopper must not change.

10.1.4.3. *Interpretation of the test*

We are looking to determine the absorption characteristic of the soil in Lugeon units, by proceeding as follows:

– we determine the volume V injected for the length of time in which each pressure has been maintained constant;

– the mean flow Q_j is measured for each of these steps;

– the loss of pressure Δp_c during the flow between the pressure measurement and the cavity using a calibration curve, the tubing for the medium flow of each step, if no direct measurement of the pressures in the cavity has been made;

– at each step, the net pressure of the flow p_j is determined in relation to the hydrostatic pressure of the water table (or to the atmospheric pressure if there is no water table at the level of the test) in the cavity using relation [10.31]:

$$p_j = p_m + 10^{-3}.(z_m - z_w).\gamma_w - \Delta p_c \qquad [10.31]$$

where p_m is the pressure measured at the surface, z_m is the altitude of the injection probe, z_w is the altitude of the water table and Δp_C is the loss of load in the tubing between the pressure conditioner and the cavity;

– the representative curve of the net injection pressure p_j is traced as a function of the injected flow Q_j;

– if neither a breakdown nor clogging has occurred, flow Q_1 is read on this curve, corresponding to a pressure of 1 MPa;

– in this case, the Lugeon unit number can be calculated using [10.32] with length of the injection cavity L:

$$U_L = {}^{Q_1}/_{L} \qquad [10.32]$$

$$\frac{\Delta Q_j}{\Delta p_j} = \frac{m.k.B}{\gamma_w} \qquad [10.33]$$

– if a breakdown or clogging has occurred, characterized by a break point on the pressure-flow curve, and an absence of reversibility, or if the 1 MPa step cannot be reached, an upper limit of the test is calculated by extrapolating the flow for a pressure of 1MPa from the tangent at the origin;

– eventually, the permeability of the rock can be deduced, using the slope of the line linking the injection pressure p_j to the flow Q_j [10.33].

The different possible curves are:

– if the flow is laminar (Figure 10.24), the flow-pressure line goes through the origin and reflects the presence of a number of fissures distributed in a homogenous manner;

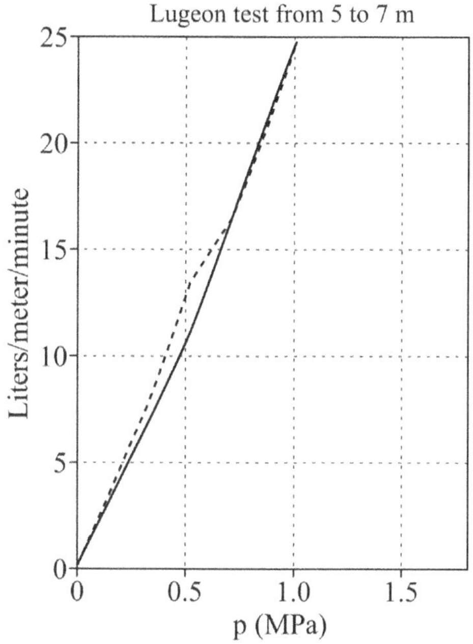

Figure 10.24. *Example of the laminar rate, loading is solid line, unloading is dotted line [CAS 80]*

– if the flow is turbulent (Figure 10.25), the flow-pressure curve is concave toward the bottom and reflects the presence of a small number of large fissures;

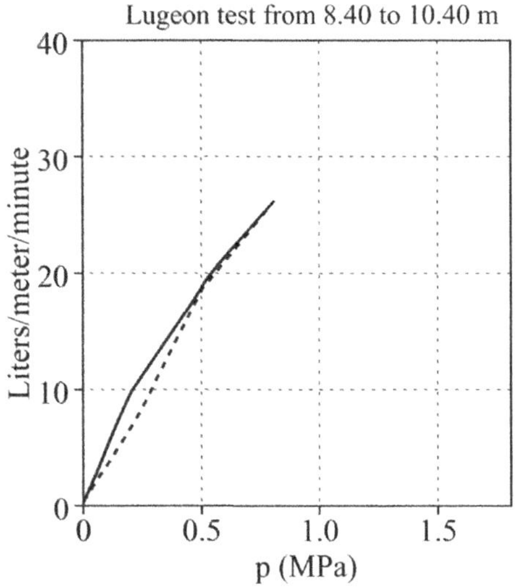

Figure 10.25. *Example of the turbulent rate, loading is solid line, unloading is dotted line [CAS 80]*

– if clogging has taken place (Figure 10.26), the slope of the flow-pressure slope decreases, and the loading and unloading slopes are no longer superimposed, despite both passing by the origin;

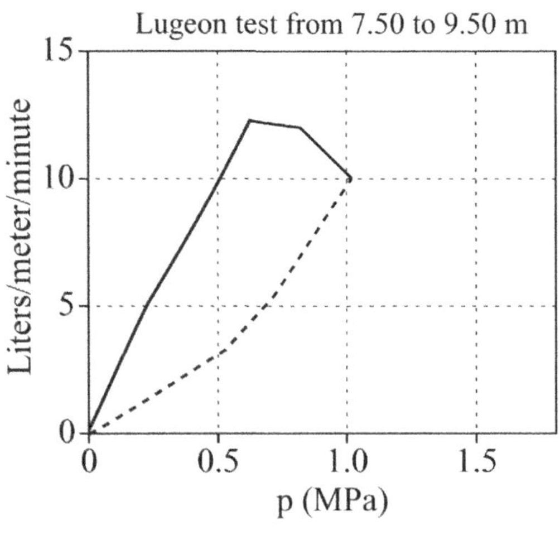

Figure 10.26. *Example of clogging, loading is solid line, unloading is dotted line [CAS 80]*

– if breakdown has taken place (Figure 10.27), the flow-pressure curve presents a point of inflexion toward the top and the permeability at unloading is lower than at the initial loading.

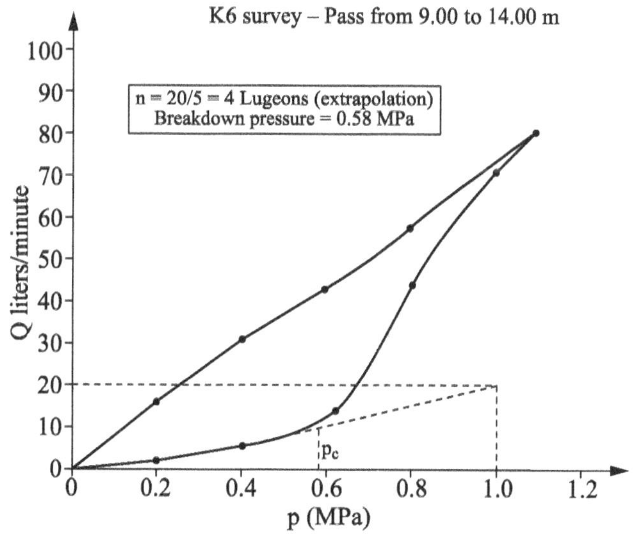

Figure 10.27. *Example of a breakdown, loading is solid line, unloading is dotted line [CAS 80]*

10.1.4.4. *Test report*

The test report contains the following information (Figure 10.28):

– reference to the standard;

– the name of the company that carried out the test;

– the site identification;

– localization of the fixed reference system and its altimetric level;

– the bore reference;

– the test date;

– the static equilibrium level of the water table at rest;

– the diameter of the drill tool, the nature of the drilling fluid and eventually the tubing used;

– the depth of the high and low levels of the cavity;

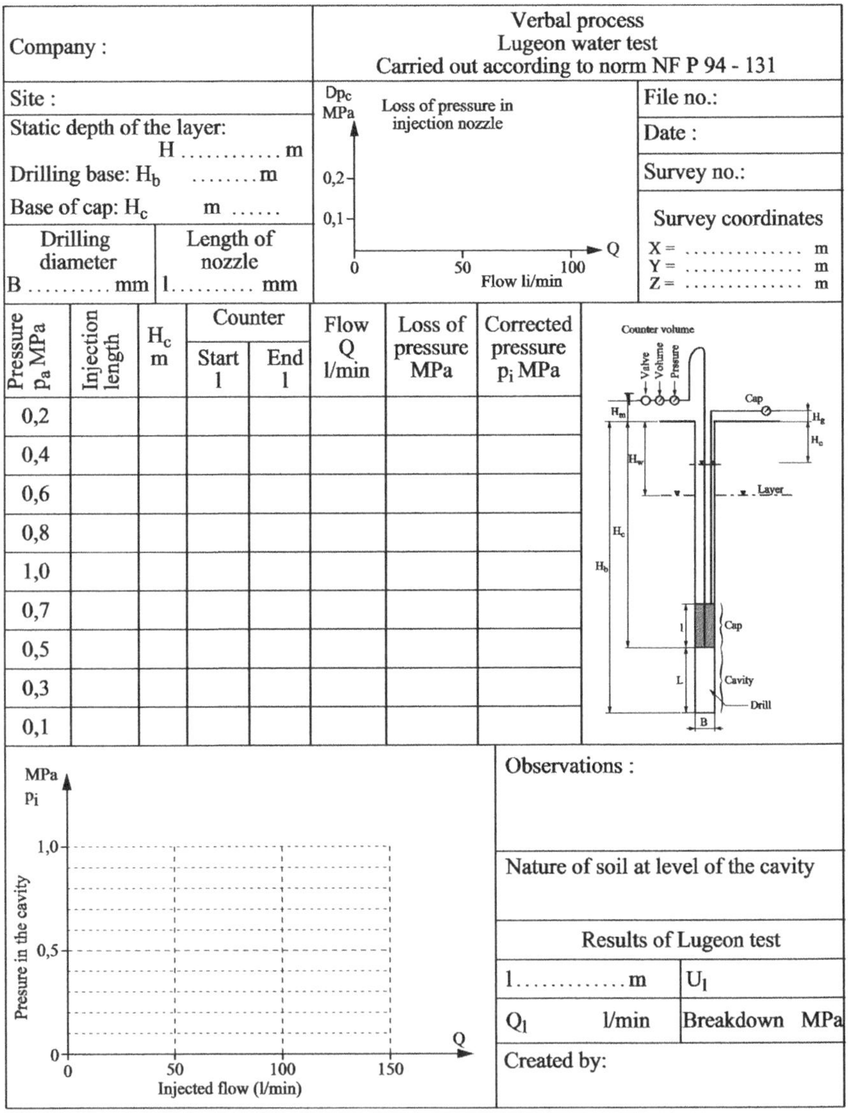

Company :	Verbal process Lugeon water test Carried out according to norm NF P 94 - 131	
Site :	Loss of pressure in injection nozzle (D_{pc} MPa: 0,1; 0,2 – Flow li/min: 0, 50, 100 – Q)	File no.:
Static depth of the layer: H .... m		Date :
Drilling base: H_b .... m		Survey no.:
Base of cap: H_c m		Survey coordinates X = ... m Y = ... m Z = ... m
Drilling diameter B .... mm / Length of nozzle l.......... mm		

Pressure P_a MPa	Injection length	H_c m	Counter Start l	Counter End l	Flow Q l/min	Loss of pressure MPa	Corrected pressure p_i MPa
0,2							
0,4							
0,6							
0,8							
1,0							
0,7							
0,5							
0,3							
0,1							

Observations :	
Nature of soil at level of the cavity	
Results of Lugeon test	
l............. m	U_l
Q_l l/min	Breakdown MPa
Created by:	

Figure 10.28. *Example of a Lugeon test report [AFN 94]*

– the pressure steps and the corresponding injection flows:

- the characteristics of the stopper: length and pressure applied,

- the nature of the soil along the length of the cavity,

- the representative curve of pressure p_j between the cavity and the level of the water table at rest as a function of the flow Q_j injected into the cavity, if the pressure is not measured directly at the level of the cavity.

– the pressure curve as a function of the net flow;

– the Lugeon unit number or eventually the breakdown number;

– the observations linked to the carrying out of the test.

10.2. Pumping or transmission tests

The pumping test is standardized under NFP 94-130 [AFN 92a].

A water table is pumped, and the depression at the surface of the table is measured, which has the shape of an inversed cone (Figures 10.35 and 10.36). This allows us to determine:

– the horizontal permeability coefficient k of the layer of soil tested;

– the dimensionless storage factor S;

– the action radius of the pumping R;

– the transmissivity T for the captive water tables of thickness H (T = k.H).

10.2.1. *Principle of the test*

The test involves pumping water into a borehole to determine the permeability coefficient of a layer of soil by measuring the extracted flow in the permanent rate, or the evolution of the water level as a function of time in the transitory rate. The level of the water table is monitored by the piezometers far from the boring to measure its drawdown level.

10.2.2. *Execution of the test*

10.2.2.1. *Test equipment*

The test material is the following:

– a strainer with an internal diameter greater than the external diameter of the pumping equipment by at least 5 cm, its height must be at least 15 times that of its internal diameter, the smallest dimension of an opening must be contained between $0.5D_{85}$ and D_{85} which is the dimension of the characteristics of the grains of the filtering material between the sieve and the borehole, the ratio between the surface of the openings and the total surface of the sieve must be greater than 10%;

– boring material;

– a pump with a great enough flow rate;

– a device for controlling the flow rate fixed onto the backflow canalization;

– flow measuring device;

– a system for measuring the water level in the wells, protected by a tube with a great enough height to protect the measurement of turbulences created by the pumping;

– piezometers with low response times;

– an emulsion device;

– a mechanism for measuring the water level in the piezometers;

– a stopwatch.

10.2.2.2. *Preparation of the test*

It contains the following steps (Figure 10.29):

– the drilling of the well, with a diameter more than 10 cm greater than the external diameter of the sieve. The boring must penetrate by more than 25 times the diameter of the sieve in the aquifer layer. Eventually, it can be carried out under the protection of provisional tubing, but cannot be consolidated with mud;

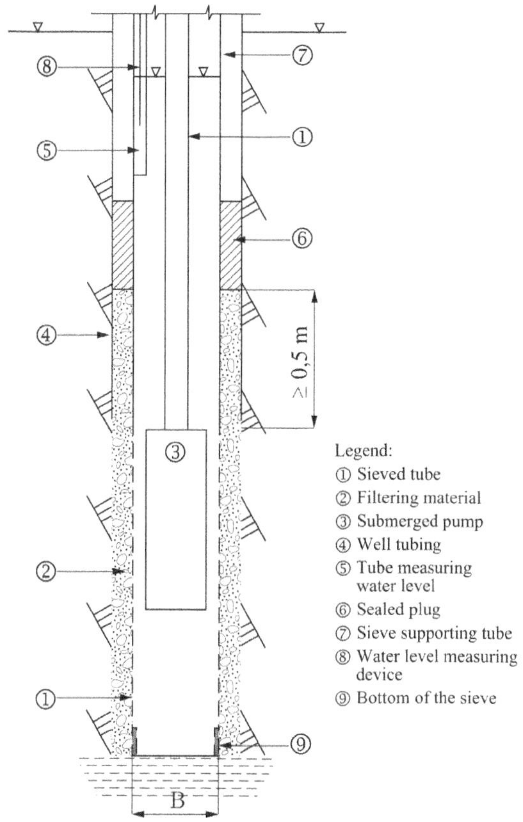

Figure 10.29. *Equipment of the well for the pumping test [AFN 92a]*

– equipping the well with a sieved tube over a minimal length of 2 m and plugged at its lower part;

– installing of a filtering material in the space between the sieve and the wall of the bore, up to 50 cm above the higher level of the sieve. If required, the realization of a sealed cap above the sieve and between the tube and the wall of the bore can be executed;

– the determination of the characteristics of the filtering material. Its role is to stop the particles carried through by the flow. The filtering material must satisfy this double inequality [10.34]:

$$5.D_{15}(soil) \leq D_{15}(filter) \leq 5.D_{85}(soil) \qquad [10.34]$$

– in the other cases, the filter is made of gravel such that the minimum diameter of the filter D_{min} satisfies equation [10.35], and that the maximum diameter D_{max} satisfies equation [10.36]:

$$2mm \leq D_{min}(filter) \leq 3mm \qquad [10.35]$$

$$5mm \leq D_{max}(filter) \leq 8mm \qquad [10.36]$$

– the number of piezometers is at least three. The implantation of piezometers over a radius is defined by (Table 10.3) when the piezometers are distributed over two perpendicular radii O_x and O_y (Figure 10.30);

Piezometers	Distance from the piezometers to the axis of the well (m)			
On one radius	$2<=X_1<=3$	$8<=X_2<=12$	$15<=X_3<=20$	
On two radii	$2<=X_1<=3$ $3<=Y_1<=6$	$8<=X_2<=12$ $12<=Y_2<=20$	$15<=X_3<=20$ $Y_3>=20$	$X_4>=25$

Table 10.3. *Distances to respect of the piezometers in relation to the well, with X_i the distance from the piezometer i to the well depending on O_x, Y_i the distance from the piezometer i to the well depending on O_y*

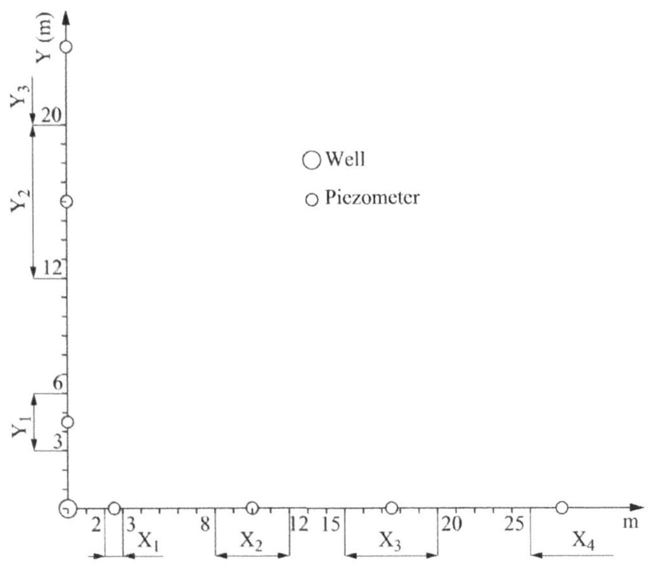

Figure 10.30. *Implantation in the horizontal plane of the piezometers [AFN 92a]*

– the piezometric tubes are sieved over a minimal length of 2 m and blocked in the inferior part;

– a filtering material is eventually placed between the sieve and the piezometer if the latter is not equipped with a filtering device;

– the length of the piezometers is such that their inferior part is located at the level of the base of the well for the closest ones;

– the piezometers are cleared by air emulsion;

– the well is cleaned with the emulsion or with fresh water until not particles are being carried through;

– the pumping flow Q_d must be such that the flow rate at the entrance of the filter, given by relation [10.37], must be less than 0.01 m/s:

$$v = \frac{Q_d}{\pi . B . L} \qquad [10.37]$$

– the water guard above the sieve must be at least 50 cm;

– the drawdown for a flow of $0.2Q_d$ must be systematically measurable;

– the piezometers are cleaned by air emulsion, their response time is verified by observing the rise of the water in the tube;

– the water levels in the piezometers are noted periodically after the end of the cleaning operations and up until stabilization;

– the water level in the well is measured after stabilization of the water levels in the piezometers.

10.2.2.3. *Execution of the test*

The test contains two phases:

– the preliminary pumping phase to determine the pumping flow rate of the test;

– the test phase.

10.2.2.3.1. Preliminary phase: determination of the characteristic curve of the well

We start off by determining the characteristic curve of the well, by carrying out the following operations:

– we pump (or inject) a constant flow to the nearest 10%;

– the water levels are measured to the nearest 2 mm;

– the measurement of the flows is done to the nearest 5%;

– the frequency of the measurements is such that the total volume must be known to 10%;

– the characteristic curve of the wells is fixed by equation [10.23], with the experimental coefficients a and b:

$$s = a.Q^2 + b.Q \qquad [10.38]$$

– the pumping is carried out with four successive steps and a flow maintained at a constant rate and over a length of time of 2 h for the first one and 1 h for the next three. Flows Q_i retained for steps I area function of flow Q_d depending on [10.39]:

$$Q_i = 0{,}8.\frac{Q_d}{5-i} \qquad [10.39]$$

– for each step, the water level is measured every 5 min for the first 30 min, and then every 10 min after that;

– during the first pumping step Q_1, the measurement of the unitary drawdown s/Q of the water in the well as a function of the time allows us to calculate:

- the slope α of the tangent to the curve of unitary drawdown as a function of the $\log_{10}$ t, for pumping maintained over 2 h (Figure 10.31),

- the unitary drawdown c = s/Q_1 at 48 h for a pumping of flow Q_1 extrapolated logarithmically at 48 h (Figure 10.31).

– determination of the coefficients a and b of the characteristic curve of the well (Figure 10.32). Coefficient a is the slope of the average line passing through the points with the coordinates (Qi , s_i/ Q_i), calculated for the four steps, with s'_i the drawdown extrapolated after 2 h for the flow Q_i (measured for the first step Q_1);

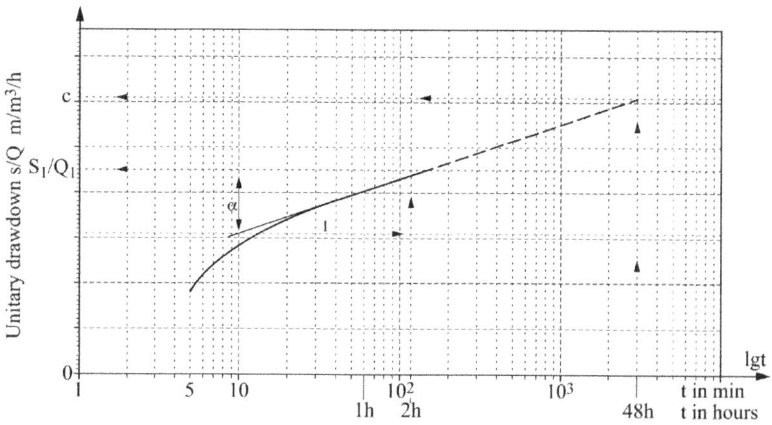

Figure 10.31. *Determination of the coefficient α and the unitary drawdown c during the preliminary test [AFN 92a]*

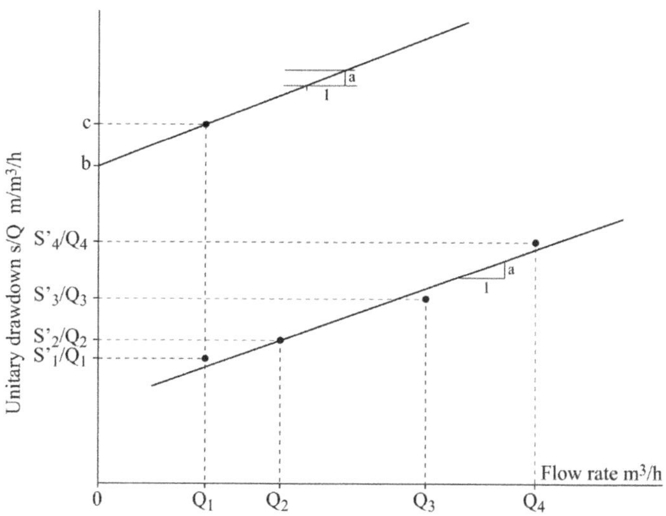

Figure 10.32. *Determination of coefficients a and b of the characteristic drawdown curve in the well during the preliminary test [AFN 92a]*

– extrapolation s'_2 of the flow Q_2 is down from the drawdown s_2 measured after 1 h using relation [10.40]:

$$s_2' = s_2 + \alpha.(0{,}3.Q_2 - 0{,}48.Q_1) \qquad [10.40]$$

– extrapolation s'_3 of flow Q_3 is done using the drawdown s_2 measured after 1 h using relation:

$$s_3' = s_3 + 0{,}3.\alpha.(Q_3 - Q_2 - Q_1) \qquad [10.41]$$

– extrapolation s'_4 of flow Q_4 is done using the drawdown s_4 measured after 1 h using relation:

$$s_4' = s_4 + \alpha.(0{,}3.Q_4 - 0{,}3.Q_3 - 0{,}18.Q_2 - 0{,}22.Q_1) \qquad [10.42]$$

– α is determined graphically in Figure 10.31.

10.2.2.3.2. Preliminary phase: determination of the test pumping flow Q_e

The test flow is the solution to the equation of the characteristic curve of the drawdown in the well [10.38], the solution for which is given by relation [10.43] with s_f the drawdown given at the end of the test, determined in the following manner:

$$Q_e = \frac{-b + \sqrt{b^2 - 4.a.s_f}}{2.a} \qquad [10.43]$$

– for a well that has penetrated an aquifer layer whose surface of the free water table is located at a height e above the base of the well, $s_f \leq e/3$;

– for a well in a captive groundwater table, s_f is the final drawdown that leads the water level in the well to 0.5 m above that of the roof the aquifer layer, at the lowest.

10.2.2.3.3. Pumping test

To carry out the pumping test, the following operations must take place:

– a new verification of the stabilization of the water levels in the well and in the piezometers is carried out;

– the pump is set to work, and the fixed flow (precision of 5%) for the test must be stabilized within the 2 min that follow. The test then starts at t = 0;

– the intervals between the noting of the drawdown (precision of 1 cm) and the time intervals (precision 1 s) are carried out at:

- 1 min for: t <= 15 min,
- 5 min for: 15 min < t <= 30 min,
- 10 min for: 30 min < t <= 1 h,
- 30 min for: 1 h <t<= 4 h,
- 1 h for: t > 4 h.

– the flow rate of the pumping is measured at the same time intervals for $t \geq 15$ min;

– the level of the water bodies located close to the test are periodically noted over the course of the test;

– the pumping is continued until three successive values are obtained for the water levels in the piezometers, separated in time by at least 1 h and not more than 1 cm apart from each other;

– the test is stopped after 48 h even if the permanent rate has not been reached;

– after stopping the pumping, the levels are measured at the following time intervals:

- 1 min for: t <= 15 min,
- 5 min for: 15 min < t <= 30 min,
- every 1 h past 30 min.

– the measurements are made until a rise of 0.75 s is obtained, with a maximal duration of 8 h;

– duration t' of the rise is counted for from the end of the pumping and reduced to t_p;

– the unitary drawdown s/Q_i is traced in m/m^3/h as a function of the $\log_{10}(t)$ (Figure 10.31);

– the slope α of the line obtained is equal to [10.44]:

$$\alpha = \frac{k.H}{0{,}183} \qquad [10.44]$$

– the transmissivity T obtained by relation [10.45]:

$$T = k.H \qquad [10.45]$$

10.2.2.3.4. Test report

The test report must contain (Figures 10.33 and 10.34):

– reference to the standard;

– a situation plan containing the implantation of the well and the piezometers as well as the water of the backflow column outlet;

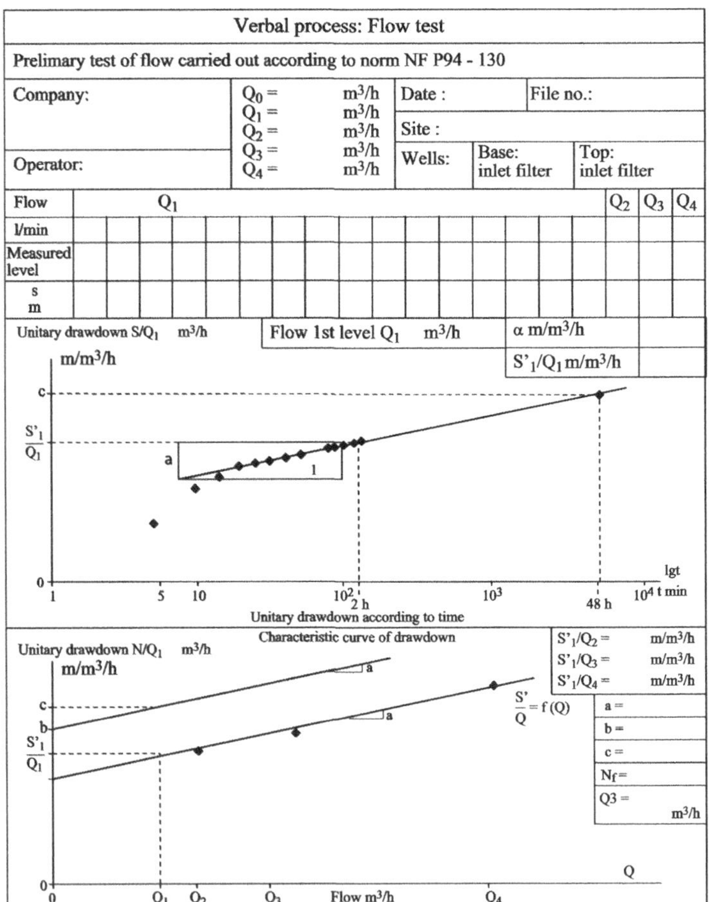

Figure 10.33. *Preliminary flow test report [AFN 92a]*

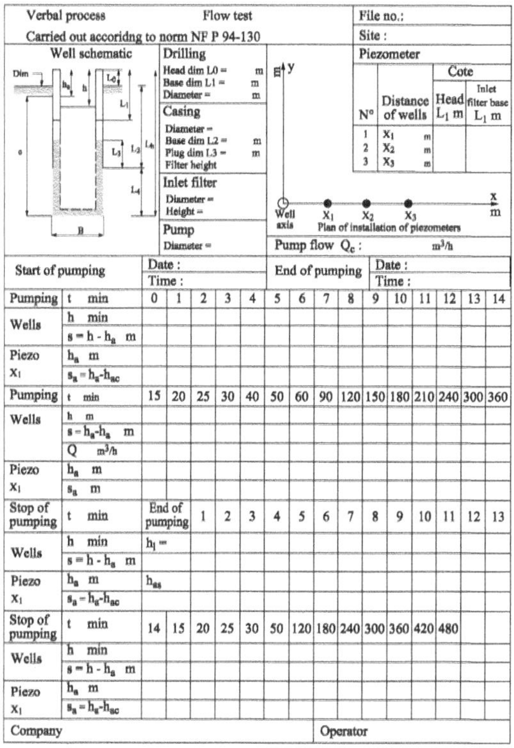
Verbal process Flow test
Carried out accoridng to norm NF P 94-130
File no.:
Site :
Well schematic

Drilling
Head dim L0 = m
Base dim L1 = m
Diameter = m
Casing
Diameter =
Base dim L2 = m
Plug dim L3 = m
Filter height
Inlet filter
Diameter =
Height =
Pump
Diameter =

Pump flow Q_c : m^3/h

Piezometer

N°	Distance of wells	Cote: Head L_1 m	Cote: Inlet filter base L_1 m
1	X_1 m		
2	X_2 m		
3	X_3 m		

Start of pumping — Date : Time : — End of pumping — Date : Time :

Pumping	t min	0	1	2	3	4	5	6	7	8	9	10	11	12	13	14
Wells	h min															
	$s = h - h_a$ m															
Piezo x_1	h_a m															
	$s_a = h_a - h_{ac}$															
Pumping	t min	15	20	25	30	40	50	60	90	120	150	180	210	240	300	360
Wells	h m															
	$s = h_a - h_a$ m															
	Q m^3/h															
Piezo x_1	h_a m															
	s_a m															
Stop of pumping	t min	End of pumping		1	2	3	4	5	6	7	8	9	10	11	12	13
Wells	h min	h_l =														
	$s = h - h_a$ m															
Piezo x_1	h_a m	h_{as}														
	$s_a = h_a - h_{ac}$															
Stop of pumping	t min	14	15	20	25	30	50	120	180	240	300	360	420	480		
Wells	h min															
	$s = h - h_a$ m															
Piezo x_1	h_a m															
	$s_a = h_a - h_{ac}$															

Company Operator

Figure 10.34. *Pumping test report [AFN 92a]*

– the main information relative to the execution of the well and its equipment:

- the diameters, depths, execution mode of the well and characteristics of the tubing,

- section of the terrains,

- geometrical characteristics of the sieve and the opening coefficient,

- granulometry of the aquifer terrains and the filter,

- composition of the bedding grout,

- levels of the head of well, the terrain, the sieve and the surface of the filter,

- characteristics of the pump and the flowmeter,

- report of the well preparation operations.

– the main information relative to each piezometer and its installation:

- distance to the well,

- diameter, depth and boring execution mode,

- total length of the piezometric tube, internal diameter, height of the sieve and distance to head of the piezometer,

- level of the head of piezometer and the surface of the line of the piezometer.

– a table which reflects the following curves:

- variation of the flow rate as a function of time,

- variation of the water level or drawdown in the well and in the piezometers (during and after pumping).

10.2.3. *Interpretation*

10.2.3.1. *Permanent rate*

The Dupuit formula [10.46] gives the value of the flow of a well as a function of the characteristics of the soil as an aquifer layer of thickness H (case of the captive water table presented in Figure 10.35), with the sample punctual flow water Q, the permeability coefficient k, the thickness of the permeable soil layer H, the drawdown of the water table in the well s_p, the radius of influence of the well R and the radius of the well r_p.

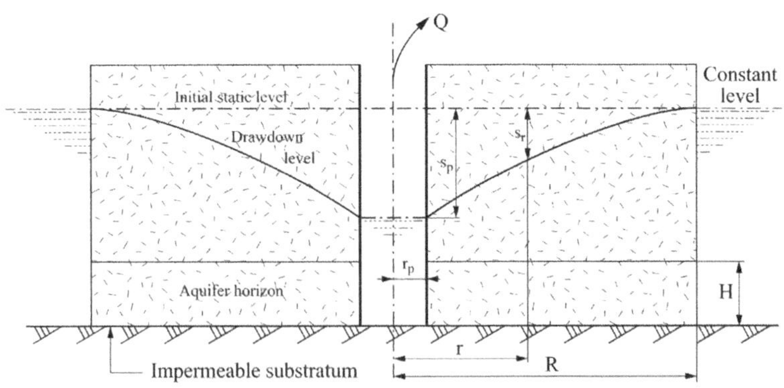

Figure 10.35. *Equilibrium rate for a well with a captive water table*

In the case of a free water table (Figure 10.36), the formula becomes [10.47], with the height of the water in the well h_p.

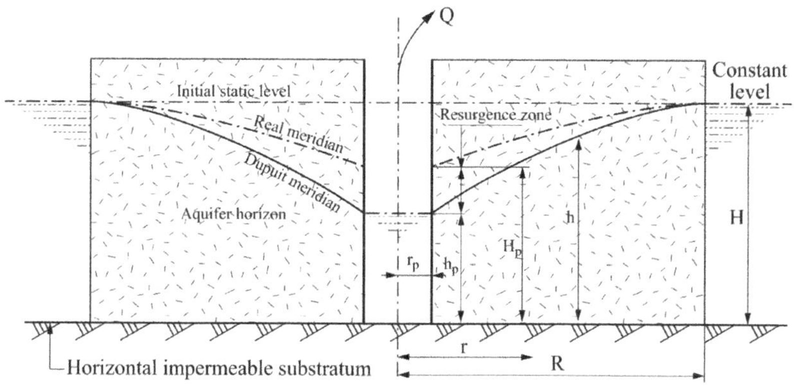

Figure 10.36. *Equilibrium rate for a well with a free water table*

In the case where resupply comes from a river, we can assume that there is an overlaying of the two states, and that the effect of the rectilinear resupply front is identical to that of a source with the same symmetrical flow rate of the well in relation to the resupply front (Figure 10.37). The overlaying of the flows for a point influenced by the pumping, located at a distance r from the pumping, and at a distance r' from its image in relation to the resupply front, results in a captive water table for pumping with a drawdown of s_2 [10.49]. The total drawdown s is the sum of s_1 and s_2, and we can deduce equation [10.50] as a result.

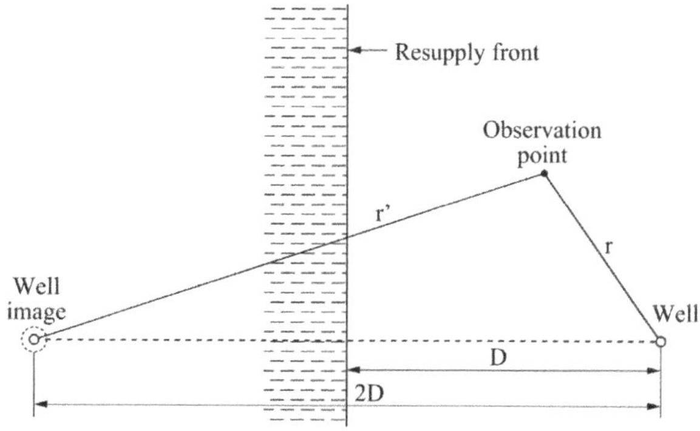

Figure 10.37. *Pumping in a water table bordered by a rectilinear resupply front*

We measure the level of the water in piezometers further and further from the well, and the drawdown s in the piezometers is traced as a function of the logarithm of the distance r (Figure 10.38). This method allows us to measure the influence radius R of the well and determine the permeability coefficient using one of the previous equations.

$$Q = \frac{2.\pi.k.H.s_p}{Ln\left(\frac{R}{r_p}\right)} \qquad [10.46]$$

$$Q = \frac{\pi.k.(H^2 - h_p^2)}{Ln\left(\frac{R}{r_p}\right)} \qquad [10.47]$$

$$Q = \frac{2.\pi.k.H.s_1}{Ln\left(\frac{R}{r}\right)} \qquad [10.48]$$

$$Q = -\frac{2.\pi.k.H.s_2}{Ln\left(\frac{R}{r\prime}\right)} \qquad [10.49]$$

$$Q = \frac{2.\pi.k.H.s}{Ln\left(\frac{r\prime}{r}\right)} \qquad [10.50]$$

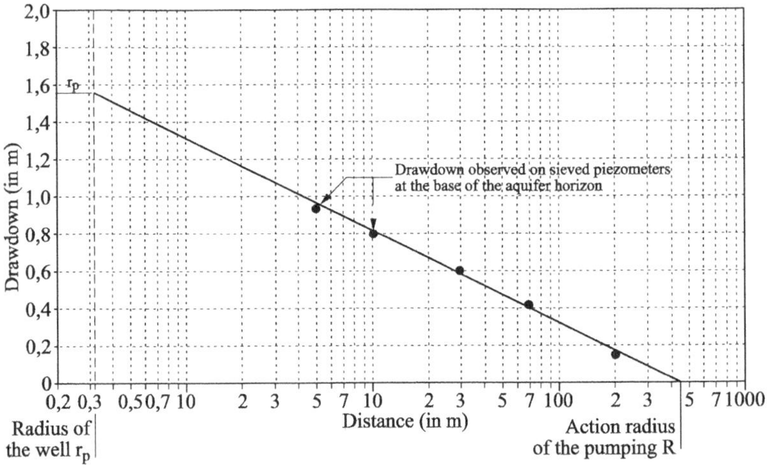

Figure 10.38. *Dupuit method. Drawdown as a function of the distance to the test well (case of the captive water table)*

10.2.3.2. *Transitory rate*

In the case of the transitory rate, the drawdown s is provided by the Theis theory. It is written in accordance with relation [10.51] as a function of the pumped flow Q, permeability k, the thickness of the aquifer H, W(u) which is itself a function [10.52] of the storage coefficient S, and of time t.

The W(u) function can be obtained using an abacus (Figure 10.39) or by using the Jacob approximation (common logarithm approximation) shown in [10.53]:

$$s = \frac{Q.W(u)}{4.\pi.k.H} \qquad [10.51]$$

$$W(u) = \int_0^{\infty} \frac{e^{-u}}{u}.du \quad with \quad u = \frac{r^2.S}{4.k.H.t} \qquad [10.52]$$

$$s = \frac{0{,}183.Q}{k.H}.Log\left(\frac{2{,}25.k.H.t}{r^2.S}\right) \qquad [10.53]$$

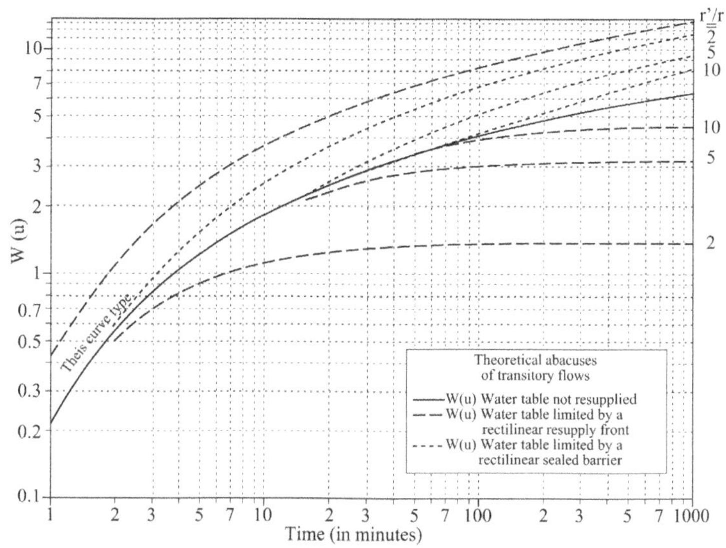

Figure 10.39. *Theoretical abacuses of transitory flows, W(u) function*

The representative points of the evolution of the drawdown as a function of time on a diagram linking the logarithm of time to the drawdown

(Figure 10.40) line up to form a line whose slope is equal to 0.183.Q/(k.H), which allows us to determine the value of the permeability k.

The intersection of this line with x-axis determines the time t_0 for which the drawdown is equal to zero (s = 0). This value allows us to determine the storage value S [10.54] knowing k by the slope of the straight part (Figure 10.40).

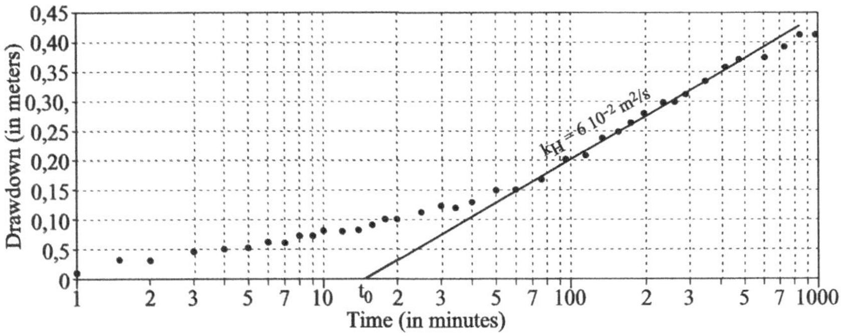

Figure 10.40. *Jacob logarithmic approximation method – evolution of the drawdown as a function of time in a piezometer*

In the case of the rising of the water table after the stopping of the pumping (flow rate of Q), we can show that the value of the drawdown is obtained using formula [10.55], with the total time passed since the start of the pumping being t (pumping time and pumping stoppage time) and the time passed since the stopping of the pumping being t'. The graphical representation of the relation between the logarithm of the t/t' ratio and the value of the drawdown results in a line whose slope is equal to 0.183.Q/(k.H), allowing us to determine the value of the permeability k.

$$S = \frac{2{,}25.k.H.t_0}{r^2} \tag{10.54}$$

$$s = \frac{0{,}183.Q}{k.H}.Log\left(\frac{t}{t'}\right) \tag{10.55}$$

11

Characterization of Sites and Soils by *In Situ* Tests

The study of soil stratification is a necessary geotechnical step for the characterization of the site through the identification of soil layers and their horizontal limits. We can assume that the mechanical characteristics are homogenous within each layer. As a result, if a material is determined to have similar mechanical characteristics at certain ground level, it can be deduced that this material is identifiable as a layer of soil.

11.1. Characterization of sites

Studying variations of the operating parameters of boring machines gives a better idea of the position and the nature of the layers that are crossed. The drilling tool can be considered a variant of the penetrometer. The use of using drilling parameters to characterising the soil is carried out along the standard [AFN 14].

11.1.1. *Analysis by drilling parameter recording*

Studying the variations of the operating parameters of the boring machines (Figure 11.1) gives a better idea of the position and the nature of the layers crossed. The drilling tool can be considered a variation of the penetrometer for which the pressure on the tool provides information with regard to the resistance to penetration, and the pressure on the drilling tool gives an indication of the permeability of the soil [REI 10].

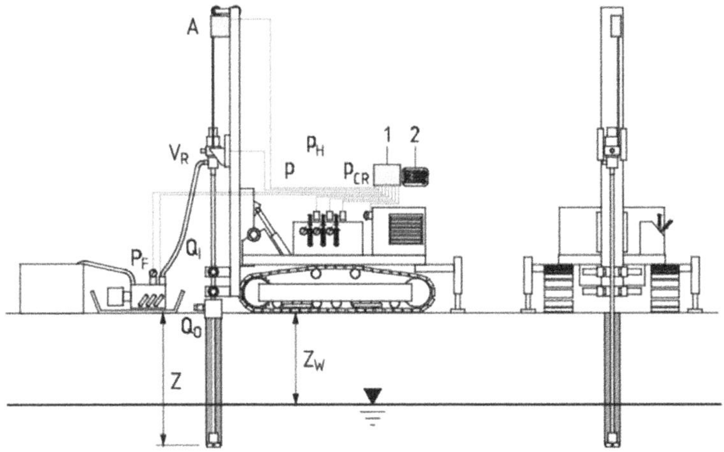

Legend

1: Distribution/branching box   2: Recorder

Figure 11.1. *Example of the placement of sensors on a drilling machine [AFN 14]*

11.1.1.1. *Measurement system*

The drilling parameters can be classified into four different categories [CAI 86]:

– the parameters imposed by the method and not recorded digitally (type of tool and diameter, nature of the injection fluid, and maximum and minimum performances of the machine and the injection system);

– the parameters set by the operator (the push exerted on the tool P_O, rotation velocity V_R and injection flow Q_I);

– the parameters that depend on the response of the terrain (advancement velocity V_A, rotational torque C_R, injection pressure P_I, return flow Q_O and back pressure P_R);

– non-controlled evolving parameters (wear of the tools and fluid composition changes).

The measurement system (Figure 11.2) is made up of [CAL 86, GUI 02]:

– a recorder with a user interface and method for transferring data;

– a conditioner supplying the sensors, multiplexing the measurement paths and conditioning the signals (sometimes integrated into the recorder);

– a set of sensors set out on the machine.

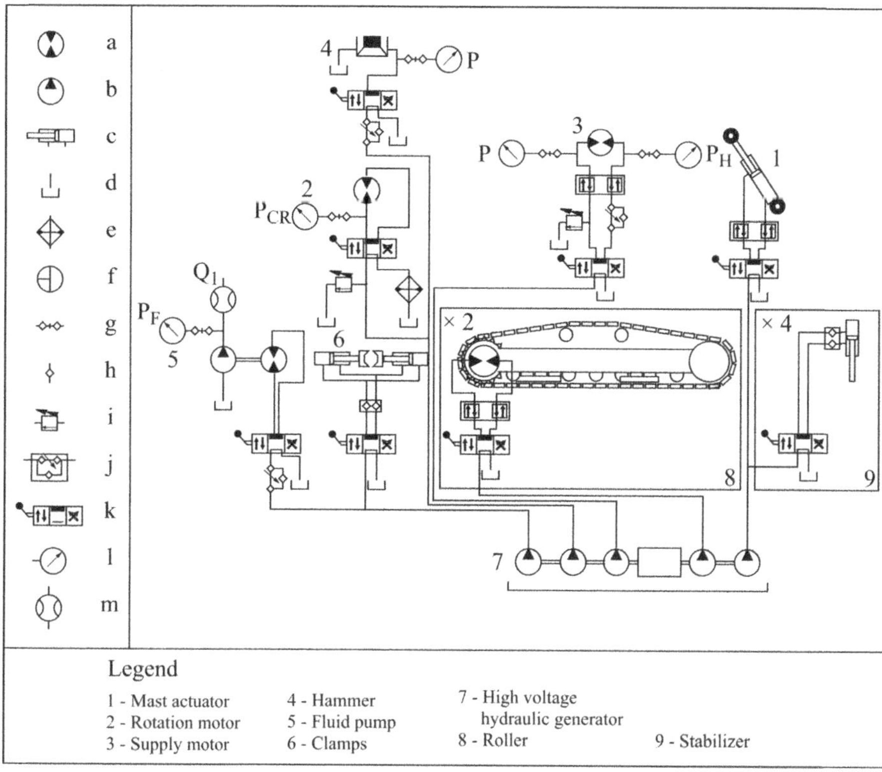

Figure 11.2. *Example of a hydraulics diagram of a drilling machine, with placement of sensors. The hydraulic symbols conform to ISO 1219-1 [AFN 14]*

Three MWD measurement categories are defined depending on the number of parameters measured during the drilling. The notations are given in Table 11.1:

– MWD A, more than four parameters are measured;

– MWD B, between two and four parameters are measured;

– MWD C, at least two parameters are measured.

11.1.1.2. *Recorded parameters*

The parameters recorded [AFN 14] are the hydraulic push pressure p, torque p_{CR} and back pressure p_R, the pressure of the injection fluid p_F and the advancement A of the tool, the rotation velocity of the head V_R as well as the flow rate of the fluid Q_I (Table 11.1). Some of these parameters require scaling before obtaining measurements that are representative of the action of the tool.

Category	Recording	Examples of parameters measured for the drilling machine	Minimal authorized precision
MWD C	Manual or automatic	Length of penetration (A) Resistance to penetration (P_R)	0,2 m or 2 % of PM' 1 s/0,2 m
MWD B	Automatic	Length of penetration (A) Pushing pressure (ρ) Backpressure (ρ_H) Drilling fluid pressure (P_F) Torque pressure (ρ_{CR})	1 % of PM 2 % of PM 2 % of PM 2 l/mn 2 % of MR
MWD A	Automatic	Length of penetration (A) Pushing pressure (ρ) Backpressure (ρ_H) Drilling fluid pressure (P_F) Rotation velocity of drill head (V_R) Drilling fluid flow rate (Q_I) Torque pressure (ρ_{CR})	5 mm or 0,1 % de PM 1 % of PM 1 % of PM 1 % of PM 1 % of PM 1 l/mn 1 % of PM

Table 11.1. *Types of systems for recording drilling parameters*

11.1.1.2.1. Advancement velocity

The advancement speed VA is obtained directly from the measurement of advancement A and the time interval dt between measurement [11.1].

$$V_A = \frac{A_{t+dt} - A_t}{dt} \qquad [11.1]$$

11.1.1.2.2. Push on the tool

Measurement of the pushing force $S_O.P_O$ corresponds to the hydraulic pressure present in the cylinder of the piston of the actuator or the hydraulic motor p scaled to the maximum pushing force F_{max} determined during the calibration of the head with a low endstop, to which the weight of the drilling head $W_{drill\ head}$ is added, as is the number of rods n, which have a unitary weight of W_{rod} [11.2]. S_o is the action surface of the tool, which has a diameter of D.

$$S_0.P_0 = \frac{p}{p_{max}}.F_{max} + W_{drill\ head} + n.W_{rod} \quad [11.2]$$

11.1.1.2.3. Back pressure

Complementing the push on the tool, the measurement of the backpressure P_H that exists in the return circuit is vital in calculating the net push applied to the tool $P_E = P_O - P_H$. Variable p_{Hmax} is determined by calibration by imposing an effort of H_{max} under which the backpressure is surpassed and the descent of the head begins [11.3]. It can be noted that the breaking, caused by the safety brake on the translation of the head, by the load valves and by load losses in the return circuit (flexible, filter, etc.), is the main cause for faulty interpretations of the combinations of parameters.

$$S_0.P_H = \frac{p_H}{p_{H\,max}}.H_{max} \quad [11.3]$$

11.1.1.2.4. Drilling fluid injection pressure

The pressure of the drilling fluid P_F is measured using a pressure sensor placed at the exit of the injection pump or the compressor. To obtain the pressure at the level of the tool P_I, it is necessary to correct, in a permeable terrain, by adding the height of the drilling fluid column (with a specific weight of γ_M) present in the string of rods, and to reduced from the height of the roof of the water table, measured at the level of the tool (Figure 11.1).

$$P_I = P_F + \gamma_M.z_w - \gamma_w.(z - z_w) \quad [11.4]$$

11.1.1.2.5. Rotational torque

The pressure sent to the rotation motor p_{CR} is measured using a pressure sensor placed on the hydraulic circuit of the drill at the entrance of the hydraulic motor. Measurement of the rotational torque C_R is, therefore, an indirect estimation of the value of the actual torque from the pressure p_{CR} and from the maximal torque C_{Rmax} determined during calibration [11.5]:

$$C_R = \frac{p_{CR} - p_{CR0}}{p_{CRmax}}.C_{Rmax} \quad [11.5]$$

11.1.1.3. *Composite parameters*

11.1.1.3.1. Resistance to penetration

Resistance to penetration corresponds to the time duration of a penetration of 0.2 m measured in seconds [11.6]. The parameter of "resistance to penetration" allows us to effectively evaluate the depth up to the rock, the relative firmness of the rock and the formation of cracks.

A similar relation is used for another composite parameter called soil-rock resistance [11.7]:

$$R_P = t_{dz=0,2\,m} \qquad [11.6]$$

$$R_{SR} = \frac{P_E}{V_A} \qquad [11.7]$$

11.1.1.3.2. Alteration index

Correlations based on the advancement velocity, the push of the drilling tool and the relative density of non-consolidated materials or the alteration index I_A in rocks can be used [11.8], with P_{max} the maximal push on the tool measured in the field; V_{max} is the theoretical maximum advancement velocity measured in the field, which allows us to normalize the real advancement velocity; k_0 and k_1 are the empirical parameters that are dependent on the drilling results of the calibration test. This relation is well adapted to rocky soils and used to qualify the rippability of the formation undergoing the test.

$$I_A = 1 + k_0.\left(\frac{P_E}{P_{max}} - k_1.\frac{V_A}{V_{max}}\right) \qquad [11.8]$$

11.1.1.3.3. Somerton index

The Somerton index S_d describes the resistance to drilling in the rock. It provides an estimate of the advancement velocity as a function of the drilling parameters. Relation [11.9] was developed based on experimental results obtained in the laboratory.

For a given site and for each geological formation, it is possible to calibrate specific relations with geotechnical parameters such as the number of blows of the core drill test (SPT) and the pressure limit of the Ménard pressuremeter test p_{LM} using relation [11.10].

For geotechnical drilling devices, even if the rotation velocity is not often recorded, the drill has a tendency to maintain a constant speed for the duration of the advancement. As a result, the Somerton index can be simplified as shown in [11.11].

$$V_A = 1{,}5.V_R.D.\left(\frac{P_E}{S_d.D^2}\right)^2 \quad [11.9]$$

$$S_d = P_E.\sqrt{\frac{V_R}{V_A}} \quad [11.10]$$

$$S_d = \frac{P_E}{\sqrt{V_A}} \quad [11.11]$$

11.1.1.3.4. Energy

Estimation of the energy E used to drill a hole using a specific tool can be obtained from the following equation [11.12]; α, β, γ are the coefficients of the efficiency of the drilling that needs to be estimated, P_M is the power of the hammer and F is the frequency.

For boreholes that are shallow, a simplification of the drilling energy composite parameter can be used, combining energy and drilling depth [11.13].

The term P_E is often neglected, which means using the work produced by the rotational torque C_R. The rotational energy per unit of penetration is obtained [11.14].

The rotation energy is useful to differentiate between variations of the hardness of the soil; however, only the alteration index allows us to better classify the most supple materials, despite a variation of only 0–2.

$$E = \frac{\alpha.P_E.V_A+\beta.C_R.2.\pi.V_R+\gamma.P_M.F}{V_A} \quad [11.12]$$

$$E = P_E + C_R.\frac{V_R}{V_A} \quad [11.13]$$

$$E = C_R.\frac{V_R}{V_A} \quad [11.14]$$

11.1.1.4. *Modus operandi*

In order to obtain drilling parameters that can be interpreted adequately, the following must be kept as constant as possible:

– the push on the tool;

– the injection flow;

– the rotation velocity or at least the ratio between the gear ratios of the motor.

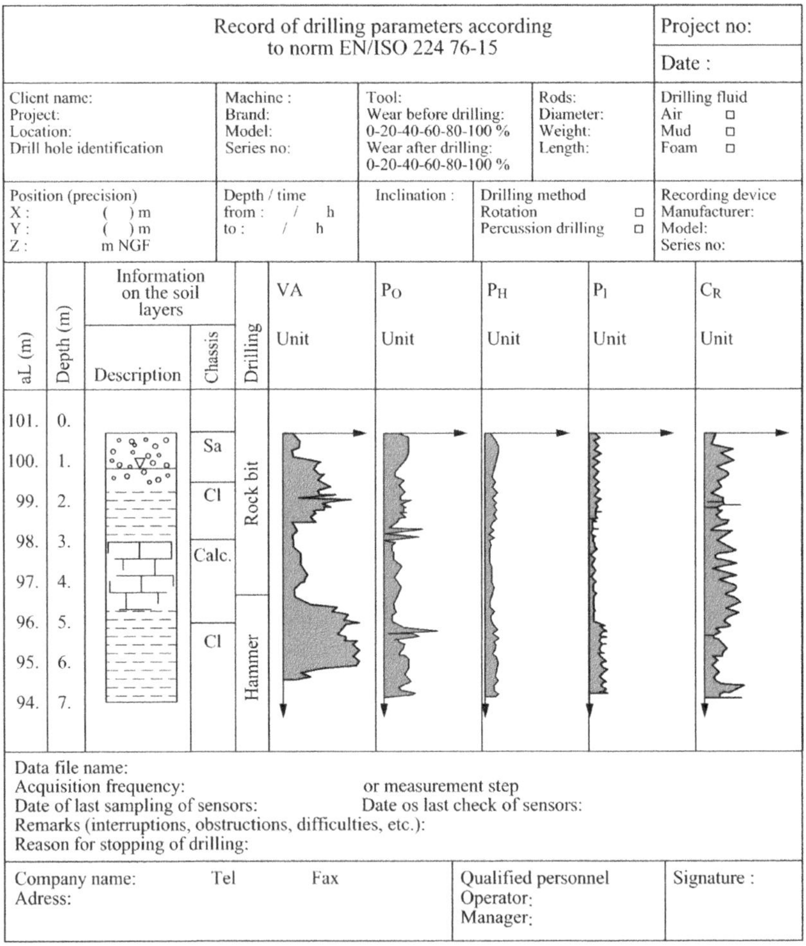

Record of drilling parameters according to norm EN/ISO 224 76-15

Project no:

Date :

Client name:
Project:
Location:
Drill hole identification

Machine :
Brand:
Model:
Series no:

Tool:
Wear before drilling:
0-20-40-60-80-100 %
Wear after drilling:
0-20-40-60-80-100 %

Rods:
Diameter:
Weight:
Length:

Drilling fluid
Air □
Mud □
Foam □

Position (precision)
X : ( ) m
Y : ( ) m
Z : m NGF

Depth / time
from : / h
to : / h

Inclination :

Drilling method
Rotation □
Percussion drilling □

Recording device
Manufacturer:
Model:
Series no:

aL (m) | Depth (m) | Information on the soil layers: Description, Chassis | Drilling | VA Unit | P_O Unit | P_H Unit | P_I Unit | C_R Unit

101. 0.
100. 1.
99. 2.
98. 3.
97. 4.
96. 5.
95. 6.
94. 7.

Sa
Cl
Calc.
Cl

Rock bit
Hammer

Data file name:
Acquisition frequency: or measurement step
Date of last sampling of sensors: Date os last check of sensors:
Remarks (interruptions, obstructions, difficulties, etc.):
Reason for stopping of drilling:

Company name: Tel Fax
Adress:

Qualified personnel
Operator:
Manager:

Signature :

Figure 11.3. *Example of a field report for the recording of drilling parameters [AFN 14]*

If any of these takes on a value that is too high, this will be detrimental to the quality of the recordings. It is important to not work the machine or the line of tools to the limit of their capacities. This is a determining factor for the success of the log. The company and the client must be aware that fine logging, using instant well logging is not always a synonym for a high yield.

11.1.1.5. *Interpretation*

Each parameter recorder can, therefore, be attributed to some qualitative significance in terms of the characteristics of the drilled soil or in terms of its function. An example of an evolution of the combined drilling parameters with the geology of the site is shown in Figure 11.4.

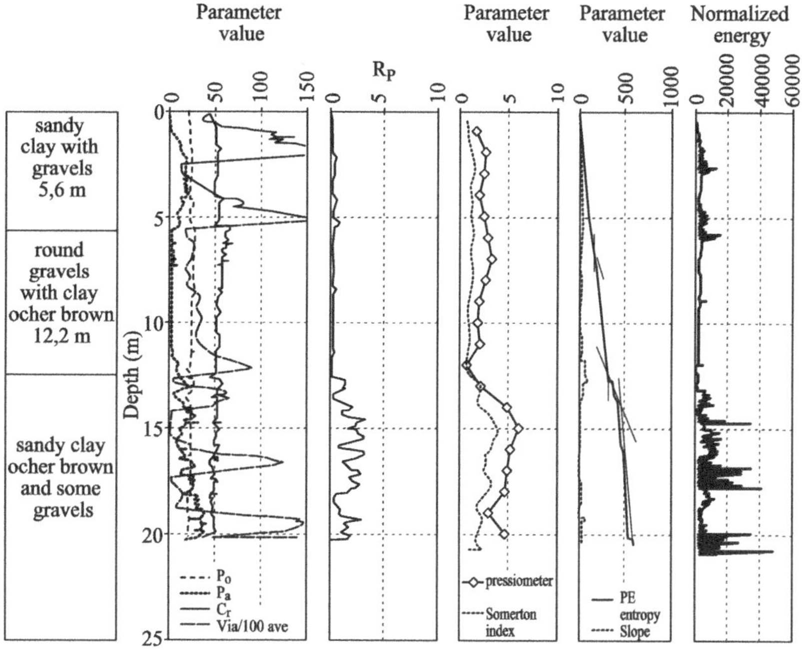

Figure 11.4. *Example of the use of drilling parameters with the geological model [REI 10]*

11.1.2. *Cluster analysis*

This method is used to visually identify the soil layers, using the penetrometer or pressuremeter, for example.

The cluster method allows us to automatically and reliably identify layers of soil. Cluster analysis [EVE 74] is a statistical method used to group data that present similar mathematical characteristics with subgroups that present common or homogenous values, and which can then be used for analyzing stratification. This method can be used for different data fields for which no division is known as of yet. It can be useful in the case of identifying ground horizons, for the characterization of the site. Cluster analyses are composed of the following six steps: normalization of the data; selection of variables; calculation of the resemblance distance matrix; choice of the grouping technique; determination of the precision of the calculation and; interpretation of the results from the cluster method.

11.1.2.1. *Choice of variables*

At a depth of I (varying from 1 to N the test depth), a series of measurement x from j variables (from 1 to p different measurements) can be written as x_{ij}.

To identify the different soil layers, the data used must be composed of different independent elements. To verify this condition, the correlation coefficient is calculated using relation [11.15] where $Cov(x_i, x_j)$ is the covariance between variables x_i and x_j. Theoretically, the covariance between two independent variables is equal to zero, and the covariance between two linked variables is 1 or –1.

From a practical point of view for pressuremeter tests, the independent variables are the Ménard Modulus E_M, the elastic modulus Ee measured in the loading-unloading cycle and the pressure limit p_{LM} according to Monnet and Broucke [MON 12c]. For penetrometer tests, the independent variables are the normalized resistance to penetration Qt [11.16] and the normalized interstitial pressure ratio Bq [11.17], according to Hegazy and Mayne [HEG 02].

$$\rho_{ij} = \frac{Cov(x_i, x_i)}{S(x_i).S(x_j)} \qquad [11.15]$$

$$Q_t = \frac{(q_c - \sigma_{v0})}{\sigma'_{v0}} \qquad [11.16]$$

$$B_q = \frac{(u_2 - u_0)}{(q_t - \sigma_{v0})} \qquad [11.17]$$

11.1.2.2. *Data standardization*

The standardization of data is a way of using and representing data in a non-dimensional space. From a theoretical point of view, standardization is not essential, but it is a logical step. At a depth of I, a series of measurements from x of j variables (from 1 to p) can be written as x_{ij}. The standardized variable z_{ij} corresponding to x_{ij} is then determined using relation [11.18] where $\overline{x_j}$ and S(xj) are, respectively, the average value and the standard deviation of variable j, whose x_{ij} measurements vary from 1 to N. After standardization, variable z_{ij} has an average value of 0 with a value of the standard deviation of 1.

$$z_{ij} = \frac{x_{ij}-\overline{x_j}}{s(x_j)} \qquad [11.18]$$

11.1.2.3. *Distance between variables*

The similarity between two different data series can be evaluated using at least seven different methods: Pearson, Manhattan, Minkowski, Puissance, Euclidien, Chebychev and Cosinus [NOR 88]. According to several researchers [HEG 02, MLY 05], the best method for estimating the distance between measurement series is cosine [11.19] rather than the Euclidian method [11.20]:

$$d_{ij} = \frac{\sum_{k=1}^{p} z_{ik}.z_{jk}}{\sqrt{\sum_{k=1}^{p}(z_{ik})^2.\sum_{k=1}^{p}(z_{jk})^2}} \qquad [11.19]$$

$$d_{ij} = \sqrt{\sum_{k=1}^{p}(z_{ik} - z_{jk})^2} \qquad [11.20]$$

The Euclidian relation [11.20] gives a result of zero for two identical data series in the same cluster, and provides a greater value for the deviation of two series corresponding to different points. The result produced by the cosine method [11.19] can be placed between –1 and +1, with 1 corresponds to identical data series and –1 to anti-correlating data series. On the contrary, with relation [11.19] there is no limit for determining non-correlated series. The cosine method, therefore, appears to be better adapted to measuring the distance between series.

11.1.2.4. *Grouping data as a cluster*

The grouping of data series into a cluster is done using the nearest neighbor method, which fulfills the mathematical conditions of continuity and deformation minimum [HEG 98]. Two data series that have the closest relative distance according to [11.19] and [11.20] are considered to be in the same cluster, independently of the altitude of the point of measurement.

The condition of homogeneity means that inside each cluster, the distance is either greater, for [11.19] or smaller for [11.20] to a value that limits the precision of the calculation. The condition of accuracy means that two clusters are different and that the average distance separating them is smaller for [11.19] or greater for [11.20] to a value that limits the precision of the calculation.

11.1.2.5. *Determination of the precision of the calculation*

The number of unclassified measurement series must be less than 5%.

11.1.2.6. *Interpretation of the results*

An example of a use of this method is presented with the piezocone test (Figure 11.5), and with the pressuremeter (Figure 11.6).

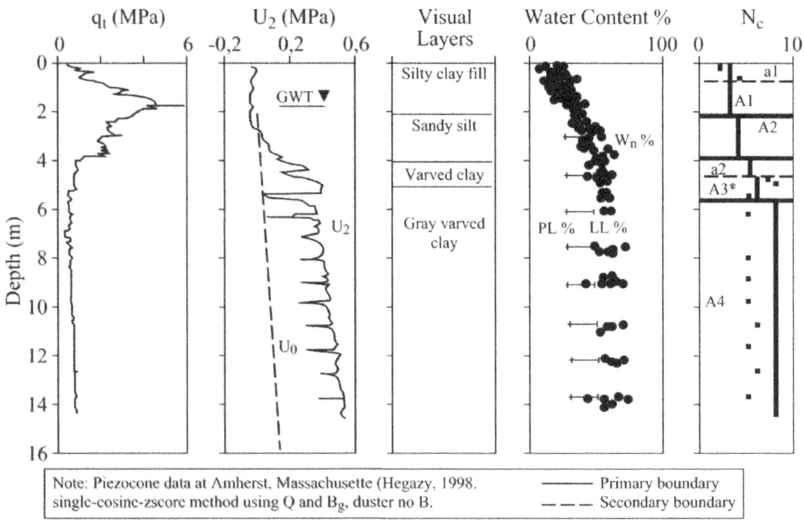

Figure 11.5. *Interpretation by cluster of the penetrometers in the Amherst site, M, Massachusetts [HEG 02]*

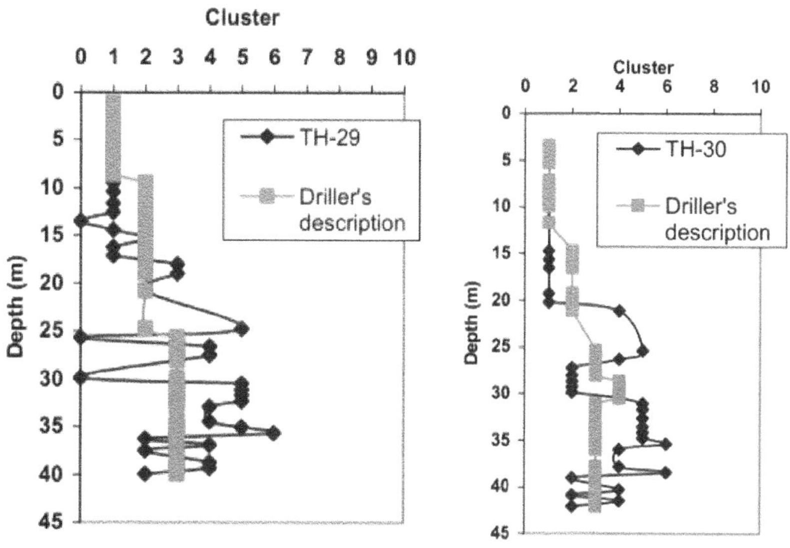

Figure 11.6. *Interpretation by cluster of the pressuremeters in the Grays Harbor site, Washington [MON 15]*

11.2. Characterization of soils

11.2.1. *Identification of the soils*

The coherent soils can be identified according to their content of calcium carbonate (Table 11.2). The classification of silts and clays to be from the size and especially the size of the fine particles (Figure 11.6).

Field testing can identify soil from experimental correlations (Table 11.2). It is also possible to use the results of the lateral sleeve of the penetrometer (Figure 11.10) with the abacus of Schertmann (1978) or this (Figure 11.11; Figure 11.12) of Robertson [ROB 90] or the results ofpressuremeter test(Figure 11.17) with the abacus of Baud and Gambin [BAU 13].

Classes of soil		I_c	p_l^* (MPa)	q_c (MPa)	$(N_{1,60})$	c_u (kPa)
Clays and silts	Very soft to soft	0,0 – 0.50	< 0,4	< 1,0		< 75
	Firm	0,5 – 0,75	0,4 to 1,2	1,0 to 2,5		75 to 150
	Stiff	0,75 – 1,00	1,2 to 2	2,5 to 4,0		150 to 300
	Very stiff	> 1,00	≥ 2	≥ 4,0		≥ 300
Intermediate soils (silty sand, clayey sand, sandy clay)	Classification to be created according to the indications of Figures A.2.1 to A.2.4					
Sands and gravels	Very loose		< 0,2	< 1,5	< 3	
	Loose		0,2 to 0,5	1,5 to 4	3 to 8	
	Reasonably dense		0,5 to 1	4 to 10	8 to 25	
	Dense		1 to 2	10 to 20	25 to 42	
	Very dense		> 2	> 20	42 to 58	
Chalk	Soft		< 0,7	< 5		
	Altered		0,7 to 3	5 to 15		
	Clean		≥ 3	≥ 15		
Marl and marl-limestone	Soft		<1	< 5		
	Stiff		1 to 4	5 to 15		
	Very stiff		> 4	> 15		
Rock	Altered		2,5 to 4			
	Fragmented		> 4			

Table 11.2. *Identification of the soils from the in situtests (AFNOR NF P94-261 [AFN 13])*

VALUE OF $CaCO_3$ [a]	CLASS OF SOIL
0 – 10%	Clay or silt
10 – 30%	Marl-clay or marl-silt
30 – 70%	Marl
70 – 90%	Marl-limestone
90 – 100%	Limestone (or chalk [b])

a value of $CaCo_3$ determined according to norm NF P 94-048

b chalk denomination designates sediment formations of light color, generally blanched or yellowed, porous and light on which certain specific laboratory tests, such as for example the Atterberg limits, can be carried out

Table 11.3. *Identification of the cohesive soils from the value of calcium carbonate (AFNOR NF P94-261 [AFN 13])*

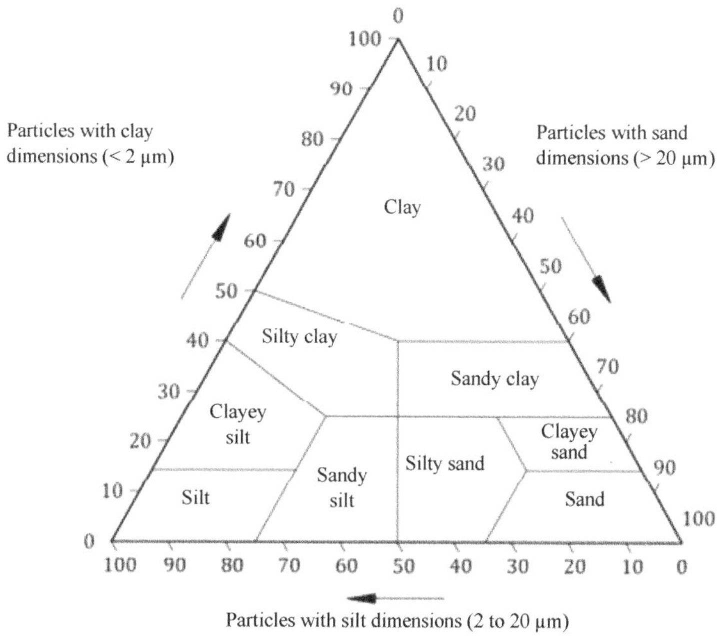

Figure 11.7. *Diagram of the soil classification from granulometry (AFNOR NF P94-261 [AFN 13])*

11.2.2. *Physical and mechanical parameters*

The parameters that describe the physical and mechanical behavior of the ground can be independent of the state of density (Table 11.4, left column) or dependent on the state of density or of deformation (Table 11.4, right column).

Physical parameters not dependent on the initial state	Physical parameters dependent on the initial state
D_{10}; D_{30}; D_{60}	
w_L ; w_p ; w_r ; I_p ;	e ; S_r ; w
γ_s ; ρ_s	γ ; γ_d; γ_{sat} ; γ_{OPN} ; ρ ; ρ_d ; ρ_{sat} ; ρ_{OPN}
	Parameters dependent on the initial state
	σ_0 ; σ'_0 ; σ_{v0} ; σ'_{v0} ; K_0 ; k_s
	Hydraulic parameters dependent on the density

	k ; C_v ; u_0 ;
	Elastic parameters dependent on the deformation
	E , ν ; E_u , ν_u ; G ; E_M ; E^e ; E_D ; K_D ; I_D
Odometric parameters not dependent on the deformation	Odometric parameters dependent on the deformation
λ ; κ ; C_s ; C_c ;	E_{oed} ; σ_p ;
Shearing parameters not dependent on the density of the soil	Shearing parameters dependent on the density of the soil
Φ_r ; Φ_μ ; Φ_{ec}	c_u ; Φ_u ; c' ; Φ' ; Ψ ; N'_{70} ; q_c ; q_d ; p_{LM} ; p^*_{LM} ;

Table 11.4. *Classification of the characteristics of the ground depending on the influence of the initial state, the density, the deformation*

11.2.3. *Correlations and relations between the characteristics measured in the laboratory*

11.2.3.1. *Correlation involving the Atterberg limits*

Evolution of the plasticity index as a function of the liquidity limit has been studied by Casagrande [CAS 36] who proposed relations [11.21] and [11.23] to separate clays and silts with low levels of plasticity (Figure 11.7). Later, Favre [FAV 80] proposed the experimental correlation [11.22] for a same class of soil.

$$I_p = 0{,}73.(w_L - 0{,}08) \quad [11.21]$$

$$I_p = 0{,}73.(w_L - 0{,}13) \quad [11.22]$$

$$I_p = 0{,}73.(w_L - 0{,}20) \quad [11.23]$$

11.2.3.2. *Correlations involving odometric parameters*

Terzaghi proposed a relation between the compression coefficient C_c and the liquidity limit.

$$C_c = 0{,}9.(w_L - 0{,}13) \quad [11.24]$$

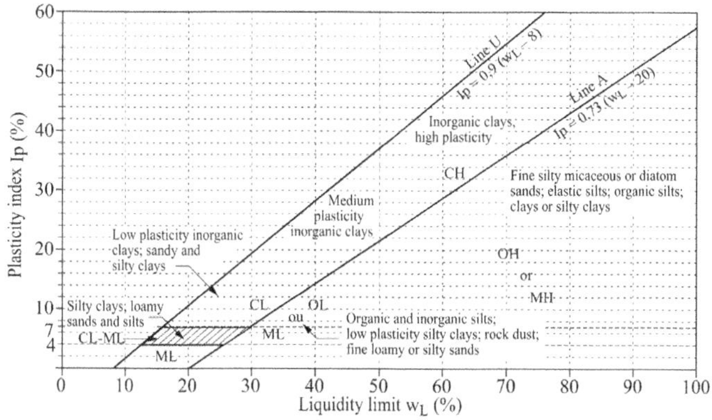

Figure 11.8. *Classification of soils [CAS 36]*

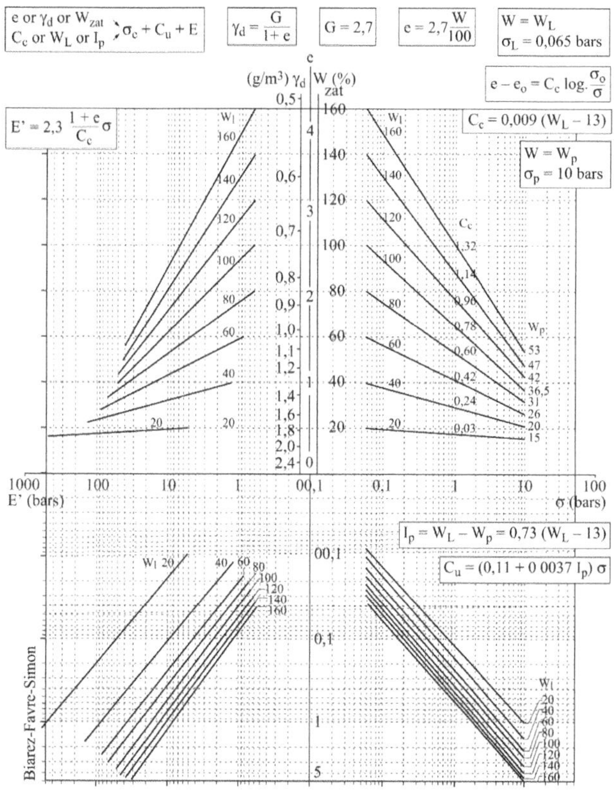

Figure 11.9. *Correlations between odometric curve and undrained cohesion for ordinary, modified and saturated clays (Biarez, Favre, Simon)*

Favre *et al.* [FAV 02] showed that the consistency index I_c [11.25] is linked to the consolidation constraint σ_c at the odometer for clays whose liquidity limit is contained between 20 and 140% through relation [11.26].

$$I_c = \frac{(w_L - w)}{I_p} \qquad [11.25]$$

$$I_c = 0.46.(Log(\sigma_c') - 0.83) \qquad [11.26]$$

The results of odometric tests [MES 77, GRA 83] show that the creep coefficient C_α, depends on the vertical constraint applied, with a correlation of the [11.27] type and with the value of c defined by [11.28]:

$$\frac{C_\alpha}{c} = constant \approx 0{,}44 \qquad [11.27]$$

$$c = \frac{\sigma_z'(1+e)}{E_{oed}} \qquad [11.28]$$

Biarez and Favre [BIA 75] proposed the following correlation [11.29] between the liquidity limit w_L and the compression index C_c of the first load. Index L corresponds to the value of the parameters at the liquidity limit. By taking into account relation [11.22], the proposed correlation [11.30] between the compression index and the liquidity limit. According to Wood [WOO 90], for a given soil or group of soils, it is better to look for a [11.31] type relation, with the experimental coefficients a and b.

$$e - e_L = -C_c.Log\left(\frac{\sigma_z'}{\sigma_{zL}'}\right) \qquad [11.29]$$

$$C_c. = 0{,}84.(w_L - 0{,}13) \qquad [11.30]$$

$$C_c. = a.(w_L - b) \qquad [11.31]$$

Favre, Biarez and Mekkaoui [FAV 02] generalize equation [11.29] to triaxial tests, finding a correlation with a slope λ of the critical state line. Index L corresponds to the value of the parameter at the liquidity limit; index s corresponds to the value of the parameter at the critical state.

$$e - e_L = -\lambda.Ln\left(\frac{p_{cs}'}{p_{csL}'}\right) \qquad [11.32]$$

Boutonnier [BOU 07] uses the modified Cam–Clay model [ROS 68] to find the equation of the isotropic compression curve [11.33] from relation e-ln(p') at the critical state

$$e = e_L - \kappa.Ln(2) - \lambda.Ln\left(\frac{p'}{p'_{0L}}\right) \qquad [11.33]$$

11.2.3.3. *Correlations involving the Proctor test*

The Proctor optimum can be estimated using the correlations shown (Table 11.5) from Boutonnier [BOU 07]. Moreover, Fleureau *et al.* [FLE 02] provide the interstitial pressure for a normal Proctor optimum (NPO) using relation [11.34] and for a modified Proctor optimum (MPO) using relation [11.35]. Acar and Nyeretse [ACA 92] propose, for mixes of sands, kaolinite and montmorillonite, in variable proportions, a relation [11.36] that depends on the plasticity index I_p.

$$u_{wOPN} = -0{,}118.(w_L)^{1{,}98}\ in\ kPa \qquad [11.34]$$

$$u_{wOPM} = -1{,}72.(w_L)^{1{,}64}\ in\ kPa \qquad [11.35]$$

$$u_{wOPM} = -69{,}7.(I_p)^{0{,}58}\ in\ kPa \qquad [11.36]$$

Authors	Compacting energy	Range w_L	Proposed correlations w_{opt} (%) and ρ_{dopt} (tons/m³) in function of w_L
Fleureau *et al.* (2002)	Normal Proctor	17 to 170	$\gamma_{dopt} = \rho_{dopt}\, g = 21.00 - 0.113\, w_L + 0.00024\, (w_L)^2$ $r^2 = 0.86$ $g = 9.81\ m/s^2$ [VI.7] $w_{opt} = 1.99 + 0.46\, w_L - 0.0012\, (w_L)^2$ $r^2 = 0.94$ [VI.8]
Fleureau *et al.* (2002)	Modified Proctor	17 to 170	$\gamma_{dopt} = \rho_{dopt}\, g = 20.56 - 0.086\, w_L + 0.00037\, (w_L)^2$ $r^2 = 0.77$ $g = 9.81\ m/s^2$ $w_{opt} = 4.55 + 0.32\, w_L - 0.0013\, (w_L)^2$ $r^2 = .88$
Gress et Autret (2002)	Normal Proctor	20 to 60	$\rho_{dopt} = 2.09 - 0.00927\, w_L$ $w_{opt} = 7.92 + 0.268\, w_L$
Popovic et Sarac (1980)	Normal Proctor	25 to 70	$\rho_{dopt} = \dfrac{2.7}{1.283 + 0.00818 \cdot w_L}$ $w_{opt} = 8.14 + 0.257\, w_L$

Table 11.5. *Correlations proposed by different authors to define the Proctoc optimum [BOU 07]*

11.2.3.4. *Correlations s between the void ratio and the friction angle*

Several relations between the laboratory characteristics can be seen below.

There is a correlation [11.37] between the void ratio and the internal friction angle for sands, provided by Caquot [CAQ 39]:

$$e.tg(\phi) = Constant \quad [11.37]$$

11.2.3.5. *Correlations involving cohesion*

The correlation between undrained cohesion and the plasticity index, during shearing of normally consolidated clay, has also been studied by Skempton, who proposes [11.38].

For overconsolidated clays, Coste *et al.* [COS 14] and Boutonnier *et al.* [BOU 13] use relation [11.39], which allows them to properly determine the stacking of LGV embankments, while odometric tests do not allow this.

$$\frac{c_u}{\sigma_p} = 0{,}11 + 0{,}38.I_p \quad [11.38]$$

$$\frac{c_u}{\sigma'_p} = 0{,}35 \quad [11.39]$$

11.2.3.6. *Correlations involving permeability*

Hazen [HAZ 95] proposed a correlation between the permeability coefficient k and the diameter D10 of a loose sand [11.40]:

$$k = 100.D_{10}^2 \quad [11.40]$$

Casagrande completed the study with a large range of soils and proposes a correlation between the permeability coefficient k and the permeability coefficient $k_{0,85}$ for the soil with a void ratio of 0.85, and the void ratio e:

$$k = 1{,}4.k_{0{,}85}.e^2 \quad [11.41]$$

11.2.3.7. *Correlations involving the coefficient of at rest pressure*

There is a correlation between the coefficient of at rest pressure and intergranular friction, which is due to Rowe [11.42] and another due to Jacky [11.43].

For fine coherent soils, Mestat [MES 98] shows [11.44].

For detrital soils, Tuna *et al.* [TUN 08] propose relation [11.45].

$$K_0 = tg\left(\frac{\pi}{4} - \frac{\phi_\mu}{2}\right) \quad [11.42]$$

$$K_0 = 1 - sin\phi \quad [11.43]$$

$$K_0 = (1 - sin\phi).\sqrt{\frac{\sigma'_p}{\sigma'_v}} \quad [11.44]$$

$$K_0 = 0.376 + 0.024.K_D - 0.00172.\frac{q_c}{\sigma'_{vo}} \quad [11.45]$$

11.2.3.8. *Relations involving intergranular friction and critical angles*

There is a relation between the intergranular friction angles Φ_μ and the critical friction angle Φ_{cv} which is obtained from the expression using the Drucker intergranular friction criterion [11.46], while the critical friction angle is defined using the Mohr–Coulomb criterion [11.47]. These two different definitions lead to relation [11.48] between Φ_μ and Φ_{cv}: these two angles depend on the plasticity of the soil, and can be correlated with the plasticity index I_p (Table 11.6).

Soil type	Ip (%)	Φ_μ (degree)	Φ_{cv} (degree)
Sand, coarse	–	29°	29,4°
Lime, class A1	0-12	28°	28,3°
Clay, little plasticity A2	12-25	20°	20°
Clay, plasticity A3	25-40	15°	15°
Clay, much plasticity A4	> 40	10°	10,1°

Table 11.6. *Normal values of intergranular friction Φ_μ and at the critical state Φ_{cv} [MON 13]*

$$tg(\phi_\mu) = \frac{(\sigma_1-\sigma_3)_{cv}.\sqrt{2}}{(\sigma_1+2.\sigma_3)_{cv}} = \frac{\left({}^{\sigma_1}/_{\sigma_3}-1\right)_{cv}.\sqrt{2}}{\left({}^{\sigma_1}/_{\sigma_3}+2\right)_{cv}} \quad [11.46]$$

$$sin\phi_{cv} = \frac{(\sigma_1 - \sigma_3)_{cv}}{(\sigma_1 + \sigma_3)_{cv}} \quad [11.47]$$

$$tg(\phi_\mu) = \frac{2.sin\phi_{cv}.\sqrt{2}}{3 - sin\phi_{cv}} \quad [11.48]$$

11.2.3.9. *Relations involving dilatancy*

There is a relation [11.49] between the intergranular friction angle Φ_μ and the dilatance angle Ψ proposed by Monnet [MON 12], and whose demonstration is presented in section 9.5.3.2, while relation [11.50] by Bolton [BOL 86] is totally experimental. The dilatance relation [11.51] according to Rowe [ROW 62] is also shown. Considering the small difference between the two angles Φ_μ and Φ_{cv} we can compare the value of the dilatance obtained with these three formulas by assuming equality between Φ_μ and Φ_{cv}. This study is shown in Figure 11.9. A difference can be seen between the value for the dilatance according to Monnet [MON 12] and according to Rowe [ROW 62] of less than 2° in the common range of friction angles (20–45°).

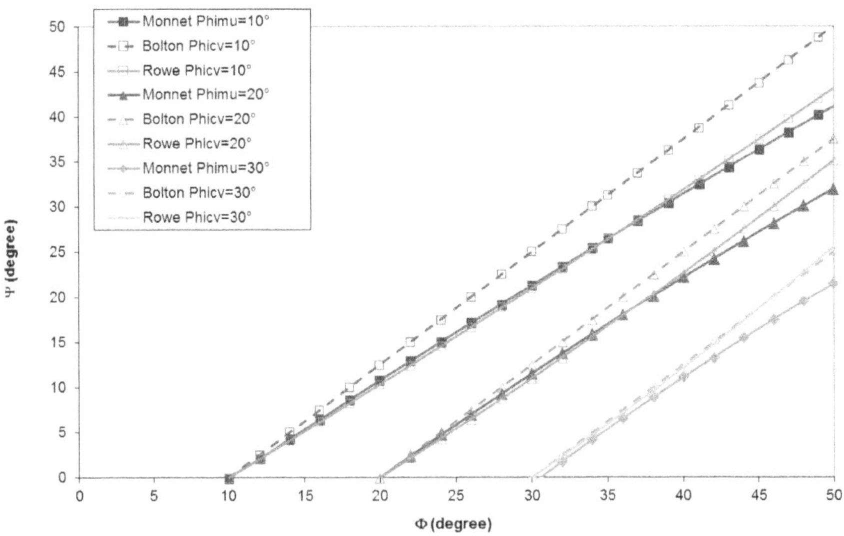

Figure 11.10. *Value of dilatance calculated according to the expressions by Monnet, Bolton and Rowe [MON 12]*

$$sin\psi = \frac{3.\sqrt{2}.\left[\frac{2.sin\phi'.\sqrt{2}}{(3-sin\phi')}-tg\phi_{\mu}\right]}{4+\sqrt{2}.\left[\frac{2.sin\phi'.\sqrt{2}}{(3-sin\phi')}-tg\phi_{\mu}\right]} \quad [11.49]$$

$$0{,}8.\psi = \phi' - \phi_{cv} \quad [11.50]$$

$$sin\psi = \frac{\frac{1+sin\phi}{1-sin\phi}\cdot\frac{1-sin\phi_{cv}}{1+sin\phi_{cv}}-1}{\frac{1+sin\phi}{1-sin\phi}\cdot\frac{1-sin\phi_{cv}}{1+sin\phi_{cv}}+1} \quad [11.51]$$

11.2.4. *Correlations involving in situ tests*

In general, correlations are built on the experience of practitioners, using measurements from different sites, and they provide an order of magnitude for the characteristics of soils that can be reached. As a result, these must be carefully used, and direct measurements of a soil characteristic should always be favored over an estimation provided by a correlation.

11.2.4.1. *The standard penetration test*

11.2.4.1.1. Correlation between the number of blows N'_{70} and the density of the soil

There is a correlation between the density of the soil and the number of blows N'_{70} in the SPT (Tables 11.7 and 11.8)

Relative density		Consistency		
Sands and Gravels	Blows/Ft (N_{SPT})	Silts and Clay	Strengh (tsf)	Blows/Ft (N_{SPT})
Very loose	0-4	Very soft	0-¼	0-2
Loose	4-10	Soft	¼-½	2-4
Medium	10-30	Firm	½-1	4-8
Dense	30-50	Stiff	1-2	8-16
Very dense	Over 50	Very stiff	2-4	16-32
		Hard	Over 4	Over 32

Table 11.7. *Correlation between the number of blows in the SPT N'70 and the density of the soil [PEC 53]*

Description*	Very loose	Loose	Medium	Dense	Very dense
Relative density D_r	0	0,15	0,35	0,65	0,85
SPT N'_{70} fine	1-2	3-6	7-15	16-30	?
medium	2-3	4-7	8-20	21-40	> 40
coarse	3-6	5-9	10-25	26-45	> 45
ϕ : fine	26-28	28-30	30-34	33-38	
medium	27-28	30-32	32-36	36-42	< 50
coarse	28-30	30-34	33-40	40-50	
γ_{wet} pcf	70-100†	90-115	110-130	110-140	130-150
(kN,m³)	(11-16)	(14-18)	(17-20)	(17-22)	(20-23)

† Excavated soil or material dumped from a truck will weigh 11 to 14 kN/m³ and must be quite dense to weigh much over 21 kN/m³. No existing soil has a D. = 0.00 nor a value of 1.00-common ranges are from 0.3 to 0.7.

Table 11.8. *Correlation between the number of blows in the SPT N'70, the density of the soil and the friction*

11.2.4.1.2. Correlation between the number of blows N'70 and the resistance to shearing characteristics

There is a correlation between the friction angle of the soil and the number of blows N'_{70} in the SPT (Table 11.6), as well as cohesion (Table 11.8).

No of Hits.N	Consistency	q_u ton/aq ft
Below 2	Very soft	Below 2.25
2 - 4	Soft	0.25 - 0.50
4 - 8	Medium	0.50 - 1.00
8 - 15	Stiff	1.00 - 2.00
15 - 30	Very stiff	2.00 - 4.00
Over 30	Hard	Over 4.00

Table 11.9. *Correlation between the number of blows in the SPT N'70, the density of the soil and cohesion [TER 48]*

11.2.4.1.3. Correlation between the number of blows N'70 and pressiometric characteristics

There are correlations between result N'_{70} of the SPT and the pressiometric Ménard values (Table 11.10).

11.2.4.2. *The penetrometer*

11.2.4.2.1. Correlation between resistance to penetration q_c and the nature of the soil

Wroth [WRO 84] proposed normalizing the penetrometer by using the normalized parameters of the cone resistance [11.52], from the normalized friction ratio F_r [11.53] is the device has a lateral friction sleeve, and from the ratio of the interstitial pressure B_q [11.54] is the device is a piezocone.

$$Q_t = \frac{q_c - \sigma'_{vo}}{\sigma'_{vo}} \qquad [11.52]$$

$$F_r = \frac{f_s}{q_c - \sigma_{vo}} \qquad [11.53]$$

$$B_q = \frac{u_2 - u_0}{q_c - \sigma_{vo}} \qquad [11.54]$$

Nature of the soil	p_{LM} (MPa)	E_M (MPa)
Silt	$N'_{70}/(30\ to\ 35)$	$N'_{70}/(2.5\ to\ 3)$
Sand	$N'_{70}/(15\ to\ 25)$	$N'_{70}/(1.5\ to\ 3)$
Clay	$N'_{70}/(15\ to\ 30)$	$N'_{70}/(1\ to\ 2.5)$
Marl	$N'_{70}/(20\ to\ 25)$	$N'_{70}/(1.5\ to\ 2.5)$
Chalk	$N'_{70}/(5\ to\ 20)$	$N'_{70}/(0.7\ to\ 1)$

Table 11.10. *Correlation between the number of blows in the SPT N'70 and the pressiometric characteristics*

By using these non-dimensional variables, Robertson [ROB 90] proposes an identification of the soils using the resistance to penetration q_c (Figure 11.10) or using the normalized resistance to penetration Q_c (Figure 11.11).

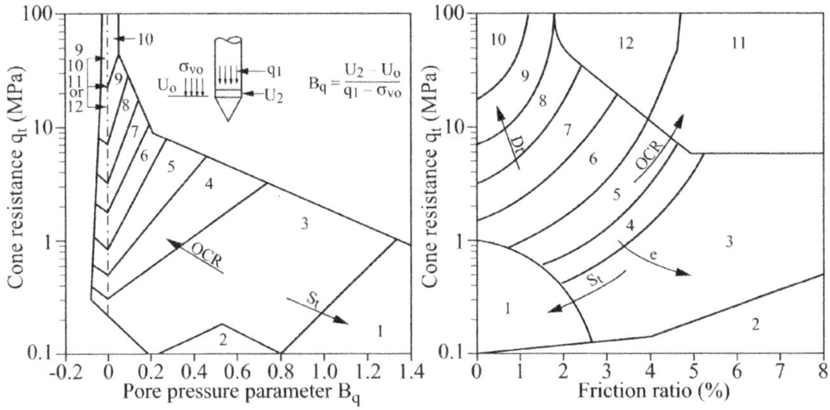

Zone: Soil Behaviour Type:

1 -	Sensitive fine grained	5 -	Clayey silt to silty clay	9 -	Sand
2 -	Organic material	6 -	Sandy silt to clayey silt	10 -	Gravelly sand to sand
3 -	Clay	7 -	Silty sand to sandy silt	11 -	Very stiff fine grained*
4 -	Silty clay to clay	8 -	Sand to silty sand	12 -	Sand to clayey sand

*Overconsolidated or cemented.

Figure 11.11. *Classification of soils according to the results of the piezocone, according to Robertson [ROB 90]*

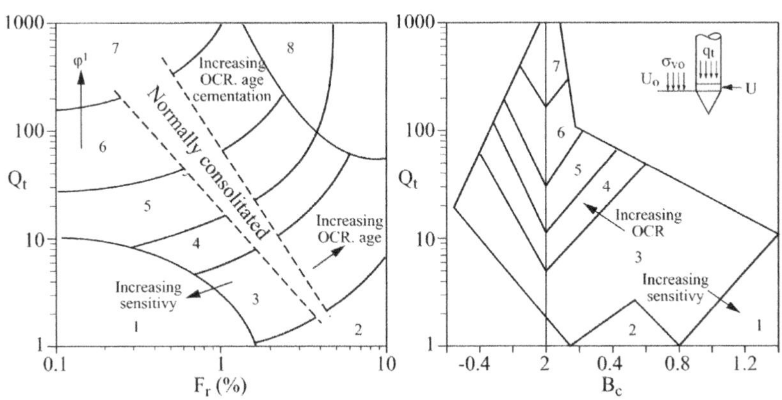

$$Q_t = \frac{q_t - \sigma_{vo}}{\sigma'_{vo}} \qquad B_c = \frac{u_2 - u_0}{q_t - \sigma_{vo}} \qquad F_r = \frac{f_s}{q_t - \sigma_{vo}} \times 100\ \%$$

Zone: Soil Behaviour Type:

1 -	Sensitive fine grained	4 -	Silt mixtures clayey silt to silty clay	7 -	Gravely sand to sand
2 -	Organic soils-peats	5 -	Sand mixtures; silty sand to sand silty	8 -	Very stiff sans to clayey sand
3 -	Clay-clay to silty clay	6 -	Sands: clean sand to silty sands	9 -	very stiff fine grained

Figure 11.12. *Classification of soils according to the normalized results of the piezocone, following [ROB 90]*

11.2.4.2.2. Correlation between the resistance to penetration q_c and deformation or stacking

An experimental relation proposed by Sanglerat [SAN 72] allows us to link static peak resistance q_c to the odometric modulus E_{oed}, following formula [11.55]. The values of the empirical coefficient α are shown in Table 11.11. In this table, q_c is written as R_p.

$$E_{od} = \alpha . q_c \qquad [11.55]$$

Class of soil	Value of q_c	Value of α
CL – low plasticity clay	$q_c < 700\ kPa$ $700\ kPa < q_c < 2\ MPa$ $2\ MPa < q_c$	$4 < \alpha < 6{,}5$ $3 < \alpha < 4$ $1{,}3 < \alpha < 2{,}2$
ML – low plasticity silt	$q_c < 2\ MPa$ $2\ MPa < q_c$	$3 < \alpha < 6$ $1 < \alpha < 3$
CH – very plastic clay MH – very plastic silt	$q_c < 2MPa$	$2{,}5 < \alpha < 5$
OL – highly organic silt	$q_c < 1{,}2\ MPa$ $1{,}2\ MPa < q_c$	$2 < \alpha < 8$ $0{,}5 < \alpha < 4$
T-OH – highly organic clay and peat	$q_c < 700\ kPa$ and $50\% < w < 100\%$ $100\% < w < 200\%$ $200\% < w$	 $1{,}5 < \alpha < 4$ $1 < \alpha < 1{,}5$ $0{,}4 < \alpha < 1$
Chalk	$q_c < 3\ MPa$ $3\ MPa < q_c$	$2 < \alpha < 4$ $1{,}5 < \alpha < 3$
Sand	$q_c < 10\ MPa$	$\alpha = 2$
Compact sandy gravel	$q_c > 10\ MPa$	$2 < \alpha < 3$

Table 11.11. *Evolution of coefficient α depending on the soil encountered [SAN 72]*

11.2.4.2.3. Correlation between the resistance to penetration q_c and the undrained cohesion c_u

In the case of purely coherent soils, peak resistance allows us to deduce an approximate value of the undrained cohesion c_u for short-term study. Sanglerat [SAN 72] proposed the following rules:

– for normally consolidated clays, when q_c is lower than 2,000 kPa and the device is equipped with a protective sleeve, we can use relation [11.56]:

$$\frac{q_c}{18} < c_u < \frac{q_c}{15}; \qquad [11.56]$$

– in the case where q_c is measured with the single cone, we can use relation [11.57]:

$$c_u < \frac{q_c}{10}; \qquad [11.57]$$

– for overconsolidated clays, when q_c is greater than 2.5 MPa, we can use this relation:

$$\frac{q_c}{26} < c_u < \frac{q_c}{22} \qquad [11.58]$$

11.2.4.2.4. Correlation between resistance to penetration q_c and resistance to liquefaction

Robertson and Wride [ROB 98] studied the liquefaction of loose sands and related the standardized results of the penetrometer to the liquefaction potential of a number of soils. They propose an abacus (Figure 11.12) to estimate the risk of liquefaction from the penetrometer test.

11.2.4.2.5. Correlation between the horizontal constraint index K_D and resistance to liquefaction

Monaco *et al.* [MON 05] studied the liquefaction of loose sands using the Marchetti flat dilatometer. This study was taken up by Viana da Fonseca [VIA 13] who validated the previous works using an abacus [9.147] to estimate the risk of liquefaction from the Marchetti test.

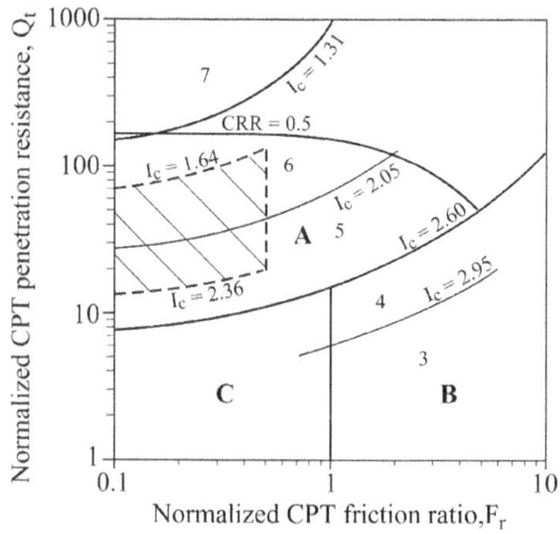

Figure 11.13. *Liquefaction potential; zone A liquefaction possible depending on the amplitude and the length of the cyclical loading; zone B No liquefaction; zone C liquefaction can be possible depending on the sensitivity of the soil as well as the amplitude and the duration of the loading cycle [ROB 98]*

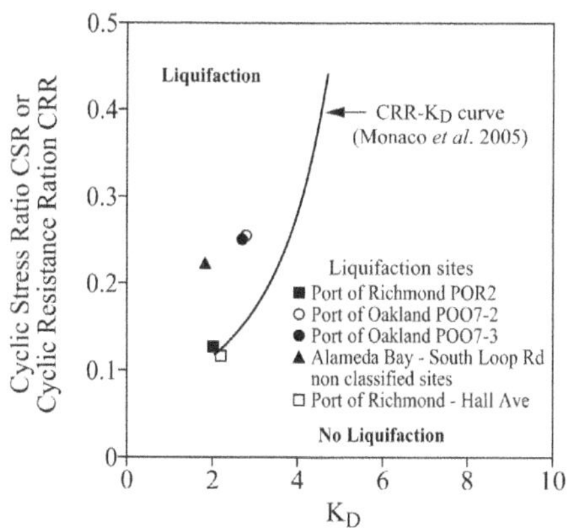

Figure 11.14. *Liquefaction potential from the Marchetti flat dilatometer test [VIA 13]*

11.2.4.2.6. Correlation between resistance to penetration q_c and q_d and the results of the other tests

There are several correlations between the static peak resistance q_c and dynamic resistance to penetration q_d shown below. The values of A are shown in Table 11.10, column 2 as well as correlations between the penetrometric qc, pressiometric p_{LM} and E_M, SPT N_{70} values shown hereafter in Tables 11.12 and 11.13.

$$q_d = A.q_c \qquad [11.59]$$

Nature of the soil	$A = q_d/q_c$	q_d/p_l	q_d/E_M
Normally consolidated clays, silts and muds, loose or medium density sands	1	1.4 to 2.5	0.1 to 0.3
Overconsolidated clays and silts	1 to 2	3 to 5	0.2 to 0.4
Sands and gravels, dense to very dense silty or loamy sands	0.5 to 1	5 to 10	0.4 to 1.5

Table 11.12. *Empirical relation between q_c and q_d; q_d and p_{LM}; q_d and E_M [WAS 82]*

Correlation	Clays	Silts	Sandy clays or loamy sands	Sands	Sands and gravel	Observations
$c_u =$	$\frac{p_{LM} - p_0}{a} + b$					$p_l - p_0$ (MPa) / a / b (MPa): ≤ 0.3 / 5.5 / 0 0.3 à 1 / 12 / 0.03 / 10 / 0.025 1 à 2.5 / 35 / 0.085
$\frac{q_c - q_0^*}{c_u} =$	15					
$\frac{q_c}{q_d} =$	1	1 1 to >3	0.5–0.9 0.4–0.1	1 1	1 1 to 2.6	Above the water table Above the water table, value of 1 seems the most frequent
$\frac{q_d}{N_{70}} =$	0.2	0.2–0.3	0.3	0.4	0.8	
$\frac{q_c}{N_{70}} =$	0.1–0.2	0.2–0.3	0.3–0.4	0.3–0.6		
$\frac{N_{70}}{p_{LM} - p_0} =$	15–20	30	20	20		MPa
$\frac{N_{70}}{E_M} =$	1–1.5	3		1,5–2		MPa
$\frac{q_c - p_0'}{p_{LM} - p_0} =$	3–4	6		8–10		
$\frac{E_M}{q_c - p_0'} =$	3–6	3		1–1.5		
$\frac{q_d}{p_{LM} - p_0} =$	3–4					Above the water table

Table 11.13. *Empirical relation between q_c and q_d; q_d and p_{LM}; q_d and E_M (Pilot)*

11.2.4.7. Correlation between the resistance to penetration q_c and the limiting constraint under a superficial foundation q_{net}

For superficial foundations, we can assume that the penetrometer represents a reduced model of the foundation. Under these conditions, the limiting constraint q_{net} that the foundation can withstand is linked to q_c through relation [11.60]. The pressiometric soil-bearing factor k_c is a function of the soil classes. In the Eurocode 7 regulations, it is possible to select a curve for the value of the soil-bearing capacity k_c (Figure 11.14), or to use the coefficients (Table 11.12) and to use formula [11.61] for a square or circular foundation and [11.62] for a rectangular foundation. The soil-bearing factor k_c is then introduced into formula [11.60] to determine q_{net}. Coefficients i_δ and i_β are dependent on the incline of the load.

$$q_{net} = k_c . q_c . i_\delta . i_\beta \quad [11.60]$$

$$k_c = k_{c0} + \left(a + b.\frac{D_e}{B}\right).\left(1 - e^{-c.\frac{D_e}{B}}\right) \quad [11.61]$$

$$k_c = k_{c0} + \left(1 - \frac{B}{L}\right) + k_c.\frac{B}{L} \quad [11.62]$$

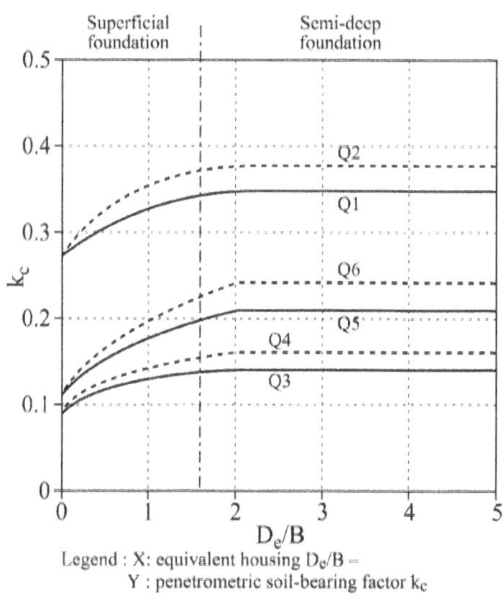

Figure 11.15. *Penetrometric soil-bearing factor k_c [AFN 13]*

<table>
<tr><th colspan="2" rowspan="2">Soil category
Variation curve of the soil-bearing factor</th><th colspan="4">Expression of k_c</th></tr>
<tr><th>a</th><th>b</th><th>c</th><th>k_{c0}</th></tr>
<tr><td rowspan="2">Clays and silts</td><td>Strip foundation - Q1</td><td>0.07</td><td>0.007</td><td>1.3</td><td>0.27</td></tr>
<tr><td>Square foundation - Q2</td><td>0.1</td><td>0.,007</td><td>1.5</td><td>0.27</td></tr>
<tr><td rowspan="2">Sands and gravel</td><td>Strip foundation - Q3</td><td>0.04</td><td>0.006</td><td>2</td><td>0.09</td></tr>
<tr><td>Square foundation - Q4</td><td>0.03</td><td>0.02</td><td>5</td><td>0.09</td></tr>
<tr><td rowspan="2">Chalk</td><td>Strip foundation - Q5</td><td>0.04</td><td>0.03</td><td>3</td><td>0.11</td></tr>
<tr><td>Square foundation - Q6</td><td>0.05</td><td>0.04</td><td>3</td><td>0.11</td></tr>
<tr><td rowspan="2">Marl and marl-limestone
Altered rocks</td><td>Strip foundation - Q5</td><td>0.04</td><td>0.03</td><td>3</td><td>0.11</td></tr>
<tr><td>Square foundation - Q6</td><td>0.05</td><td>0.04</td><td>3</td><td>0.11</td></tr>
</table>

Note 1 Intermediate soils defined in clause A.2 (4) are assimilated to either powdery soils or coherent soils in the following manner:
- "clay and silt" type soils also include silty clays, loamy silts and sandy clays (from Figures A.2.1 and A.2.3 and from zones 1, 3, 4 and 9 of Figure A.2.3);
- "sand and gravel" type soils also include loamy sands, silty sands and sandy silts (from Figure A.2.1 and A.2.3 and from zones 5 to 8 of Figure A.2.3).

Table 11.14. *Penetrometric soil-bearing factor k_c [AFN 13]*

11.2.4.2.8. Correlation between the resistance to penetration q_c and the soil-bearing capacity of a deep foundation

Deep foundations put up resistance through two terms [11.63], peak effort at the inferior extremity of the pile R_b, and the lateral friction effort along the pile R_s. The correlation between penetrometric resistance and the soil-bearing capacity of the piles is described in the regulatory standard of Eurocodes 7 [AFN 12b].

11.2.4.2.8.1. Peak term

The mobilizable effort at the peak of the pile [11.64] is linked to the terrain's rupture pressure under the peak of the pile q_b and at the straight section of the pile A_b. The following formula is used [11.65] to calculate the ultimate constraint q_b under the peak of the pile from the smoothed penetrometric peak q_{ce} and possibly also from the initial vertical constraint q_0 in the absence of foundation. The penetrometric soil-bearing coefficient k_c is dependent on the nature of the soil (Table 11.15).

$$R_c = R_b + R_s \quad [11.63]$$

$$R_b = A_b . q_b \quad [11.64]$$

$$q_b = q_0 + k_c . q_{ce} \quad [11.65]$$

The penetrometric soil-bearing coefficient is dependent on the effective housing length of the pile D_{ef} divided by the diameter of the pile B. Ehen D_{ef}/B greater than 5, then relation [11.66] is used with values indicated in Table 11.13.

$$k_c = k_{cmax} \quad [11.66]$$

When D_{ef}/B has a value of less than 5, then we use relation [11.67] for clays and silts; [11.68] for intermediate soils; [11.69] for sands and gravels and; [11.70] for chalk, marls and altered or fragmented rocks.

$$k_c = 0{,}3 + (k_{cmax} - 0{,}3) . \frac{(D_{ef}/B)}{5} \quad [11.67]$$

$$k_c = 0{,}2 + (k_{cmax} - 0{,}2) . \frac{(D_{ef}/B)}{5} \quad [11.68]$$

$$k_c = 0{,}1 + (k_{cmax} - 0{,}1) . \frac{(D_{ef}/B)}{5} \quad [11.69]$$

$$k_c = 0{,}15 + (k_{cmax} - 0{,}15) . \frac{(D_{ef}/B)}{5} \quad [11.70]$$

The resistance to equivalent penetration q_{ec} helps smooth the pressuremeter curve and obtains an average value for the resistance to penetration between 0.5 m and B around the pile peak according to relation [11.71]. D is the depth of the head of the pile around which is calculated the average resistance above the pile with distance b [11.73], below the pile with distance a [11.72] and h the height of the housing of the pile in the soil-bearing formation.

The height of the effective housing D_{ef} is given in relation [11.74] in which hD denotes a length equal to 10.B and D refers to the length of the pile.

$$q_{ce} = \frac{1}{b+3.a} \cdot \int_{D-b}^{D+3a} q_{cc}(z).dz \qquad [11.71]$$

$$a = max\left\{\frac{B}{2}; 0{,}5m\right\} \qquad [11.72]$$

$$b = min\{a; h\} \qquad [11.73]$$

$$D_{ef} = \frac{1}{q_{ce}} \cdot \int_{D-h_D}^{D} q_c(z).dz \qquad [11.74]$$

Terrain / Pile class (c)	Clay % $CaCo_3$ < 30 % Silt	Intermediate soil	Grave sand	Chalk	Marl and marl-limestone	Altered or fragmented rock (a)
1	0.4 (b)	0.3 (b)	0.2 (b)	0.3 (b)	0.3 (b)	0.3 (b)
2	0.45	0.3	0.25	0.3	0.3	0.3
3	0.5	0.5	0.5	0.4	0.35	0.35
4	0.45	0.4	0.4	0.4	0.4	0.4
5#	0.35	0.3	0.25	0.15	0.15	0.15
6#	0.4	0.4	0.4	0.35	0.2	0.2
7#	0.35	0.25	0.15	0.15	0.15	0.15
8	0.45 (b)	0.3 (b)	0.2 (b)	0.3 (b)	0.3 (b)	0.25 (b)

(#) for BAO, HB and PP type piles, installed by vibro-sinking instead of beating, a reduction of 50% must be made on factor k_c

Table 11.15. *Values of the maximum soil-bearing coefficient k_{cmax} for a relative housing D_{ef}/B greater than 5 [AFN 12b]*

11.2.4.2.8.2. Term of lateral friction

The effort that can be mobilized by friction along the pile [11.75] is linked to the perimeter of the pile P_s, to the length D of the pile and to the value of the unitary lateral friction q_s, which evolves as a function of the edge z along the pile between 0 and D. The limit value of the unitary lateral friction q_s at a depth of z is determined in terms of the penetrometric resistance to penetration q_c using the following expression [11.76].

Coefficient $\alpha_{pile\text{-}soil}$, which depends on the type of pile and on the type of soil, is shown in Table 11.16.

The friction f_{soil} is determined in terms of the nature of the soil and the resistance to penetration q_c .The function f_{soil} can be read graphically (Figure 11.15) or calculated analytically using [11.77]. The values of coefficients a, b and c are read in Table 11.17:

$$R_s = P_s . \int_0^D q_s(z). dz \quad [11.75]$$

$$q_s(z) = \alpha_{pile-soil} \cdot f_{soil}[q_c(z)] \quad [11.76]$$

$$f_{soil}[q_c(z)] = (a. q_c + b). (1 - e^{-c.q_c}) \quad [11.77]$$

No	Abbreviation	Application technique	Clay % $CaCo_3$ < 30% Silt	Intermediate soils	Grave sands	Chalk	Marl and Marl-Limestone	Altered or fragmented rock
1	FS##	Simple bore (piles and strips)	0.55	0.65	0.70	0.80	1.40	1.50
2	FB##	Mud bore (piles and strips)	0.65	0.80	1.00	0.80	1.40	1.50
3	FTP	Tubed bore (Lost ferrule)	0.35	0.40	0.40	0.25	0.85	–
4	FTR	Tubed bore (recovered ferrule)	0.65	0.80	1.00	0.75	0.13	–
5	FSR, FBR, PU##	Simple or mud bore with grooving or well	0.70	0.85	–	–	–	–
6	FTC, FTCD	Continuous auger bore with single or double rotation	0.75	0.90	1.25	0.95	1.50	1.50
7	VM	Formed auger	0.95	1.15	1.45	0.75	1.60	–
8	VT	Tubed auger	0.30	0.35	0.40	0.45	0.65	–
9	BPF++, BPR	Beating concrete, pre-made or pre-constrained	0.55	0.65	1.00	0.45	0.85	–
10	BE++	Coated beating (concrete, mortar, grout)	1.00	1.20	1.45	0.85	1.50	–
11	BM++	Formed beating	0.60	0.70	1.00	0.95	0.95	–
12	BAF++	Closed steel beating	0.40	0.50	0.85	0.20	0.85	–
13	BAD++#	Open steel beating	0.60	0.70	0.50	0.25	0.95	0.95
14	HB++#	Beaten H	0.55	0.65	0.70	0.20	0.95	0.85
15	HBi++	IGU or IRS injected beaten H	1.35	1.60	2.00	1.10	2.25	2.25
16	PP++#	Beaten sheet piling	0.45	0.55	0.55	0.20	1.25	1.15
17	M1	Type I micro-pile	–	–	–	–	–	–
18	M2	Type II micro-pile	–	–	–	–	–	–
19	PIGU, MIGU	Injected pile or micro-pile (type III)	1.35	1.60	2.00	1.10	2.25	2.25
20	PIRS, MIRS	Injected pile or micro-pile (type IV)	1.70	2.05	2.65	1.40	2.90	2.90

++ Refer to article A.100 for the calculation of the perimeter

\# For BAO, HB and PP type piles, installed by vibro-sinking, and not by beating, a reduction of 30% must be made for the values of q_s.

\## For longer piles, clause (6) of this section must be applied.

Table 11.16. *Coefficient $\alpha_{pile-soil}$ for the estimated lateral friction of the pile, penetrometric method [AFN 12b]*

Soil type	Clays	Intermediate soils Sand Grave	Sands	Chalk	Marl and marl-limestone	Altered or fragmented rock
Choice of the curve	Q1	Q2	Q3	Q2	Q2	Q2
a	0.0018	0.0015	0.0012	0.0015	0.0015	0.0015
b	0.1	0.1	0.1	0.1	0.1	0.1
c	0.4	0.25	0.15	0.25	0.25	0.25

Table 11.17. *Coefficients a, b and c for the expression of f_{soil} of the lateral friction of the pile, penetrometric method [AFN 12b]*

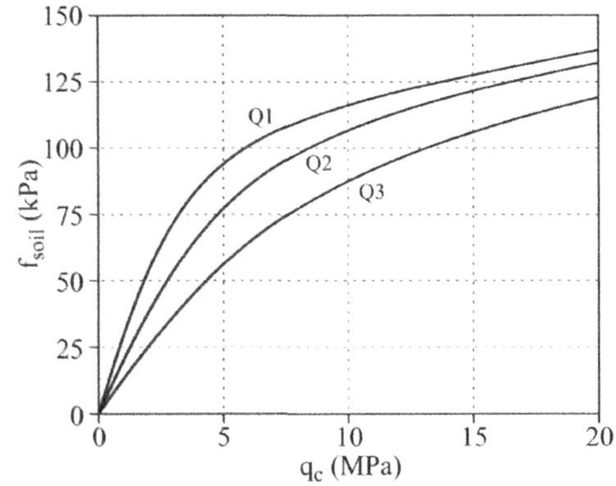

Figure 11.16. *Function f_{sol} of the lateral soil-pile friction, penetrometric method [AFN 12b]*

11.2.4.3. *The pressuremeter*

11.2.4.3.1. Correlation between the pressiometric modulus E_M and the evaluation of the stacking of superficial foundations

Ménard proposes separating stacking s_c caused by the spherical component of the constraints, from stacking s_d caused by the deviator component following relation [11.78].

Spherical stacking s_c is applied to a volume of soil corresponding to a sphere with a diameter equal to that of the foundation. It is provided through relation [11.79]. Constraint q' corresponds to the constraint transmitted to the soil through the foundation, constraint s'v0 is the effective constraint to the foundation plane, B_0 is equal to 0.6 m, B is the width of the foundation and L is its length, with E_c equal to the pressiometric modulus E_M.

Deviator stacking s_d applied to a layer of soil with a thickness of 8.B. It is provided by equation [11.80], with E_d equal to the pressiometric modulus E_M.

$$s_f = s_c + s_d \quad [11.78]$$

$$s_c = \frac{\alpha}{9.E_c}.(q' - \sigma'_{v0}).\lambda_c \quad [11.79]$$

$$s_d = \frac{2}{9.E_d}.(q' - \sigma'_{v0}).B_0.\left(\lambda_d.\frac{B}{B_0}\right)^{\alpha} \quad [11.80]$$

The value of the rheological coefficient α is given in Table 11.18 for soils and in Table 11.19 for rocks. The shape coefficients λ_c and λ_d are given in Table 11.20.

	Peat	Clay		Silt		Sand		Grave	
Type	α	E_M/p_l	α	E_M/p_l	α		α	E_M/p_l	α
Overconsolidated or very tight	–	>	1	> 14	2/3	> 12	1/2	> 10	1/3
Normally consolidated or normally tight	1	9 - 16	2/3	8 - 14	1/2	7 - 12	1/3	6 - 10	1/4
Altered overconsolidated and modified or loose	1	9 - 16	2/3	8 - 14	1/2	5 - 7	1/3	–	–

Table 11.18. *Rheological coefficient α for the stacking of superficial foundations in the soil [FAS 93] Titre V*

Rock	
Type	α
Very slight cracking	2/3
Normal cracking	1/2
Very cracked	1/3
Very altered	2/3

Table 11.19. *Rheological coefficient α for the stacking of superficial foundation in rocks [FAS 93]*

L/B	Circle	Square	2	3	5	≥ 20
	1.00	1.10	1.20	1.30	1.40	1.50
λ_d	1.00	1.12	1.53	1.78	2.14	2.65

Table 11.20. *Shape coefficient for the stacking of superficial foundations [FAS 93]*

For the case of heterogeneous soils, the soil under the foundation is divided into 16 layers of thickness B/2, and E_c from formula [11.79] is taken as equal to the pressiometric modulus E_M of the first layer, and Ed from formula [11.80] determined by relation[11.81], in which E_1 represents the pressiometric modulus of the first layer, E_2 represents the pressiometric modulus of the second layer, and so on and so forth. Moduli $E_{i\text{-}j}$ are calculated using relation [11.82]:

$$\frac{1}{E_d} = \frac{0{,}25}{E_1} + \frac{0{,}35}{E_2} + \frac{0{,}25}{E_{3-5}} + \frac{0{,}1}{E_{6-8}} + \frac{0{,}1}{E_{9-16}} \qquad [11.81]$$

$$\frac{1}{E_{i-j}} = \frac{1}{j-1+1} \cdot \sum_{k=i}^{k=j} \frac{1}{E_k} \qquad [11.82]$$

11.2.4.3.2. Correlation between the pressure limit p_{LM} and the evaluation of the ultimate constraint under superficial foundations

Ménard linked the conventional pressure limit, measured using the pressuremeter, to the limiting constraint q_{net} under a foundation. He classified soils into different categories depending on their types and the difference between the conventional pressure limit p_{LM} and the horizontal pressure of grounds at rest p_0. Regulations have since been updated and his work is reflected in the rules of Eurocode 7. The pressiometric soil-bearing factor k_p is a function of the soil classes. To define it, we can choose a soil-bearing curve (Figure 11.16), or using the coefficients (Table 11.21) and formula [11.84] for a square or circular foundation and [11.85] for a rectangular foundation. The soil-bearing factor k_p is introduced in formula [11.83] to determine q_{net}. Coefficients i_δ and i_β depend on the incline of the load.

$$q_{net} = k_p . p_{le}^* . i_\delta . i_\beta \qquad [11.83]$$

$$k_p = k_{p0} + \left(a + b.\frac{D_e}{B}\right).\left(1 - e^{-c.\frac{D_e}{B}}\right) \qquad [11.84]$$

$$k_p = k_{p0} + \left(1 - \frac{B}{L}\right) + k_p.\frac{B}{L} \qquad [11.85]$$

11.2.4.3.3. Correlation between the pressure limit p_{LM} and evaluation of the soil-bearing capacity of a deep foundation

Deep foundations put up resistance through two terms [11.63]: peak effort at the inferior extremity of the pile R_b, and the lateral friction effort along the pile R_s. The correlation between pressiometric resistance and the soil-bearing capacity of the piles is described in the regulatory standard of Eurocodes 7 [AFN 12b].

Soil category Variation curve of the soil-bearing factor		Expression of k_c			
		a	b	c	k_{c0}
Clays and silts	Strip foundation - Q1	0.2	0.02	1.3	0.8
	Square foundation - Q2	0.3	0.02	1.5	0.8
Sands and gravel	Strip foundation- Q3	0.3	0.05	2	1
	Square foundation - Q4	0.22	0.18	5	1
Chalk	Strip foundation - Q5	0.28	0.22	2.8	0.8
	Square foundation - Q6	0.35	0.31	3	0.8
Marl and marl-limestone Altered rocks	Strip foundation - Q7	0.2	0.2	3	0.8
	Square foundation - Q8	0.2	0.3	3	0.8

Note 1 Intermediate soils defined in clause A.2 (4) are assimilated to either powdery soils or coherent soils in the following manner:
- "clay and silt" type coherent soils also include silty clays, loamy silts and sandy clays (from Figure A.2.1);
- "sand and gravel" type soils also include loamy sands, silty sands and sandy silts (from Figure A.2.1).

Table 11.21. *Pressiometric soil-bearing factor k_p [AFN 13]*

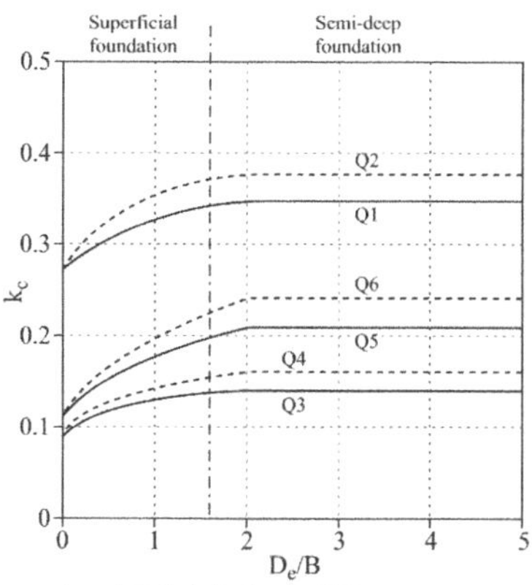

Legend: X: Equivalent housing D_e/B –
Y: Pressiometric soil-bearing factor k_c

Figure 11.17. *Pressiometric soil-bearing factor k_p [AFN 13]*

11.2.4.3.3.1. Peak term

Ménard linked the conventional pressure limit, measured with a pressuremeter, to the constraint under a pile. He classified soils into the same categories as previously. The regulations have taken up this method and they now calculate ultimate resistance under the peak of the pile q_b [11.86] as a function of the initial vertical constraint under the pile q_0 and proportionally to the equivalent net pressure limit p^*_{LMe}. The soil-bearing factor k_p depends on a maximal value, given in Table 11.22.

$$q_b = q_0 + k_p.p^*_{LMe} \quad [11.86]$$

Terrain / Pile class (c)	Clay % $CaCo_3$ < 30% Silt Intermediate soils	Intermediate soil	Chalk	Marl and marl-limestone	Altered or fragmented rock (a)
1	1.15 (b)	1.1 (b)	1.45 (b)	1.45 (b)	1.45 (b)
2	1.3	1.65	1.6	1.6	2.0
3	1.55	3.2	2.35	2.10	2.10
4	1.35	3.1	2.30	2.30	2.30
5#	1.0	1.9	1.4	1.4	1.2
6#	1.20	3.10	1.7	2.2	1.5
7#	1.0	1.0	1.0	1.0	1.2
8	1.15 (b)	1.1 (b)	1.45 (b)	1.45 (b)	1.45 (b)

(#) for BAO, HB and PP type piles, installed by vibro-sinking instead of beating, a reduction of 50% must be made on factor k_c

Table 11.22. *Soil-bearing factor k_{pmax} for a pile with a relative housing D_{ef}/B greater than 5, pressuremeter method [AFN 12b]*

When D_{ef}/B is greater than 5 for the calculation of k_p, the maximal possible value is used [11.87].

When D_{ef}/B is less than 5, then relation [11.88] is used:

$$k_p = k_{pmax} \quad [11.87]$$

$$k_p = 1 + (k_{pmax} - 1).\frac{(D_{ef}/B)}{5} \quad [11.88]$$

$$D_{ef} = \frac{1}{p^*_{LMe}}.\int_{D-h_D}^{D} p^*_{LM}(z).dz \quad [11.89]$$

The height of the effective housing D_{ef} is given by relation [11.89] in which h_D refers to a length equal to 10.B and D refers to the length of the pile.

11.2.4.3.3.2. Lateral friction term

The effort that can be mobilized by friction along the pile [11.90] is linked to the perimeter of the pile P_s, to the length D of the pile and to the unitary lateral friction q_s, which evolves as a function of the edge z along the pile between 0 and D. The limit value of the unitary lateral friction q_s at a depth of z is determined in terms of the net pressure limit p_{LM}^* using the following expression [11.91].

Coefficient $\alpha_{pile\text{-}soil}$, which depends on the type of pile and on the type of soil, is shown in Table 11.23.

No.	Abb.	Implementation technique	*Values in kPa*					
			Clay % $CaCO_3$<30% Loam	Intermediate soils	Sand Gravel	Chalk	Marl & Limestone Marl	Weathered/ fragmented rock
1	SD##	Simple drilled pile (piles and diaphragm wall)	90	90	90	200	170	200
2	MP##	Bored mud pile (piles and diaphragm wall)	90	90	90	200	170	200
3	BCL	Bored cased pile (lost shell)	50	50	50	50	90	–
4	BCR	Bored cased pile (recovered shell)	90	90	90	170	170	–
5	SDG, MDG, WE##	Drilled single or mud with grooving or wells	90	90	–	–	–	–
6	CAS, CAD	Continuous flight auger single rotation or double rotation	90	90	170	200	200	200
7	SCT	Screwedcast	130	130	200	170	170	–
8	SCD	Screwedcased	50	50	90	90	90	–
9	BCP** BCC**	Beaten precast or prestressed concrete	130	130	130	90	90	–
10	BS**	Beaten surrounded (concrete, mortar, grout)	170	170	260	200	200	–
11	BM**	Beatenmolded	90	90	130	260	200	–
12	BC**	Beatenclosedsteel	90	90	90	50	90	–
13	BO**#	Beaten open steel	90	90	50	50	90	90
14	BH**#	Beaten H	90	90	130	50	90	90
15	BHi**	H beaten injected IGU or IRS	200	200	380	320	320	320
16	BP**#	Beaten sheet piles	90	90	50	50	90	90
17	M1	Micropile type I	–	–	–	–	–	–
18	M2	Micropile type II	–	–	–	–	–	–
19	IP MIP	Injected pile or micropile (III)	200	200	380	320	320	320
20	IP MIP	Injected pile or micropile (IV)	200	200	440	440	440	500

Table 11.23. *Coefficient $\alpha_{pile\text{-}soil}$ for the estimated lateral friction of the pile, pressuremeter method [AFN 12b]*

The friction f_{soil} is determined in terms of the nature of the soil and the resistance at the pressure limit p_{LM}^*. The function f_{soil} can be read graphically in Figure 11.17 or calculated analytically using [11.92]. The values of coefficients a, b and c can be read in Table 11.22:

$$R_S = P_S . \int_0^D q_s(z). dz \quad [11.90]$$

$$q_s(z) = \alpha_{pile-soil} . f_{soil}[p_{LM}^*(z)] \quad [11.91]$$

$$f_{soil}[p_{LM}^*(z)] = (a. p_{LM}^* + b). \left(1 - e^{-c.p_{LM}^*}\right) \quad [11.92]$$

Soil type	Clay % $CaCo_3 < 30$ % Silt Intermediate soils	Intermediate soil Grave sand	Chalk	Marl and marl-limestone	Altered or fragmented rock
Choice of the curve	Q1	Q2	Q3	Q4	Q5
a	0.003	0.01	0.007	0.008	0.01
b	0.04	0.06	0.07	0.08	0.08
c	3.5	1.2	1.3	3	3

Table 11.24. *Coefficients a, b and c for the expression of f_{soil} of the lateral friction of the pile, penetrometric method [AFN 12b]*

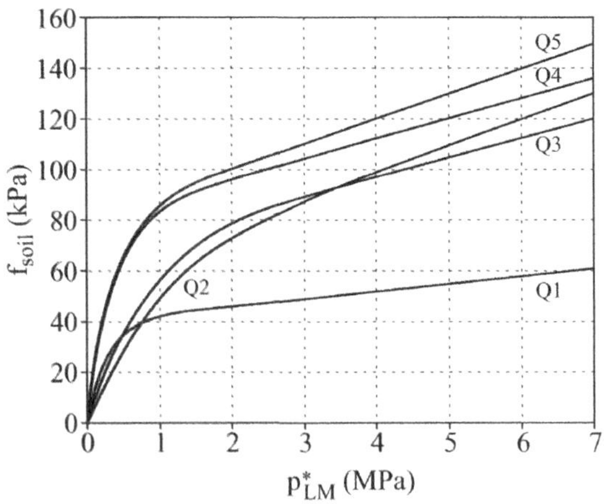

Figure 11.18. *Function f_{soil} of lateral soil-pile friction, pressuremeter method [AFN 12b]*

11.2.5. *Relations involving in situ tests*

11.2.5.1. *Measurement of the elastic modulus E^e and the shearing modulus G^e*

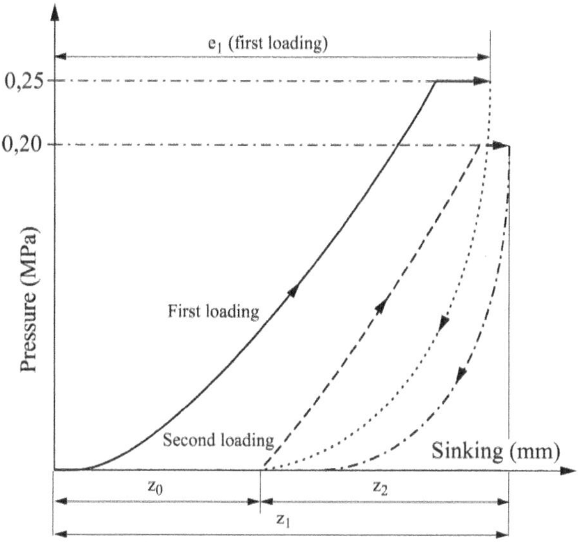

Figure 11.19. *Interpretation of the plate test*

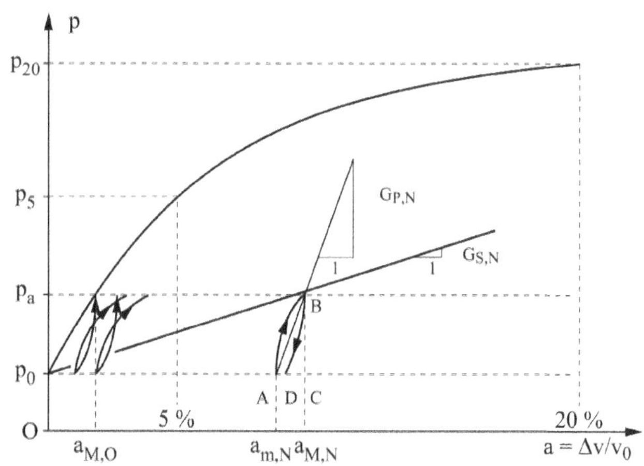

Figure 11.20. *Cyclical self-drilling pressuremeter test [BAG 81]*

11.2.5.1.1. Use of the test plate

The elastic modulus of the soil is measured using the plate test [AFN 00a], using a plate with a diameter of 60 cm and a loading–unloading cycle (Figure 11.18). Determination is done using relation [11.93] for a thickness of the investigated layer of 1.5.D, which is 0.9 m.

$$E_{v2} = E^e = \frac{\pi.(1-v^2)}{2}.\frac{\Delta\sigma_i}{\Delta s_i}.R = \frac{90}{z_2} \qquad [11.93]$$

11.2.5.1.2. Use of the self-drilling pressuremeter test

The shearing modulus G is obtained by considering the elasticity equation [11.94], dividing the pressure loading increment by the volume change corresponding to the sensor in the loading-unloading cycle (Figure 11.19). It is determined at the test depth in the borehole and for a radius equal to the length of the measurement cell, which is 0.6 m.

$$G^e = \frac{p_a - p_0}{\Delta V / V_0} \qquad [11.94]$$

11.2.5.1.3. Use of the Ménard pressuremeter test

The elastic modulus E^e is evaluated, linked to the shearing modulus G, by the pressuremeter over loading-unloading cycle (Figure 11.20) using the formula obtained from the value of the elasticity [11.95] by assuming a Poisson coefficient ν equal to 0.3, and average volume of the probe V_s, the initial volume of the cycle being the lowest V_{r3}, the final volume of the cycle being the greatest V_{r4}, the initial pressure of the cycle being the lowest p_{r3}, the final pressure of the cycle being the greatest p_{r4}. The elastic shearing modulus G^e is determined using relation [11.96].

$$E^e = 2.(1+\vartheta).\left(V_s + \frac{V_{r3}+V_{r4}}{2}\right).\frac{p_{r4}-p_{r3}}{V_{r4}-V_{r3}} \qquad [11.95]$$

$$G^e = \left(V_s + \frac{V_{r3}+V_{r4}}{2}\right).\frac{p_{r4}-p_{r3}}{V_{r4}-V_{r3}} \qquad [11.96]$$

11.2.5.2. *Measurement of the undrained cohesion c_u*

11.2.5.2.1. Use of the vane tester

For clays, the vane tester helps determine the value of the undrained cohesion through relation [11.97], with the value of the moment M_{mes}

measured at the maximum of the shearing curve (Figure 11.22). Coefficient K is equal to 1,310 cm^3 and the Pilot coefficient k is read in Figure 11.23:

$$c_u = k.\frac{M_{mes}}{K} \quad [11.97]$$

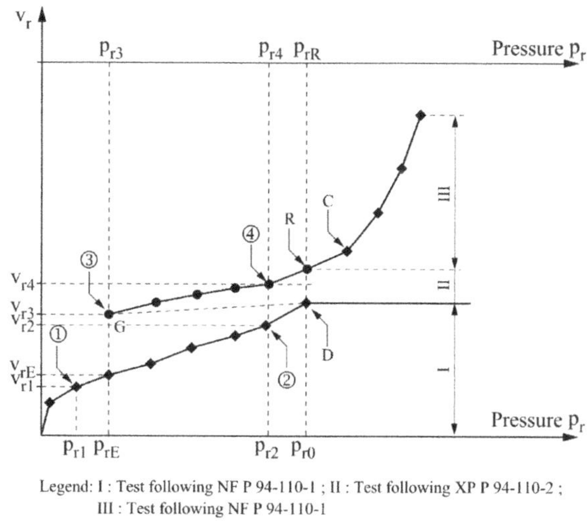

Figure 11.21. *Example of a cyclical pressiometric curve, function of p [AFN 99b]*

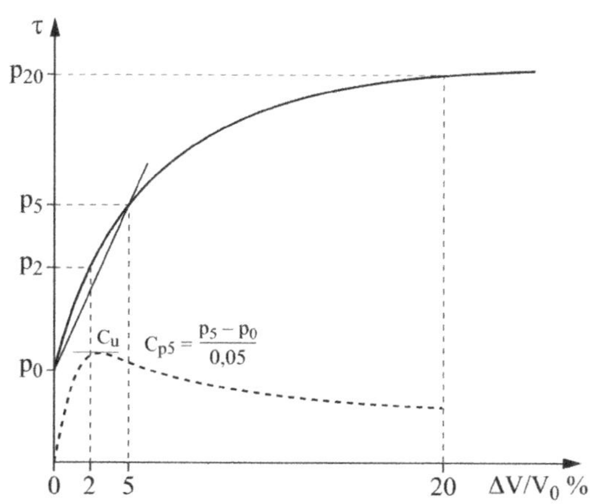

Figure 11.22. *Interpretation of the self-drilling pressuremeter test and determination of the shearing curve*

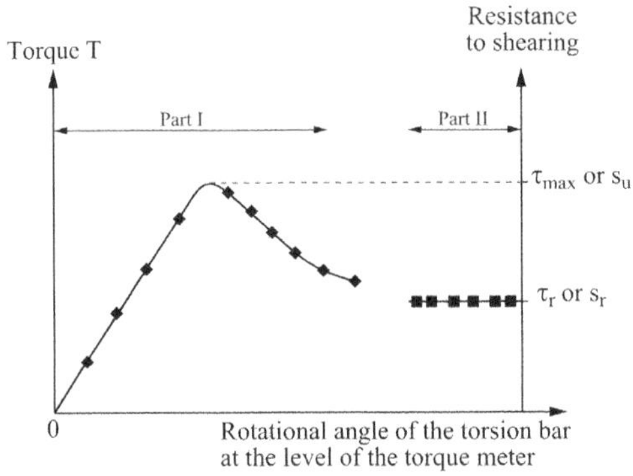

Figure 11.23. *Determination of τ_{max} [AFN 91]*

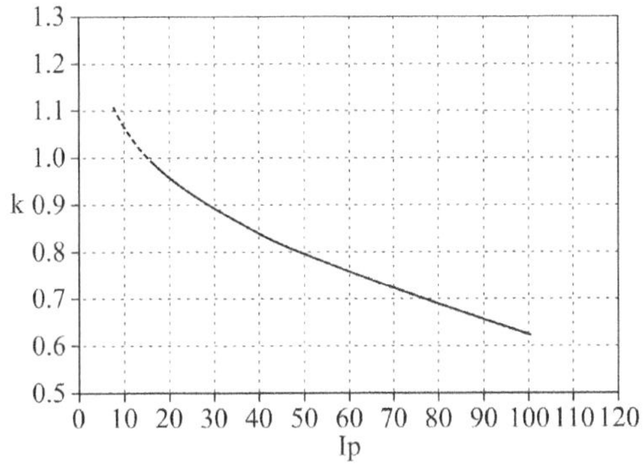

Figure 11.24. *Determination of coefficient k in function of I_p (Pilot)*

11.2.5.2.2. Use of the self-drilling pressuremeter test

For clays, and impermeable soils in general, the undrained cohesion can be evaluated using the Baguelin *et al.* [BAG 72] method, using formula [11.98], with the circumferential deformation ε_θ being equal to the deformation of the bore dR/R_0 t. The following are measured in Figure 11.21:

– the peak undrained cohesion at the maximum of this shearing curve;

– the residual undrained cohesion at the plateau after the shearing peak.

$$f(\varepsilon_\theta) = \left[\varepsilon_\theta.(1+\varepsilon_\theta/2).(1+\varepsilon_\theta).\frac{dF(\varepsilon_\theta)}{d\varepsilon_\theta}\right] = \tau \qquad [11.98]$$

This method is suitable for determining the residual undrained cohesion, but it overestimates the peak cohesion. It is also sensitive to the pressuremeter curve derivation method.

11.2.5.2.3. Use of the Ménard pressuremeter test

Monnet and Chemaa's theory [MON 95] shows that the pressuremeter curve – with semi-logarithmic coordinates – is transformed into a line linking Ln(dR/R0) to the pressure applied to the bore (see Figure 11.8). The slope of this line (Figure 11.24) is inversely proportional to the value of the undrained cohesion following formula [11.99] for the case with two plastic zones:

$$Ln\left[\frac{u_a}{a}+\frac{c_u}{2.\mu}\right]=\frac{p}{c_u}-\frac{\gamma.z}{c_u}+Ln\left[\frac{\gamma.z.(1-K_0)}{2.\mu}+\frac{c_u}{2.\mu}\right] \qquad [11.99]$$

The value of the conventional pressure limit corresponds to a deformation of the boring of (?). (?) is then, respectively, equal to $\sqrt{2}-1$, allowing us to determine the expression of the pressure limit p_{LM} [11.100] in the case where two plastic zones develop. This relation [11.100] can be transformed into an abacus (Fgure 11.25) providing the undrained cohesion as a function of the elastic shearing modulus G^e and the net pressure limit p_{LM}-p_0.

$$p_{LM}=\gamma.z+c_u.Ln\left[\frac{c_u+2.\mu.(\sqrt{2}-1)}{c_u+\gamma.z.(1-K_0)}\right] \qquad [11.100]$$

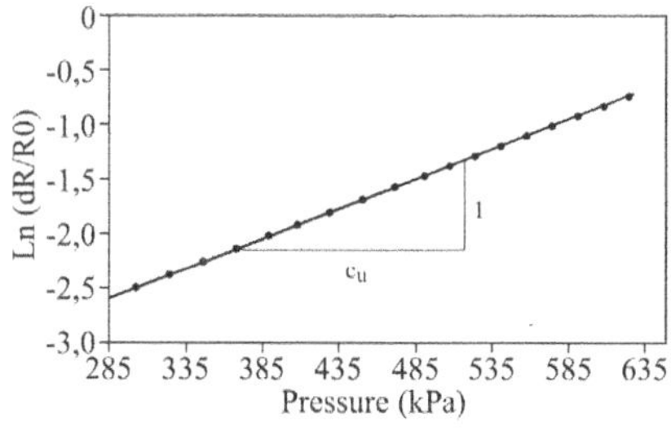

Figure 11.25. *Linear transformation of the pressuremeter curve for a test at 13.20 m in Geneva clay [MON 95]*

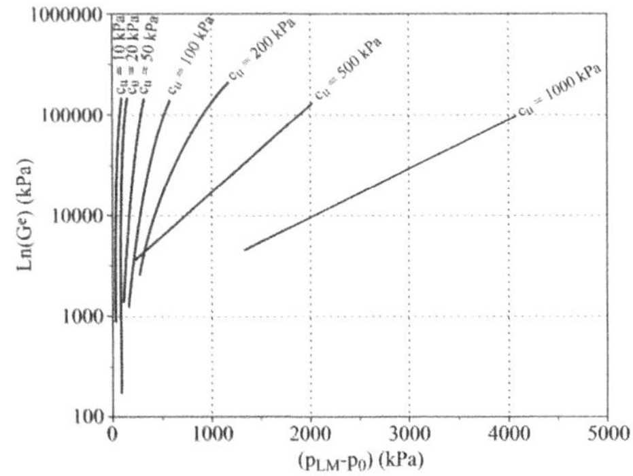

Figure 11.26. *Determination of the undrained cohesion as a function of the elastic modulus and the pressure limit [MON 13]*

11.2.5.3. *Measurement of the friction angle Φ'*

11.2.5.3.1. Use of the direct shearing test

The direct *in situ* shearing test allows us to determine the direct value of the shearing as a function of the displacement (Figure 11.26). The relation between the shearing and normal constraint to the shearing plane allows us to measure both the cohesion and the friction angle of the soil.

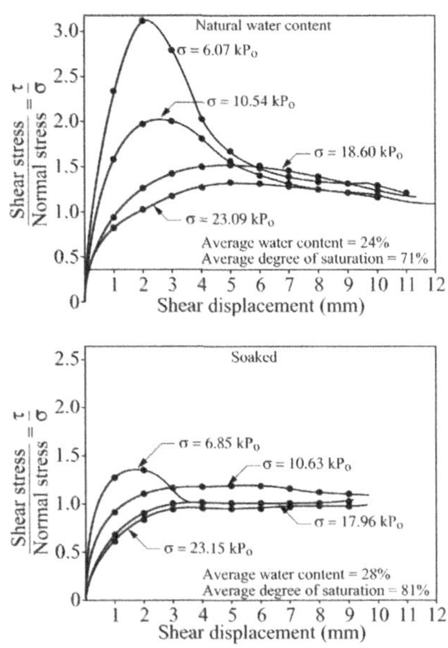

Figure 11.27. *Evolution of the shear during the direct shear box test in a soil*

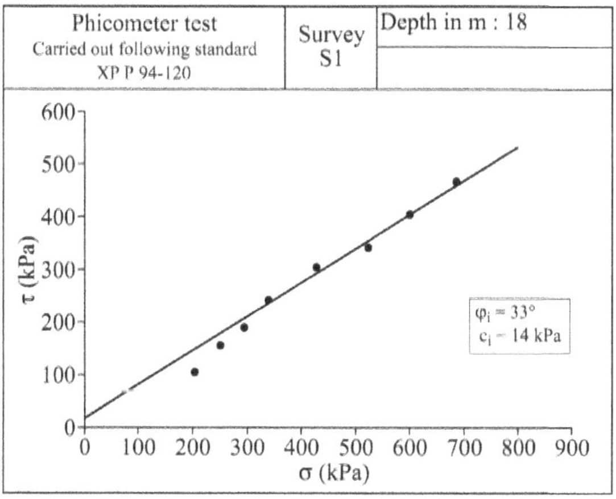

Figure 11.28. *Example of a phicomètre test result, determination of c and Φ in situ*

11.2.5.3.2. Use of the phicomètre

The phicometer test allows us to determine the different torques, of value (τ_{maxi} , σ_{ni}), on the envelop external to the vanes of the probe, correspond to each radial constraint applied to the boring. The torques allow us to draw (Figure 11.27) the linear Coulomb curve, and determine the value of the cohesion c and the internal friction angle of the soil Φ.

11.2.5.3.3. Use of the Ménard pressuremeter

For powdery and permeable soils, the theoretical analysis [MON 94] of the expansion of the pressuremeter probe, while taking dilatance into account, allows us to prove relation [11.101] for the case where a plastic zone develops. Relation [11.101] allows us to transform the pressuremeter curve into a straight line by drawing the graph that links the logarithm of the borehole deformation u_a/a to the logarithm of the pressure applied to the borehole (Figure 11.28) following the relations below. The slope of the line obtained in this way takes on the value of δ shown in formula [11.102]. The value of δ is known graphically. It is a function of two parameters Φ_μ and Φ through relations [11.103]–[11.405]. As the interganular friction angle Φ_μ does not depend on the state of density, but is rather only a function of the geologically origin of the soil (Table 11.23), knowing δ as well as the nature of the soil allows us to determine the friction angle Φ.

$$Ln\left[\left({}^{u_a}/_{a}\right).(1+n)-C_1\right]=\delta.Ln(p)-\delta.Ln\left(\frac{2.K_0.\gamma'.z.}{(1+N)}\right)+$$
$$Ln\left[\frac{K_0.\gamma'.z.(1-N).(1+n)}{2.\mu.(1+N)}.-C_1\right] \quad [11.101]$$

$$\delta=\frac{(1+n)}{(1-N)} \quad [11.102]$$

$$n=\frac{(1-sin\psi)}{(1+sin\psi)} \quad [11.103]$$

$$sin\psi=\frac{3.\sqrt{2}.\left[\frac{2.sin\phi'.\sqrt{2}}{(3-sin\phi')}-tg\phi_\mu\right]}{4+\sqrt{2}.\left[\frac{2.sin\phi'.\sqrt{2}}{(3-sin\phi')}-tg\phi_\mu\right]} \quad [11.104]$$

$$N=\frac{(1-sin\phi)}{(1+sin\phi)} \quad [11.105]$$

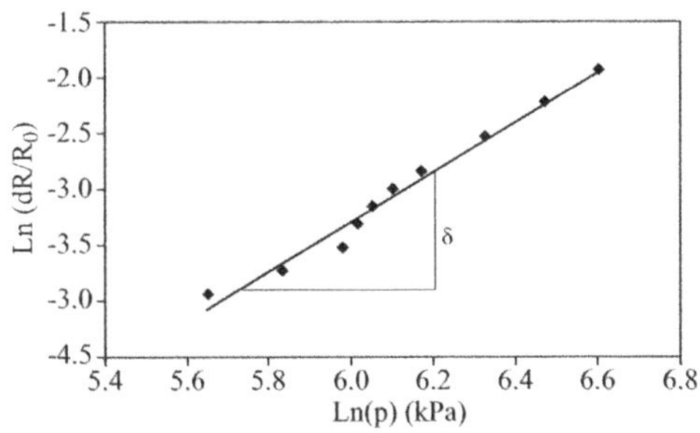

Figure 11.29. *Linear transformation of the pressuremeter curve; test at 9 m depth in the Lyon grave [MON 94]*

Class of soil	Ip (%)	Φ_μ (degree)	Φ_{cv} (degree)
Sand, grave	–	29°	29,4°
Silt, class A1	0 - 12	28°	28,3°
Low plasticity clay, A2	12 - 25	20°	20°
Plastic clay, A3	25 - 40	15°	15°
Very plastic clay, A4	> 40	10°	10.1°

Table 11.25. *Normal values of the intergranular friction [MON 13]*

If we now consider a drilling deformation of $(\sqrt{2}-1)$ introduced in equation [11.101], we can then find the theoretical pressure limit [11.106] in the case where a single plastic zone develops. This expression can be transformed into an abacus that will allow us to determine the friction angle Φ as a function of variables G^e/p'_0 and p'_{LM}/p'_0 (Figure 11.29).

$$p'_{LM} = \frac{2.K_0.\gamma'.z.}{(1+N)}\sqrt[\delta]{\frac{\left[\left((\sqrt{2}-1).(1+n)-C_1\right).2.\mu.(1+N)\right]}{[K_0.\gamma'.z.(1-N).(1+n)-2.\mu.C_1.(1+N)]}} \qquad [11.106]$$

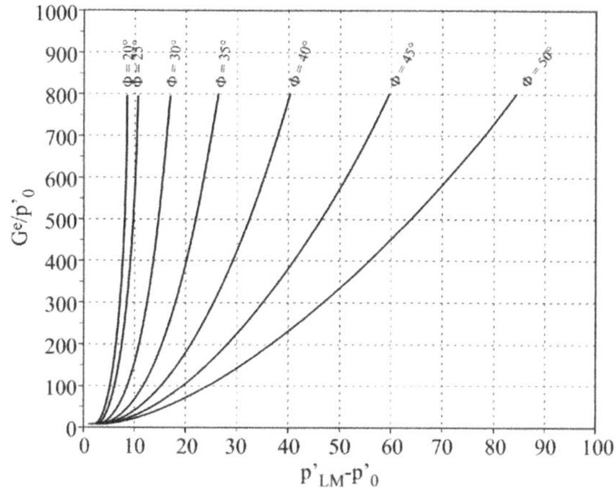

Figure 11.30. *Determination of the friction as a function of the shearing elastic modulus and the pressure limit [MON 13]*

11.2.5.3.4. *Measurement of the permeability of the soil*

11.2.5.3.4.1. Use of the infiltrometer

For the open infiltrometer, a constant load h is imposed on the terrain under the device, and the flow rate Q that percolates the soil is measured for a straight section of the infiltrometer A. This results in relation [11.107]. In the field, the depth of the saturated zone is observed as being equal to z_w. As the load is equal to zero at the saturation front, we can deduce the hydraulic gradient using relation [11.108], as well as the surface permeability using relation [11.109]:

$$Q = k.A.i \qquad [11.107]$$

$$i = \frac{h+z_w}{z_w} \qquad [11.108]$$

$$k = \frac{Q.z_w}{A.(h+z_w)} \qquad [11.109]$$

For the closed infiltrometer, increasing hydraulic gradients are imposed (Figure 11.30). The permeability coefficient k is found for a hydraulic gradient to be equal to 1.

In both cases, the permeability corresponds to the layer of soil with a thickness equal to 30 cm at the surface.

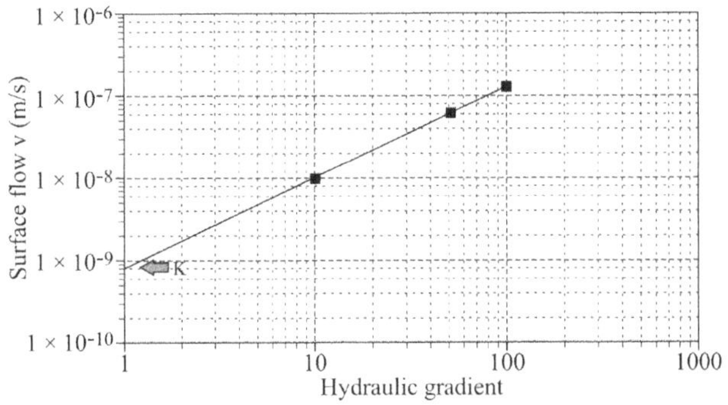

Figure 11.31. *Determination of the permeability coefficient for a gradient of 1 [AFN 99a]*

11.2.5.3.4.2. Use of the Lefranc test

A cavity of any shape (but unchanging) is created within the permeable soil and under the water table. The dimensions of the cavity must be known. It is created at the bottom of the borehole and built so that filtration only takes place through the edges of the cavity and not through the borehole. Water is removed (for depths of less than 9 m) or injected (preferable) into the cavity. The load is fixed at a set amount.

The shape coefficients can take on the values shown in Table 10.1. If the location is close to the limit of an aquifer domain (surface of the water table and impermeable wall), the m is replaced in formula [11.110] by m_0 shown in formula [11.111], and in Table 10.2, with H_w the distance of the cavity under the level of the water table, H the distance from the cavity to the impermeable wall and L the length of the cavity.

The value of the permeability k is obtained using relation [11.110]. It corresponds to a layer of soil around the borehole with a diameter of 3.D (D is the diameter of the borehole)

$$Q = m.k.h.B \qquad [11.110]$$

$$\frac{1}{m_0} = \frac{1}{m} - \frac{B}{8.\pi.H_w} + \frac{B}{8.\pi.H} \qquad [11.111]$$

11.2.5.3.4.3. Use of the drilling permeameter test

Water is injected into the terrain at a chosen level of the borehole using the drilling permeameter. This test allows us to determine the permeability of the soil at a chosen level of the borehole. The current lines are assumed to be horizontal in the region of the injection cell (Figure 10.18). In this case, the flow revolves around the bore axis. In the area close to the borehole, the equipotential surfaces are revolving cylinders. However, at larger distances from the probe, the flow seems to come from a punctual source. Measurement of the flow Q injected at a pressure p allows us to obtain the value of the permeability k using relation [11.112].

$$Q = \left[\frac{4.\pi.\lambda}{2.Ln(2.\lambda)+1}\right].\left(\frac{p-p_0}{\gamma_w}\right).k.B \qquad [11.112]$$

$$m = \frac{4.\pi.\lambda}{2.Ln(2.\lambda)+1} \qquad [11.113]$$

11.2.5.3.4.4. Use of the "géomécamètre" test

A measured flow of Q is injected into the terrain at an injection pressure of p. The device then allows us to determine the permeability of the soil k using relation [11.114].

$$k = \frac{Q.\gamma_w}{2.\pi.l.(p-p_0)}.\left[Ln\frac{l}{r_0}+\frac{1}{2}\right] \qquad [11.114]$$

Bibliography

[ACA 92] ACAR Y.B., NYERETSE P., "Total suction of artificial mixtures of soil compacted at optimum water content", *Geotechnical Testing Journal*, vol. 15, no. 1, pp. 65–73, 1992.

[ADE 05] ADEME, *Guide méthodologique pour le suivi des tassements des Centres de Stockage de Classe II*, Publication ADEME, 2005.

[AFN 88] AFNOR DTU 13-12 P11-711, Règles pour le calcul des fondations superficielles, CSTB, Cahier no. 2225, 1988.

[AFN 90a] AFNOR – NF P 94-114, Sols: Reconnaissance et essais, Essai de pénétration dynamique de type A, 1990.

[AFN 90b] AFNOR – NF P 94-115, Sols: Reconnaissance et essais, Essai de pénétration dynamique de type B, 1990.

[AFN 90c] AFNOR – NFP 94-150-1, Sols: Reconnaissance et Essais – Essai statique de pieu isolé sous compression axiale, partie 1: en compression, 1990.

[AFN 91a] AFNOR – NF P 94-110-1, Sols: Reconnaissance et essais, Essai pressiométrique Ménard, partie 1: essai sans cycle, 1991.

[AFN 91b] AFNOR – NF P 94-112, Sols: Reconnaissance et essais, Essai scissométrique en place, 1991.

[AFN 91c] AFNOR – NF P 94-116, Sols: Reconnaissance et essais, Essai de pénétration au carottier, 1991.

[AFN 92a] AFNOR – NF P 94-130, Sols: Reconnaissance et essais, Essai de pompage, 1992.

[AFN 92b] AFNOR – NFP 11-300, Classification des matériaux utilisables dans la construction des remblais et des couches de forme d'infrastructures routières, 1992.

[AFN 93a] AFNOR – NF P 94-151, Sols: Reconnaissance et essais, Essai statique de pieu isolé sous effort transversal, 1993.

[AFN 93b] AFNOR – NF P 94-153, Sols: Reconnaissance et essais, Essai statique de tirant d'ancrage, 1993.

[AFN 93c] AFNOR – NF P 94-242, Renforcement des sols, Essai statique d'arrachement de clou soumis à effort axial de traction, essai à vitesse de déplacement constante, 1993.

[AFN 94] AFNOR – NF P 94-131, Sols: Reconnaissance et essais, Essai d'eau Lugeon, 1994.

[AFN 95a] AFNOR NF P94-119, Sols: reconnaissance et essais – Essai au piézocone, 1995.

[AFN 95b] AFNOR NF P94-156, Sols: reconnaissance et essais – Mesures à l'inclinomètre, 1995.

[AFN 95c] AFNOR – NF P 94-050, Sols: Reconnaissance et Essais – Détermination de la teneur en eau pondérale des matériaux – Méthode par étuvage, 1995.

[AFN 95d] AFNOR – NF P 94-202, Sols: Reconnaissance et Essais – Prélèvement des sols et des roches, Méthodes et procédures, 1995.

[AFN 96a] AFNOR – NF P 94-010, Glossaire géotechnique, définitions, notations, symboles, 1996.

[AFN 96b] AFNOR – NF P 94-049-1, Sols: Reconnaissance et Essais – Détermination de la teneur en eau pondérale des matériaux – Partie 1: Méthode de la dessiccation au four à micro-ondes, 1996.

[AFN 96c] AFNOR – NF P 94-049-2, Sols: Reconnaissance et Essais – Détermination de la teneur en eau pondérale des matériaux – Partie 2: Méthode à la plaque chauffante ou panneaux rayonnants, 1996.

[AFN 96d] AFNOR – NF P 94-113, Sols: Reconnaissance et essais, Essai de pénétration statique, 1996.

[AFN 97] AFNOR NF P94-120, Sols: Reconnaissance et essais – Essai de cisaillement au phicometre, 1997.

[AFN 98] AFNOR – NFX 30-418, Détermination de la perméabilité d'une formation géologique en place de matériaux rapportés, ou artificiellement reconstitués: infiltromètre à double anneau, de type ouvert, 1998.

[AFN 99a] AFNOR NFX 30-420, Détermination de la perméabilité d'une formation géologique en place de matériaux rapportés, ou artificiellement reconstitués: infiltromètre à double ou simple anneau, de type fermé, 1999.

[AFN 99b] AFNOR – NF P 94-110-2, Sols: Reconnaissance et essais, Essai pressiométrque Ménard, partie 2: essai avec un cycle, 1999.

[AFN 00a] AFNOR NF P94-117-1, Sols: Reconnaissance et essais – Portance des plates-formes – Partie 1: module sous chargement statique à la plaque (EV2), 2000.

[AFN 00b] AFNOR – NF P 94-132, Sols: Reconnaissance et essais, Essai d'eau Lefranc, 2000.

[AFN 08] AFNOR NF P94-117-3, Sols: Reconnaissance et essais – Portance des plates-formes – Partie 3: coefficient de réaction de WESTERGAARD sous chargement statique d'une plaque, 2008.

[AFN 12a] AFNOR – NF P 94-105, Sols: Reconnaissance et essais – Contrôle de la qualité du compactage – Méthode au pénétromètre dynamique à énergie variable – Principe et méthode d'étalonnage du pénétromètre – Exploitation des résultats – Interprétation, 2012.

[AFN 12b] AFNOR – NF P 94-262, Justification des ouvrages géotechniques, Normes d'application nationale de l'Eurocode 7, Fondations profondes, 2012.

[AFN 13] AFNOR – NF P 94-261, Justification des ouvrages géotechniques, Normes d'application nationale de l'Eurocode 7, Fondations superficielles, 2013.

[AFN 14] AFNOR NFP 94-521-15PR, Reconnaissance et essais géotechniques – Essais en place – Partie 15: Enregistrement des paramètres de forages, 2014.

[AGE 03] AGENCE INTERNATIONALE DE L'ÉNERGIE ATOMIQUE, Les sondes à neutrons et à rayons gamma: leurs applications en agronomie, 2003.

[AGG 01] AGGOUR M.S., RADDING W.R., Standard penetration test (SPT) correction, Maryland Department of Transport State Highway Administration, 2001.

[AGU 78] AGUELIN F., JEZEQUEL J.F., SHIELDS D.H., *The Pressuremeter and Foundation Engineering*, Trans Tech Publication, 1978.

[ALL 56] ALLWOOD R.J., An experimental investigation of the distribution of pressure in sand, PhD Thesis, University of Birmingham, 1956.

[AMA 91] AMAR S., CLARKE B.G.F., GAMBIN M. *et al.*, *The Application of Pressuremeter Test Results to Foundation Design in Europe*, Comité Technique Régional Européen, Pressiomètre, Balkema, pp. 1–24, 1991.

[ASH 03] ASHBY N., "Relativity in the global positioning system", *Living Reviews in Relativity*, vol. 6, pp. 4–42, 2003.

[AST 78] ASTM – D2573-72, Standard test method for filed vane shear test in cohesive soil, ASTM International, 1978.

[AST 94a] ASTM – D1143-81, Standard test method for piles under static axial compressive load, 1994.

[AST 94b] ASTM – D4719-87, Standard test method for pre-bored pressuremeter testing in soils, 1994.

[AST 07a] ASTM D4714, Standard test method for pre-bored pressuremeter testing in soils, ASTM international, ASTM International, West Conshohocken, PA, 2007.

[AST 07b] ASTM D6635, Standard test method for performing the flat plate dilatometer, Book of Standards, p. 14, 2007.

[AST 11] ASTM – D1586-11, Standard test method for standard penetration test (SPT) and split-barrel sampling of soils, ASTM International, West Conshohocken, PA, 2011.

[AST 12] ASTM – D5778-12, Standard test for electronic friction core and piezocore peretration testing of soils, ASTM international, West Conshohocken, PA, 2012.

[AST 14] ASTM – D3441-05, Standard test method for mechanical core penetration tests in soils, ASTM International, 2014.

[BAG 72] BAGUELIN F., JEZEQUEL J.F., LE MEE E. *et al.*, "Expansion de sondes cylindriques dans les sols cohérents", *Bulletin de liaison des Ponts et Chaussées*, no. 61, pp. 189–202, 1972.

[BAG 78] BAGUELIN F., JEZEQUEL J.F., SHIELDS D.H., *The Pressuremeter and Foundation Engineering*, Trans Tech Publication, 1978.

[BAG 81] BAGUELIN F., JEZEQUEL J.F., LE MEHAUTÉ A., "Essais cycliques au pressiomètre autoforeur", *Bulletin de liaison des Ponts et Chaussées*, pp. 19–23, 1981.

[BAU 13] BAUD J.P., GAMBIN M., "Determination du coefficient rhéologique de Ménard dans le diagramme pressionama", *Proceedings of the 18th International Conference on Soil Mechanics and Geotechnical Engineering*, pp. 487–490, 2013.

[BER 02] BERTRAND L., "Mesures de faibles perméabilités (in-situ et en laboratoire)", *Journée Hydraulique*, p. 65, CFMS, 2002.

[BIA 75] BIAREZ J., FAVRE J.L., "Parameters filing and statistical analysis of data in soil mechanics", *Proceedings of the 2nd International Conference on Applications of Statistics and Probability in Soil and Structural Engineering*, Aachen, Germany, vol. 2, pp. 249–264, 1975.

[BIA 94] BIAREZ J., HICHER P.Y., *Elementary Mechanics of Soils Behaviour, Saturated Remoulded Soils*, Balkema, Rotterdam, 1994.

[BIS 45] BISHOP R.F., HILL R., MOTT N.F., "Theory of indentation and hardness tests", *Proceedings of the Physical Society*, vol. 57, p. 47, 1945.

[BJE 60] BJERRUM L., SIMONS N.E., "Comparison of shear strength characteristics of normally consolidated clays", *Research Conference of Shear Strength of Cohesive Soils*, Boulder, Colorado, ASCE, pp. 711–727, June 1960.

[BJE 72] BJERRUM L., "Embankments on soft ground", *Proceedings Speciality Conference on Performance of Earth and Earth Supported Structures*, Purdue University, Lafayette, Indiana, ASCE, pp. 1–54, June 1972.

[BOL 86] BOLTON M.D., "The strength and dilatancy of sand", *Geotechnique*, vol. 36, pp. 65–77, 1986.

[BON 72] BONNEVAL H., *Photogrammétrie générale*, 4th ed., Eyrolles, Paris, 1972.

[BOU 07] BOUTONNIER L., Comportement hydromécanique des sols fins proches de la saturation cas des ouvrages en terre: coefficient B, déformations instantanées et différées, retrait/gonflement, PhD Thesis, INP Grenoble, 2007.

[BOU 13] BOUTONNIER L., HAJOUAI F., BACAR FADHULI N. *et al.*, "LGV Est Lot 41: tassements calculés puis mesurés sous remblais", *Proc. 18th Int. Conf. on Soil Mech. and Found. Eng.*, Paris, pp. 1233–1236, 2013.

[CAI 86] CAILLEUX J.-B., Étude des diagraphies instantanées en forage, Rapport des laboratoires, GT12, p. 97, 1986.

[CAM 61] CAMBEFORT H., CHADEISSON R., "Critère pour l'évaluation de la force portante d'un pieu", *Congrès International de la Mécanique des Sols et des Travaux de Fondations*, Paris, 1961.

[CAM 72] CAMBEFORT H., *Géotechnique de l'ingénieur et reconnaissance des sols*, Eyrolles, 1972.

[CAM 93] CAMBOU B., BAHAR R., "L'utilisation de l'essai pressiométrique pour l'identification de paramétres intrinsèques du comportement du sol", *Revue Française Géotechnique*, vol. 63, pp. 39–53, 1993.

[CAS 36] CASAGRANDE A., Characteristics of cohesionless soils affecting the stability of slopes and earth fills, PhD Thesis, Harvard University, 1936.

[CAS 78] CASSAN M., *Les essais in situ en mécanique des sols*, vol. 1, Eyrolles, 1978.

[CAS 80] CASSAN M., *Les essais d'eau dans la reconnaissance des sols*, Eyrolles, 1980.

[CAS 81] CAST, Les essais in situ, Document de stage INSA de Lyon, 1981.

[CAS 05] CASSAN M., *Les essais de perméabilité sur site dans la reconnaissance des sols*, Presse des Ponts et Chaussées, 2005.

[CHA 87] CHAFFOIS S., MONNET J., *Model of Sand Behaviour Towards Shearing and Compressibility in 3 Dimensional Condition of Stress and Strain, Constitutive Equations for Granular and Non-Cohesive Soil*, Balkema, Cleveland, 1987.

[CHA 91] CHAFFOIS S., Modélisation numérique des sols pulvérulents sollicités au cisaillement et en compression triaxiale vraie – Application aux tunnels et aux pieux, Thèse de Docteur de l'INSA de Lyon, p. 454, 1991.

[COM 01] COMBARIEU O., CANÉPA Y., "L'essai cyclique au pressiomètre", *Bul. Liais. Pots et Chaussées*, vol. 233, pp. 37–65, 2001.

[COS 81] COSTET J., SANGLERAT G., *Cours pratique de Mécanique des sols*, Dunod, 1981.

[COS 14] COSTE P., DELUZARCHE R., BOUTONNIER L. *et al.*, "Tassements et temps de consolidation mesurés et calculés sur projets LGV", *Journ. Nat. de Géot. et de Géol. Ing. (JNGG)*, Beauvais, 2014.

[DOR 10] DORBANI B., HAZA-ROZIER E., REIFFSTECK P., "Appareillage pour diagraphie de perméabilité incluant un piezocone", *Jour. Nat. de Géotech. et de Géol. de Ing. JNGG*, Grenoble, pp. 65–72, 2010.

[DTU 83] DTU 13-12, "Etude Géotechnique et reconnaissance des sols", *Revue Française de Géotechnique*, no. 22, p. 132, February 1983.

[EUR 97] EUROCODE 7, Geotechnical design, Part 3: design assisted by field tests, Flat dilatometer test (DMT), p. 8, 1997.

[EUR 07] EUROCODE 7 EN 1997-2, Geotechnical design, Part 2: ground investigation and testing, 2007.

[EVE 94] EVERITT B., *Cluster analysis*, Halsted – Wiley, New York, 1994.

[FAH 93] FAHEY M., CARTER J.P., "A finite element study of the pressuremeter test in sand using a non-linear elastic plastic model", *Canadian Geotechnical Journal*, vol. 30, no. 2, pp. 348–362, 1993.

[FAR 13] FARGIER, DURAND E., ROUSSELET R., "Le Perméafor, un outil de diagnostic des ouvrages hydrauliques en terre et de leur fondation", *Colloque Modernisation des barrages*, Chambéry, pp. 133–141, 2013.

[FAS 93] FASCICULE 62 TITRE V, Règles techniques de conception et de calcul des fondations des ouvrages de génie civil, Ministère de l'Equipement et du Logement et des Transports, 1993.

[FAV 80] FAVRE J.L., Milieu continu et milieu discontinu, Mesure statistique indirecte des paramètres rhéologiques et approche probabiliste de la sécurité, PhD Thesis, University of Paris VI, 1980.

[FAV 02] FAVRE J.L., BIAREZ J., MEKKAOUI S., "Modèles de comportement en grandes déformations des sables et argiles remaniées à l'oedomètre et au triaxial", PARAM, Paris, pp. 369–384, 2002.

[FLE 02] FLEUREAU J.M., VERBRUGGE J.-C., HUERGO P.J. *et al.*, "Aspects of the behavior of compacted clayey soils on drying and wetting paths", *Canadian Geotechnical Journal*, vol. 39, pp. 1341–1357, 2002.

[FRA 86] FRANK R., NAHRA R., Contribution numérique et analytique à l'étude de la consolidation autour du pressiomètre, Rapport recherche LCPC (No. 137), 1986.

[FRY 73] FRYDMAN S., ZEITLEN J.G., ALPAN I., "The yielding behaviour of particulate media", *Canadian Geotechnical Journal*, vol. 10, pp. 341–362, 1973.

[GAI 12] GAIATECH, available at: www.gaiatech.fr, 2012.

[GIB 61] GIBSON R.E., ANDERSON W.F., "In-situ measurement of soils properties with the pressuremeter", *Civ. Eng. Public Works Rev.*, vol. 56, no. 658, pp. 615–618, 1961.

[GRA 83] GRAHAM J., CROOKS J.H.A., BELL A.L., "Time effects on the stress-strain behaviour of natural soft clays", *Géotechnique*, vol. 33, no. 3, pp. 327–340, 1983.

[GRE 02] GRESS J.C., AUTRET A., "Correlations between standard Proctor parameters, Californian bearing ratio and blue methylene value", in MAGNAN (ed.), *Paramètres de calcul géotechnique*, Presses de l'ENPC/LCPC, Paris, 2002.

[GUI 00] GUIDE TECHNIQUE REFERENCE, Réalisation des remblais et des couches de forme – Vol. 1 principles généraux, Laboratoire central des Ponts et Chaussées, July 2000.

[GUI 02] GUI M.W., SOGA K., BOLTON M.D. *et al.*, "Instrumented borehole drilling for subsurface investigation", *Journal of Geotechnical and Geoenvironmental Engineering ASCE*, vol. 128, no. 4, pp. 283–291, 2002.

[HAR 74] HARA A., OHTA T., NIWA M. *et al.*, "Shear modulus and shear strength of cohesive soils", *Soils and Foundations*, vol. 14, no. 3, pp. 1–12, 1974.

[HAT 96] HATANAKA F.H., UCHIDA A., "Empirical correlation between penetration resistance and internal friction angle of sandy soils", *Soils and Foundations*, vol. 36, no. 4, pp. 1–9, 1996.

[HEG 98] HEGAZY Y.A., Delineating geostratigraphy by cluster analysis of piezocone data, PhD Thesis, School of Civil and Environmental Engineering Department, Georgia Institute of Technology, Atlanta, 1998.

[HEG 02] HEGAZY Y.A., MAYNE P.W., "Objective site characterization using clustering of piezocone data", *Journal of Geotechnical and Geoenvironmental Engineering*, vol. 128, no. 12, pp. 986–996, 2002.

[HOL 81] HOLTZ R.D., KOVACS W.D., *An Introduction to Geotechnical Engineering*, Prentice Hall, New Jersey, 1981.

[HUG 74] HUGHES J., WITHERS N.J., "Reinforcing soft cohesive soil with stone columns", *Ground Engineering*, vol. 7, no. 3, pp. 42–49, May 1974.

[HUG 77] HUGHES J., WROTH C.P., WINDLE D., "Pressuremeter test in sand", *Geotechnique*, vol. 27, no. 4, pp. 455–477, 1977.

[HVO 76] HVORSLEV M.J., The changeable interaction between soils and pressure cells – tests and reviews at the waterways experiment station, Technical Report no. S.76.7, 1976.

[JEZ 68] JEZEQUEL J., TOUZÉ J., Sonde pressiométrique autoforeuse, Brevet français d'invention, numéro 1.596.747, Paris, p. 6, 2 December 1968.

[JON 82] JONES G.A., RUST E., "Piezometer penetration testing CPTU", *Proc. 2nd Eur. Symp. on Penetration Testing*, Amsterdam, pp. 607–614, 1982.

[KAL 56] KALLSTENIUS T., BERGAU W., *Investigation of Soil Pressure Measuring by Means of Cells*, Proceedings Royal Swedish Geotechnical Institute, Stockholm, no. 12, p. 50, 1956.

[KOG 38] KOGLER F., SCHEIDIG J., *Baugrunde und Bauwerk*, Ernst, Berlin, 1938.

[KOV 82] KOVACS W.D., SALOMONE L., "SPT hammer energy measurement", *Journal of Geotechnical Engineering*, vol. 110, no. 4, 1982.

[KUL 90] KULHAWY F.H., MAYNE P.W., *Manual on Estimating Soil Properties for Foundation Design*, Electric Power Research Institute, Palo Alto, 1990.

[LAD 63] LADANYI, "Expansion of a cavity in a saturated clay medium", *Journal of the Soil Mechanics and Foundations Division, Proc. of ASCE*, vol. 89, no. SM 4, pp. 127–161, 1963.

[LAD 72] LADANYI, "In-situ determination of undrained stress-strain behavior of sensitive clays with the pressuremeter", *Can. Geot. J.*, vol. 9, pp. 313–319, 1972.

[MAG 81] MAGNAN J.P., MIEUSSENS C., "Remblais sur sol compressibles, Instrumentation et exploitation des mesures", *Bull. liaison des P.*, vol. 115, pp. 61–70, 1981.

[MAL 10] MALLET C., Lidar aéroporté topographiques et bathymétriques, Ecole été Lidar géophysique, available at: http://recherche.ign.fr/labos/matis clement.mallet@ign.fr, 2010.

[MAR 80] MARCHETTI S., "In situ test by flat dilatometer", *ASCE Journal of the Geotechnical Engineering Division*, vol. 106, pp. 299–321, 1980.

[MEM 80] MÉMIER B., Contribution à l'étude des capteurs de pression totale dans les sols, PhD Thesis, INSA de LYON, 1980.

[MEN 55] MÉNARD L.F.A., Pressiométre, Brevet Français d'Invention, Paris, no. 1.117.983, pp. 3–19, 19 January 1955.

[MEN 59] MÉNARD L.F.A., Dispositif d'étude de la déformation sous charge d'un milieu homogène, Brevet Français d'Invention, Paris, no. 1.234.756, p. 5–15, 15 May 1959.

[MES 77] MESRI G., GODLEWSKI P.M., "Time and stress-compressibility interrelationship", *Proc. ASCE, Journal of the Geotechnical Engineering Division*, vol. 103, no. GT5, pp. 417–430, 1977.

[MES 98] MESTAT P., "Etat des contraintes initiames dans les sols et calcul par éléments finis", *Bul. Des Lab. Des Ponts et Chaus.*, vol. 215, pp. 15–32, 1998.

[MLY 05] MLYNAREK Z., WIERZBICKI J., WOLYNSKI W., "Use of cluster method for in situ tests", *Studia Geotechnica et Mechanica*, vol. 28, nos. 3–4, pp. 15–27, 2005.

[MON 50] MONFORE G.E., An analysis of the stress distribution in and near stress gages embedded in elastic soils, Structural Laboratory Report no. SP 26, U.S. Bureau of Reclamation, Research and Geology Division, Denver, Colorado, 1950.

[MON 69] MONNET J., GIELLY J., "Détermination d'une loi de comportement pour les sols pulvérulents", *Revue Française de Géotechnique*, no. 7, pp. 45–56, 1969.

[MON 90] MONNET J., "Theoretical study of elasto-plastic equilibrium around pressuremeter in sands", *Proc. 3rd Int. Symposium on Pressuremeter*, Oxford, pp. 137–148, 1990.

[MON 94] MONNET J., KHLIF J., "Etude théorique de l'équilibre élasto-plastique d'un sol pulvérulent autour du pressiomètre", *Revue Française de Géotechnique*, Paris, no. 67, pp. 3–12, 1994.

[MON 95] MONNET J., CHEMAA T., "Etude théorique de l'équilibre élasto-plastique d'un sol cohérent autour du pressiomètre", *Revue Française de Géotechnique*, Paris, no. 73, pp. 15–26, 1975.

[MON 99] MONNET J., Brevet français Gaiatech, Procédé et appareil d'essai de terrain à partir d'un forage, No. 99 12020, Lyon, 22 September 1999.

[MON 02] MONNET J., "Calibration of the parameters used in a theoretical model for several natural soils", *Symposium International Param 2002*, Paris, pp. 413–422, 02 September 2002.

[MON 05] MONACO P., MARCHETTI S., TOTANI G. *et al.*, "Sand liquefaction assessment by flat dilatometer test (DMT)", *Proc. 16th Int. Conf. Soil Mech. Geot. Eng.*, Osaka, vol. 4, pp. 2693–2697, 2005.

[MON 12a] MONNET J., "An elasto-plastic analysis of the pressuremeter test in granular soil – part 1: theory", *European Journal of Environmental and Civil Engineering*, vol. 16, no. 6, pp. 699–714, 2012.

[MON 12b] MONNET J., "An elasto-plastic analysis of the pressuremeter test in granular soil – part 2: numerical study", *European Journal of Environmental and Civil Engineering*, vol. 16, no. 6, pp. 715–729, 2012.

[MON 12c] MONNET J., BROUCKE M., "The use of a cluster analysis in a Ménard pressuremeter survey", *ICE – Geotechnical Engineering*, London, vol. 165, no. 6, pp. 367–377, 2012.

[MON 13] MONNET J., Caractérisation mécanique des sols par l'essai pressiométrique, ISP6 – PRESSIO 2013, Session 4–5, Paris, 2013.

[MON 15] MONNET J., HUGHES J., DAFNI J., "The use of cluster analysis for identification of soil layer – the Grays harbor pontoon construction project", *Symposium International ISP7*, pp. 249–260, 2015.

[NEG 88] NEGUSSEY D., WIJEWICKREME W.K.D., VAID Y.P., "Constant-volume friction angle of granular materials", *Canadian Geotechnical Journal*, vol. 25, pp. 50–55, 1988.

[NOR 88] NORUSIS M.J., The SPSS guide to data analysis for SPSS, SPSS Inc., Chicago, vol. 402, 1988.

[PEA 54] PEATTLE K.R., SPARROW R.R., "The fundamental action of earth pressure cells" *Journal of the Mechanics and Physics of Solids*, London, vol. 2, pp. 141–155, 1954.

[PEC 53] PECK R.B., HANSON W.E., THORNBURN T.H., *Foundation Engineering*, John Wiley & Sons, New York, 1953.

[PHI 85] PHILIPPONNAT G., Procédé et dispositif de mesure des caractéristiques de cisaillement du sol, Brevet Français d'Invention, Paris, no. 85 09466, p. 9, 1985.

[PIL 82] PILOT G., *Foundation Engineering*, Presse de l'E.N.P.C., Paris, 1982.

[PIR 12] PIRON A., CHULLIAT O., MOREL LOT G., Suivi de la tension résiduelle par auscultation des tirants d'ancrage du parc hydraulique EDF, Colloque CFBR: Auscultation des barrages et des Digues, Chambéry, 2012.

[POP 80] POPOVIC M., SARAC D., "Some properties of compacted clay materials used in earth dams and embankments", *International Conference on Compaction*, Editions Anciens ENPC, Paris, vol. 1, pp. 189–193, 22–23–24 April 1980.

[RAT 68] RAT M., LAVIRON F., JOREZ J.C., "Essai Lefranc", *Bulletin des Laboratoires des Ponts et Chaussées*, pp. 56–66, 1968.

[REI 05] REIFFSTECK P., REVERDY G., VINCELAS G. *et al.*, "Pressiomètre auto-foreur de nouvelle generation", *Symp. Int. ISP5 – Pressio 2005*, vol. 1, pp. 113–126, 2005.

[REI 10] REIFFSTECK P., "Utilisation des paramètres de forage en reconnaissance géotechnique", *Journ. Nat. Géot. Et Géol. Ing. (JNGG)*, Grenoble, pp. 157–164, 2010.

[ROB 90] ROBERTSON P.K., "Soil classification using the cone penetration test", *Can. Geotch. J.*, vol. 27, pp. 151–158, 1990.

[ROB 97] ROBERTSON P.K., WRIDE C.E., "Cyclic liquefaction and its evaluation based on the SPT and CPT", *Proc. of NCEER*, Buffalo, NY, pp. 41–87, 1997.

[ROB 98] ROBERTSON P.K., WRIDE C.E., "Evaluating cyclic liquefaction potential using the cone penetration test", *Can. Geotech. J.*, vol. 35, pp. 442–459, 1998.

[ROS 68] ROSCOE K.H., BURLAND J.B., "On the generalised stress-strain behaviour of wet clay", in HEYMAN J., LECKIE F.A. (eds), *Engineering Plasticity*, Cambridge University Press, 1968.

[ROW 62] ROWE P.W., "The stress dilatancy relation for static equilibrium", *Proceedings of the Royal Society London A*, vol. 269, pp. 500–527, 1962.

[ROW 69] ROWE P.W., "The relation between the shear strength of sands in triaxial compression in plane strain and direct shear", *Geotechnique*, vol. 19, pp. 75–86, 1969.

[RUE 67] RUELLAN F., *Photogrammétrie, interprétation de photographies stéréoscopiques 8terrestres et aériennes*, Premier fascicule: Initiation, Masson and Cie, Paris, 1967.

[SAL 66] SALENÇON J., "Expansion quasi statique d'une cavité à symétrie sphérique ou cylindrique dans un milieu élasto-plastique", *Annales des Ponts et Chaussées*, no. 3, pp. 175–187, 1966.

[SAN 72] SANGLERAT G., *The Penetrometer and Soil Exploration*, Elsevier, 1972.

[SED 85] SEDITECH, Catalogue de matériel de sols, bétons, bitumes, Géotechnique, Géophysique, p. 120, 1985.

[SEN 99] SENOUCI S.M., MONNET J., “Modélisation numérique du Géomécamètre”, *Revue Française de Géotechnique*, no. 88, pp. 21–35, 1999.

[TAY 45] TAYLOR D.W., Review of pressure distribution theories and earth pressure cell investigation, Chapter 2 in report to the U.S. Army Engineer Waterways Experiment Station, 1945.

[TAY 15] TAYLOR D.W., Pressure cell considerations, Triaxial shear research and pressure distribution studies on soils, Chapter 2, U.S. Army Engineer Waterways Experiment Station, CE, Vicksburg, Mississippi, pp. 211–236, 2015.

[TEC 01] TECHNICAL COMMITTEE 16, “The flat dilatometer test (DMT) in soil investigations”, *ISSMGE, Int. Conf. on In Situ Measurement of Soil Properties*, Bali, Indonesia, 2001.

[TER 48] TERZAGHI K., PECK R.B., *Soil Mechanics in Engineering Practice*, 1st ed., John Wiley & Sons, New York, p. 566, 1948.

[TER 68] TERZAGHI K., PECK R.B., *Soil Mechanics in Engineering Practice*, Wiley, 1968.

[TIM 51] TIMOSHENKO K., GOODIER J.N., *Theory of Elasticity*, 2nd ed., Mac Graw-Hill, New York, 1951.

[TRO 61] TROLOPPE D.H., LEE B.C.E., “The measurement of soil pressures”, *5th International Conference on Soil Mechanics and Foundation Engineering*, Paris, vol. 2, pp. 403–409, 1961.

[VIA 13] VIANA DA FONSECA A., “Application of in situ testing in tailing dams, emphasis on liquefaction: case history”, *4th International Conference on Site Characterisation*, Poito de Galinhas, pp. 181–203, 2013.

[VIA 15] VIANA DA FONSECA A., TOPA GOMES A., COUTINHO R.Q., “The use of pressuremeter tests for modelling residual soils geomechanic and foundation behavior”, *Int. Symposium 60 Years of Pressuremeters, ISP7*, Hammamet, pp. 33–71, 2015.

[WAS 82] WASCHKOWSKI E., “Le pénétromètre dynamique”, *Bul. Liaison Labo P et Ch.*, vol. 125, pp. 95–103, May–June 1982.

[WOL 89] WOLFF T.F., “Pile capacity prediction using parameter functions”, *ASCE Geotechnical Special Publication*, vol. 23, pp. 96–107, 1989.

[WOO 77] WOOD D.M., WROTH C.P., “Some laboratory experiments related to the results of pressuremeter tests”, *Geotechnique*, vol. 27, no. 2, pp. 181–201, 1977.

[WOO 90] WOOD D.M., *Soil Behaviour and Critical State Soil Mechanics*, Cambridge University Press, 1990.

[WRO 72] WROTH C.P., HUGHES J.M.O.-, An instrument for the in situ measurement of the properties of soft clays, Report of Department of Engineering, University of Cambridge, CUED/C, Soils TR13, 1972.

Index

A, B

C

D, E, F

G, H, I

J, L, M

N, O, P

R, S, T

U, V, W

Other titles from

iSTE

in

Civil Engineering and Environmental Geomechanics

2015

AÏT-MOKHTAR Abdelkarim, MILLET Olivier
Structure Design and Degradation Mechanisms in Coastal Environments

2014

DAÏAN Jean-François
Equilibrium and Transfer in Porous Media – 3-volume series
Equilibrium States – Volume 1
Transfer Laws – Volume 2
Applications, Isothermal Transport, Coupled Transfers – Volume 3

2013

AMZIANE Sofiane, ARNAUD Laurent
Bio-aggregate-based Building Materials: Applications to Hemp Concretes

BONELLI Stéphane
Erosion in Geomechanics Applied to Dams and Levees

CASANDJIAN Charles, CHALLAMEL Noël, LANOS Christophe, HELLESLAND Jostein
Reinforced Concrete Beams, Columns and Frames: Mechanics and Design

GUÉGUEN Philippe
Seismic Vulnerability of Structures

HELLESLAND Jostein, CHALLAMEL Noël, CASANDJIAN Charles, LANOS Christophe
Reinforced Concrete Beams, Columns and Frames: Section and Slender Member Analysis

LALOUI Lyesse, DI DONNA Alice
Energy Geostructures: Innovation in Underground Engineering

LEGCHENKO Anatoly
Magnetic Resonance Imaging for Groundwater

2012

BONELLI Stéphane
Erosion of Geomaterials

JACOB Bernard *et al.*
ICWIM6 – Proceedings of the International Conference on Weigh-In-Motion

OLLIVIER Jean-Pierre, TORRENTI Jean-Marc, CARCASSES Myriam
Physical Properties of Concrete and Concrete Constituents

PIJAUDIER-CABOT Gilles, PEREIRA Jean-Michel
Geomechanics in CO_2 Storage Facilities

2011

BAROTH Julien, BREYSSE Denys, SCHOEFS Franck
Construction Reliability: Safety, Variability and Sustainability

CREMONA Christian
Structural Performance: Probability-based Assessment

HICHER Pierre-Yves
Multiscales Geomechanics: From Soil to Engineering Projects

IONESCU Ioan R. *et al.*
Plasticity of Crystalline Materials: from Dislocations to Continuum

LOUKILI Ahmed
Self Compacting Concrete

MOUTON Yves
Organic Materials for Sustainable Construction

NICOT François, LAMBERT Stéphane
Rockfall Engineering

PENSE-LHERITIER Anne-Marie
Formulation

PIJAUDIER-CABOT Gilles, DUFOUR Frédéric
Damage Mechanics of Cementitious Materials and Structures

RADJAI Farhang, DUBOIS Frédéric
Discrete-element Modeling of Granular Materials

RESPLENDINO Jacques, TOULEMONDE François
Designing and Building with UHPFRC

2010

ALSHIBLI A. Khalid
Advances in Computed Tomography for Geomechanics

BUZAUD Eric, IONESCU Ioan R., VOYIADJIS Georges
Materials under Extreme Loadings / Application to Penetration and Impact

LALOUI Lyesse
Mechanics of Unsaturated Geomechanics

NOVA Roberto
Soil Mechanics

SCHREFLER Bernard, DELAGE Pierre
Environmental Geomechanics

TORRENTI Jean-Michel, REYNOUARD Jean-Marie, PIJAUDIER-CABOT Gilles
Mechanical Behavior of Concrete

2009

AURIAULT Jean-Louis, BOUTIN Claude, GEINDREAU Christian
Homogenization of Coupled Phenomena in Heterogenous Media

CAMBOU Bernard, JEAN Michel, RADJAI Fahrang
Micromechanics of Granular Materials

MAZARS Jacky, MILLARD Alain
Dynamic Behavior of Concrete and Seismic Engineering

NICOT François, WAN Richard
Micromechanics of Failure in Granular Geomechanics

2008

BETBEDER-MATIBET Jacques
Seismic Engineering

CAZACU Oana
Multiscale Modeling of Heterogenous Materials

HICHER Pierre-Yves, SHAO Jian-Fu
Soil and Rock Elastoplasticity

JACOB Bernard *et al.*
HVTT 10

JACOB Bernard *et al.*
ICWIM 5

SHAO Jian-Fu, BURLION Nicolas
GeoProc2008

2006

DESRUES Jacques *et al.*
Advances in X-ray Tomography for Geomaterials

FSTT
Microtunneling and Horizontal Drilling

MOUTON Yves
Organic Materials in Civil Engineering

2005

PIJAUDIER-CABOT Gilles *et al.*
Creep Shrinkage and Durability of Concrete and Concrete Structures CONCREEP – 7

Printed and bound by CPI Group (UK) Ltd, Croydon, CR0 4YY
13/09/2026
14961020-0007